K.R. Venugopal, K.G. Srinivasa and L.M. Patnaik

Soft Computing for Data Mining Applications

Studies in Computational Intelligence, Volume 190

Further volumes of this series can be found on our homepage: springer.com

Vol. 168. Andreas Tolk and Lakhmi C. Jain (Eds.)
Complex Systems in Knowledge-based Environments: Theory, Models and Applications, 2009
ISBN 978-3-540-88074-5

Vol. 169. Nadia Nedjah, Luiza de Macedo Mourelle and Janusz Kacprzyk (Eds.)
Innovative Applications in Data Mining, 2009
ISBN 978-3-540-88044-8

Vol. 170. Lakhmi C. Jain and Ngoc Thanh Nguyen (Eds.)
Knowledge Processing and Decision Making in Agent-Based Systems, 2009
ISBN 978-3-540-88048-6

Vol. 171. Chi-Keong Goh, Yew-Soon Ong and Kay Chen Tan (Eds.)
Multi-Objective Memetic Algorithms, 2009
ISBN 978-3-540-88050-9

Vol. 172. I-Hsien Ting and Hui-Ju Wu (Eds.)
Web Mining Applications in E-Commerce and E-Services, 2009
ISBN 978-3-540-88080-6

Vol. 173. Tobias Grosche
Computational Intelligence in Integrated Airline Scheduling, 2009
ISBN 978-3-540-89886-3

Vol. 174. Ajith Abraham, Rafael Falcón and Rafael Bello (Eds.)
Rough Set Theory: A True Landmark in Data Analysis, 2009
ISBN 978-3-540-89886-3

Vol. 175. Godfrey C. Onwubolu and Donald Davendra (Eds.)
Differential Evolution: A Handbook for Global Permutation-Based Combinatorial Optimization, 2009
ISBN 978-3-540-92150-9

Vol. 176. Beniamino Murgante, Giuseppe Borruso and Alessandra Lapucci (Eds.)
Geocomputation and Urban Planning, 2009
ISBN 978-3-540-89929-7

Vol. 177. Dikai Liu, Lingfeng Wang and Kay Chen Tan (Eds.)
Design and Control of Intelligent Robotic Systems, 2009
ISBN 978-3-540-89932-7

Vol. 178. Swagatam Das, Ajith Abraham and Amit Konar
Metaheuristic Clustering, 2009
ISBN 978-3-540-92172-1

Vol. 179. Mircea Gh. Negoita and Sorin Hintea
Bio-Inspired Technologies for the Hardware of Adaptive Systems, 2009
ISBN 978-3-540-76994-1

Vol. 180. Wojciech Mitkowski and Janusz Kacprzyk (Eds.)
Modelling Dynamics in Processes and Systems, 2009
ISBN 978-3-540-92202-5

Vol. 181. Georgios Miaoulis and Dimitri Plemenos (Eds.)
Intelligent Scene Modelling Information Systems, 2009
ISBN 978-3-540-92901-7

Vol. 182. Andrzej Bargiela and Witold Pedrycz (Eds.)
Human-Centric Information Processing Through Granular Modelling, 2009
ISBN 978-3-540-92915-4

Vol. 183. Marco A.C. Pacheco and Marley M.B.R. Vellasco (Eds.)
Intelligent Systems in Oil Field Development under Uncertainty, 2009
ISBN 978-3-540-92999-4

Vol. 184. Ljupco Kocarev, Zbigniew Galias and Shiguo Lian (Eds.)
Intelligent Computing Based on Chaos, 2009
ISBN 978-3-540-95971-7

Vol. 185. Anthony Brabazon and Michael O'Neill (Eds.)
Natural Computing in Computational Finance, 2009
ISBN 978-3-540-95973-1

Vol. 186. Chi-Keong Goh and Kay Chen Tan
Evolutionary Multi-objective Optimization in Uncertain Environments, 2009
ISBN 978-3-540-95975-5

Vol. 187. Mitsuo Gen, David Green, Osamu Katai, Bob McKay, Akira Namatame, Ruhul A. Sarker and Byoung-Tak Zhang (Eds.)
Intelligent and Evolutionary Systems, 2009
ISBN 978-3-540-95977-9

Vol. 188. Agustín Gutiérrez and Santiago Marco (Eds.)
Biologically Inspired Signal Processing for Chemical Sensing, 2009
ISBN 978-3-642-00175-8

Vol. 189. Sally McClean, Peter Millard, Elia El-Darzi and Chris Nugent (Eds.)
Intelligent Patient Management, 2009
ISBN 978-3-642-00178-9

Vol. 190. K.R. Venugopal, K.G. Srinivasa and L.M. Patnaik
Soft Computing for Data Mining Applications, 2009
ISBN 978-3-642-00192-5

Tejaswi

Foreword

The authors have consolidated their research work in this volume titled Soft Computing for Data Mining Applications. The monograph gives an insight into the research in the fields of Data Mining in combination with Soft Computing methodologies. In these days, the data continues to grow exponentially. Much of the data is implicitly or explicitly imprecise. Database discovery seeks to discover noteworthy, unrecognized associations between the data items in the existing database. The potential of discovery comes from the realization that alternate contexts may reveal additional valuable information. The rate at which the data is stored is growing at a phenomenal rate. As a result, traditional ad hoc mixtures of statistical techniques and data management tools are no longer adequate for analyzing this vast collection of data. Several domains where large volumes of data are stored in centralized or distributed databases includes applications like in electronic commerce, bioinformatics, computer security, Web intelligence, intelligent learning database systems, finance, marketing, healthcare, telecommunications, and other fields.

Efficient tools and algorithms for knowledge discovery in large data sets have been devised during the recent years. These methods exploit the capability of computers to search huge amounts of data in a fast and effective manner. However, the data to be analyzed is imprecise and afflicted with uncertainty. In the case of heterogeneous data sources such as text and video, the data might moreover be ambiguous and partly conflicting. Besides, patterns and relationships of interest are usually approximate. Thus, in order to make the information mining process more robust it requires tolerance toward imprecision, uncertainty and exceptions.

With the importance of soft computing applied in data mining applications in recent years, this monograph gives a valuable research directions in the field of specialization. As the authors are well known writers in the field of Computer Science and Engineering, the book presents state of the art technology in data mining. The book is very useful to researchers in the field of data mining.

Bangalore,
November 2008

N.R. Shetty
President, ISTE, India

Preface

In today's digital age, there is huge amount of data generated everyday. Deriving meaningful information from this data is a huge problem for humans. Therefore, techniques such as data mining whose primary objective is to unearth hithero unknown relationship from data becomes important. The application of such techniques varies from business areas (Stock Market Prediction, Content Based Image Retrieval), Proteomics (Motif Discovery) to Internet (XML Data Mining, Web Personalization). The traditional computational techniques find it difficult to accomplish this task of Knowledge Discovery in Databases (KDD). Soft computing techniques like Genetic Algorithms, Artificial Neural Networks, Fuzzy Logic, Rough Sets and Support Vector Machines when used in combination is found to be more effective. Therefore, soft computing algorithms are used to accomplish data mining across different applications.

Chapter one presents introduction to the book. Chapter two gives details of self adaptive genetic algorithms. An iterative merge based genetic algorithms for data mining applications is given in chapter three. Dynamic association rule mining using genetic algorithms is described in chapter four. An evolutionary approach for XML data mining is presented in chapter five. Chapter six, gives a neural network based relevance feedback algorithm for content based image retrieval. An hybrid algorithm for predicting share values is addressed in chapter seven. The usage of rough sets and genetic algorithms for data mining based query processing is discussed in chapter eight. An effective web access sequencing algorithm using hashing techniques for better web reorganization is presented in chapter nine. An efficient data structure for personalizing the Google search results is mentioned in chapter ten. Classification based clustering algorithms using naive Bayesian probabilistic models are discussed in chapter eleven. The effective usage of simulated annealing and genetic algorithms for mining top-k ranked webpages from Google is presented in chapter twelve. The concept of mining bioXML databases is introduced in chapter thirteen. Chapter fourteen and fifteen discusses algorithms for DNA compression. An efficient algorithm for motif discovery in protein sequences is presented in chapter sixteen. Finally, matching techniques for genome sequences and genetic algorithms for motif discovery are given in chapter seventeen and eighteen respectively.

The authors appreciate the suggestions from the readers and users of this book. Kindly communicate the errors, if any, to the following email address: venugopalkr@gmail.com.

Bangalore,
November 2008

K.R. Venugopal
K.G. Srinivasa
L.M. Patnaik

Acknowledgements

We wish to place on record our deep debt of gratitude to Shri M C Jayadeva, who has been a constant source of inspiration. His gentle encouragement have been the key for the growth and success in our career. We are indebted to Prof. K Venkatagiri Gowda for his inspiration, encouragement and guidance throughout our lives. We thank Prof. N R Shetty, President, ISTE and Former Vice Chancellor, Bangalore University, Bangalore for his foreword to this book. We owe debt of gratitude to Sri K Narahari, Sri V Nagaraj, Prof. S Lakshmana Reddy, Prof. K Mallikarjuna Chetty, Prof. H N Shivashankar, Prof. P Sreenivas Kumar, Prof. Kamala Krithivasan, Prof. C Sivarama Murthy, Prof. T Basavaraju, Prof. M Channa Reddy, Prof. N Srinivasan, Prof. M Venkatachalappa for encouraging us to bring out this book in the present form. We sincerely thank Sri K P Jayarama Reddy, T G Girikumar, P Palani, M G Muniyappa for their support in the preparation of this book.

We are grateful to Justice M Rama Jois, Sri N Krishnappa for their encouragement. We express our gratitude to Sri Y K Raghavendra Rao, Sri P R Ananda Rao, Justice T Venkataswamy, Prof. V Y Somayajulu, Sri Sreedhar Sagar, Sri N Nagabhusan, Sri Prabhakar Bhat, Prof. K V Acharya, Prof. Khajampadi Subramanya Bhat, Sri Dinesh Kamath, Sri D M Ravindra, Sri Jagadeesh Karanath, Sri N Thippeswamy, Sri Sudhir, Sri V Manjunath, Sri N Dinesh Hegde, Sri Nagendra Prasad, Sri Sripad, Sri K Thyagaraj, Smt. Savithri Venkatagiri Gowda, Smt. Karthyayini V and Smt. Rukmini T, our well wishers for inspiring us to write this book.

We thank Prof. K S Ramanatha, Prof. K Rajanikanth, V K Ananthashayana and T V Suresh Kumar for their support. We thank Smt. P Deepa Shenoy, Sri K B Raja, Sri K Suresh Babu, Smt. J Triveni, Smt. S H Manjula, Smt. D N Sujatha, Sri Prakash G L, Smt. Vibha Lakshmikantha, Sri K Girish, Smt. Anita Kanavalli, Smt. Alice Abraham, Smt. Shaila K, for their suggestions and support in bringing out this book.

We are indebted to Tejaswi Venugopal, T Shivaprakash, T Krishnaprasad and Lakshmi Priya K for their help. Special thanks to Nalini L and Hemalatha for their invaluable time and neat desktop composition of the book.

About the Authors

K.R. Venugopal is Principal and Dean, Faculty of Engineering, University Visvesvaraya College of Engineering, Bangalore University, Bangalore. He obtained his Bachelor of Technology from University Visvesvaraya College of Engineering in 1979. He received his Masters degree in Computer Science and Automation from Indian Institute of Science Bangalore He was awarded Ph.D. in Economics from Bangalore University and Ph.D. in Computer Science from Indian Institute of Technology, Madras. He has a distinguished academic career and has degrees in Electronics, Economics, Law, Business Finance, Public Relations, Communications, Industrial Relations, Computer Science and Journalism. He has authored and edited twenty seven books on Computer Science and Economics, which include Petrodollar and the World Economy, Programming with Pascal, Programming with FORTRAN, Programming with C, Microprocessor Programming, Mastering C++ etc. He has been serving as the Professor and Chairman, Department of Computer Science and Engineering, UVCE. He has over two hundred research papers in refereed International Journals and Conferences to his credit. His research interests include computer networks, parallel and distributed systems and database systems.

K.G. Srinivasa obtained his a Ph.D. in Computer Science and Engineering from Bangalore University. Currently he is working as an Assistant Professor in the Department of Computer Science and Engineering, M S Ramaiah Institute of Technology, Bangalore. He received Bachelors and Masters degree in Computer Science and Engineering from the Bangalore University in the year 2000 and 2002 respectively. He is a member of IEEE, IETE, and ISTE. He has authored more than fifty research papers in refereed International Journals and Conferences. His research interests are Soft Computing, Data Mining and Bioinformatics.

L.M. Patnaik is Vice Chancellor of Defence Institute of Advanced Studies, Pune, India. He was the Professor since 1986 with the Department of

Computer Science and Automation, Indian Institute of Science, Bangalore. During the past 35 years of his service at the Institute. He has over 400 research publications in in refereed International Journals and Conference Proceedings. He is a Fellow of all the four leading Science and Engineering Academies in India; Fellow of the IEEE and the Academy of Science for the Developing World. He has received twenty national and international awards; notable among them is the IEEE Technical Achievement Award for his significant contributions to high performance computing and soft computing. His areas of research interest have been parallel and distributed computing, mobile computing, CAD for VLSI circuits, soft computing, and computational neuroscience.

Contents

Acronyms

GA	Genetic Algorithms
ANN	Artificial Neural Networks
AI	Artificial Intelligence
SVM	Support Vector Machines
KDD	Knowledge Discovery in Databases
OLAP	On-Line Analytical Processing
MIQ	Machine Intelligence Quotient
FL	Fuzzy Logic
RS	Rough Sets
XML	eXtended Markup Language
HTML	Hyper Text Markup Language
SQL	Structured Query Language
PCA	Principal Component Analysis
SDI	Selective Dissemination of Information
SOM	Self Organizing Map
CBIR	Content Based Image Retrieval
WWW	World Wide Web
DNA	Deoxyribo Nucleic Acid
IGA	Island model Genetic Algorithms
SGA	Simple Genetic Algorithms
PID	Pima Indian Diabetes
Wisc	Wisconsin Breast Cancer Database
Hep	Hepatitis Database
Ion	Ionosphere Database
LVQ	Learning Vector Quantization
BPNN	Backpropagation Neural Network
RBF	Radial Basis Function
ITI	Incremental Decision Tree Induction
LMDT	Linear Machine Decision Tree
DTD	Document Type Definition
MFI	Most Frequently used Index

LFI	Less Frequently used Index
hvi	Hierarchical Vector Identification
UIC	User Interest Categories
KNN	*k* Nearest Neighborhood
DMQL	Data Mining Query Languages
TSP	Travelling Salesman Problem
MAD	Mean Absolute Deviation
SSE	Sum of Squared Error
MSE	Mean Squared Error
RMSE	Root Mean Squared Error
MAPE	Mean Absolute Percentage Error
STI	Shape Texture Intensity
HIS	Hue, Intensity and Saturation
DCT	Discrete Cosine Transform
PWM	Position Weight Matrix
PSSM	Position Specific Scoring Matrix
PRDM	Pairwise Relative Distance Matrix
DSSP	Secondary Structure of Proteins
LSI	Latent Semantic Indexing
GIS	Geographical Information Systems
CAD	Computer Aided Design
FS	Free Search
BGA	Breeder Genetic Algorithm
STIRF	Shape, Texture, Intensity-distribution features with Relevance Feedback

Chapter 1
Introduction

Database mining seeks to extract previously unrecognized information from data stored in conventional databases. Database mining has also been called *database exploration* and *Knowledge Discovery in Databases*(KDD). Databases have significant amount of stored data. This data continues to grow exponentially. Much of the data is implicitly or explicitly imprecise. The data is valuable because it is collected to explicitly support particular enterprise activities. There could be valuable, undiscovered relationships in the data. A human analyst can be overwhelmed by the glut of digital information. New technologies and their application are required to overcome information overload. Database discovery seeks to discover noteworthy, unrecognized associations between data items in an existing database. The potential of discovery comes from the realization that alternate contexts may reveal additional valuable information. A metaphor for database discovery is mining. Database mining elicits knowledge that is implicit in the databases. The rate at which the data is stored is growing at a phenomenal rate. As a result, traditional ad hoc mixtures of statistical techniques and data management tools are no longer adequate for analyzing this vast collection of data [1]. Several domains where large volumes of data are stored in centralized or distributed databases include the following applications in electronic commerce, bioinformatics, computer security, Web intelligence, intelligent learning database systems, finance, marketing, healthcare, telecommunications, and other fields, which can be broadly classified as,

1. *Financial Investment*: Stock indexes and prices, interest rates, credit card data, fraud detection.
2. *Health Care*: Several diagnostic information stored by hospital management systems.
3. *Manufacturing and Production*: Process optimization and trouble shooting.
4. *Telecommunication Network*: Calling patterns and fault management systems.
5. *Scientific Domain*: Astronomical observations, genomic data, biological data.
6. *The World Wide Web*.

The area of Data Mining encompasses techniques facilitating the extraction of knowledge from large amount of data. These techniques include topics such as

K.R. Venugopal, K.G. Srinivasa, L.M. Patnaik: Soft Comput. for Data Min. Appl., SCI 190, pp. 1–17.
springerlink.com

pattern recognition, machine learning, statistics, database tools and On-Line Analytical Processing (OLAP). Data mining is one part of a larger process referred to as Knowledge Discovery in Database (KDD). The KDD process is comprised of the following steps: (i) Data Cleaning (ii) Data Integration (iii) Data Selection (iv) Data Transformation (v) Data Mining (vi) Pattern Evaluation (vii) Knowledge Presentation.

The term *data mining* often is used in discussions to describe the whole KDD process, when the data preparation steps leading up to data mining are typically more involved and time consuming than the actual mining steps. Data mining can be performed on various types of data, to include: Relational Database, Transactional Database, Flat File, Data Warehouse, Images (Satellite, Medical), GIS, CAD, Text, Documentation, Newspaper Articles, Web Sites, Video/Audio, Temporal Databases/Time Series(Stock Market Data, Global Change Data), etc. The steps in KDD process are briefly explained below.

- Data cleaning to remove noise and inconsistent data
- Data integration which involves combining of multiple data sources
- Data selection where data relevant to analysis task is retrieved from the database
- Data transformation where consolidated data is stored to be used by mining processes
- Data mining which is essential where intelligent methods are applied in order to extract data patterns
- Pattern evaluation where interestingness measures of discovered patterns are measured
- Knowledge presentation where user understandable forms of mined knowledge is presented

Data mining also involves an integration of different techniques from multiple disciplines such as database technology, statistics, machine learning, neural networks, image and signal processing, etc.. Data mining can be performed on a variety of data such as relational databases, data warehouses, transactional databases, object oriented databases, spatial databases, legacy databases, World Wide Web, etc.. The kind of patterns found in data mining tasks are two important ones that are descriptive and predictive. Descriptive patterns characterize the general properties of databases while predictive mining tasks perform inference on the current data in order to make predictions.

Data Mining is a step in the KDD process that consists of applying data analysis and discovery algorithms which, under acceptable computational limitations, produce a particular enumeration of patterns over data. It uses historical information to discover regularities and improve future decisions. The overall KDD process is outlined in Figure 1.1. It is interactive and iterative involving the following steps [3].

1. *Data cleaning*: which removes noise, inconsistency from data.
2. *Data Integration*: which combines multiple and heterogeneous data sources to form an integrated database.
3. *Data selection*: where data appropriate for the mining task is taken from the databases.

4. *Data transformation*: where data is transformed or consolidated into forms appropriate for mining by performing summary or aggregation operations.
5. *Data mining*: where the different mining methods like association rule generation, clustering or classification is applied to discover the patterns.
6. *Pattern evaluation*: where patterns are identified using some constraints like support, confidence.
7. *Knowledge presentation*: where visualization or knowledge presentation techniques are used to present the knowledge.
8. The updations to the database like increment/decrement is handled if any, and the steps from 1 to 7 is repeated.

Data mining involves fitting models to determine patterns from observed data. The fitted models play the role of inferred knowledge. Deciding whether the model reflects useful knowledge or not is a part of the overall KDD process. Typically, a data mining algorithm constitutes some combination of the following three components,

- *The model*: The function of the model(e.g., classification, clustering) and its representational form (e.g., linear discriminants, neural networks). A model contains parameters that are to be determined from the data.
- *The preference criterion*: A basis for preference of one model or set of parameters over another, depending on the given data. The criterion is usually some form of *goodness-of-fit* function of the model to the data, perhaps tempered by a smoothing term to avoid overfitting, or generating a model with too many degrees of freedom to be constrained by the given data.
- *The search algorithm*: The specification of an algorithm for finding particular models and parameters, given the data, models, and a preference criterion.

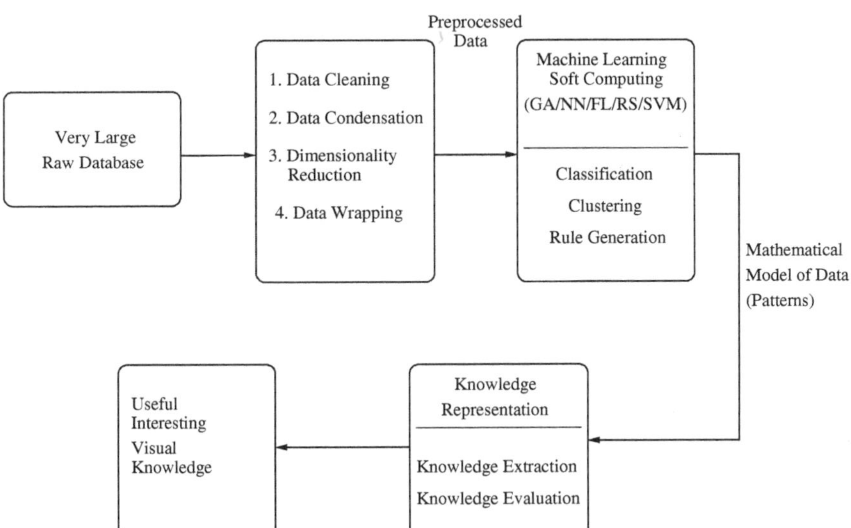

Fig. 1.1 Overall KDD Process with Soft Computing

In general, mining operations are performed to figure out characteristics of the existing data or to figure out ways to infer from current data some prediction of the future. Below are the main types of mining [4,5].

1. *Association Rule Mining* - Often used for market basket or transactional data analysis, it involves the discovery of rules used to describe the conditions where items occur together - are associated.
2. *Classification and Prediction* - involves identifying data characteristics that can be used to generate a model for prediction of similar occurrences in future data.
3. *Cluster Analysis* - attempts to look for groups (clusters) of data items that have a strong similarity to other objects in the group, but are the most dissimilar to objects in other groups.
4. *Outlier Mining* - uses statistical, distance and deviation-based methods to look for rare events (or outliers) in datasets, things that are not normal.
5. *Concept/Class Description* - uses data characterization and/or data discrimination to summarize and compare data with target concepts or classes. This is a technique to provide useful knowledge in support of data warehousing.
6. *Time Series Analysis* - can include analysis of similarity, periodicity, sequential patterns, trends and deviations. This is useful for modeling data events that change with time.

1.1 Data Mining

In general data mining tasks can be broadly classified into two categories: *descriptive data mining* and *predictive data mining*. Descriptive data mining describes the data in a concise and summary fashion and gives interesting general properties of the data whereas predictive data mining attempts to predict the behavior of the data from a set of previously built data models. A data mining system can be classified according to the type of database that has to be handled. Different kinds of databases are, relational databases, transaction databases, object oriented databases, deductive databases, spatial databases, mobile databases, stream databases and temporal databases. Depending on the kind of knowledge discovered from the database, mining can be classified as association rules, characteristic rules, classification rules, clustering, discrimination rules, deviation analysis and evolution. A survey of data mining tasks gives the following methods.

1.1.1 Association Rule Mining (ARM)

One of the strategies of data mining is association rule discovery which correlates the occurrence of certain attributes in the database leading to the identification of large data itemsets. It is a simple and natural class of database regularities, useful in various analysis and prediction tasks. ARM is a undirected or unsupervised data mining method which can handle variable length data and can produce clear, understandable and useful results. Association rule mining is computationally and I/O intensive. The problem of mining association rules over market basket data is referred so, due to its origins in the study of consumer purchasing patterns in retail shops. Mining association rules is the process of discovering expressions of the form $X \longrightarrow Y$. For example, customers usually buy coke(Y) along with cheese(X). These rules provide valuable insights to customer buying behavior, vital to business analysis.

New association rules, which reflect the changes in the customer buying pattern, are generated by mining the updations in the database. This concept is called incremental mining. This problem is very popular due to its simple statement, wide applications in finding hidden patterns in large data and paradigmatic nature. The process of discovering association rules can be split into two steps, first finding all itemsets with appreciable support and next is the generation of the desired rules.

Various applications of association rule mining are super market shelf management, Inventory management, Sequential pattern discovery, Market basket analysis including cross marketing, Catalog design, loss-leader analysis, Product pricing and Promotion. Association rules are also used in online sites to evaluate page views associated in a session to improve the store layout of the site and to recommend associated products to visitors.

Mining association rules at multiple concept levels may lead to the discovery of more specific and concrete knowledge from data. A top down progressive deepening method is developed for mining Multiple Level Association Rules(MLAR) for large databases. MLAR uses a hierarchy information encoded table instead of the original transaction table. Encoding can be performed during the collection of task-relevant data and thus there is no extra pass required for encoding. Large support is more likely to exist at high concept level, such as milk and bread, rather than at low concept level such as particular brand of milk and bread. To find strong associations at relatively low concept levels, the *min_support* threshold must be reduced substantially. One of the problems with this data mining technique is the generation of large number of rules. As the rules generated increases, it becomes very difficult to understand them and take appropriate decisions. Hence pruning and grouping the rules to improve the understandability, is an important issue.

Inter-transaction association rules break the barrier of Intra-transaction association and are mainly used for prediction. They try to relate items from the different transactions, due to which the computations become exhaustive. Hence the concept of sliding window is used to limit the search space. A frequent inter-transaction itemsets must be made up of frequent intra-transaction itemsets.

Intra-transaction association rules is a special case of inter-transaction association rules. Some of the applications are (i) to discover traffic jam association patterns among different highways to predict traffic jams, (ii) from weather database to predict flood and drought for a particular period.

1.1.2 Incremental Mining

One of the important problems of the data mining problem is to maintain the discovered patterns when the database is updated regularly. In several applications new data is added continuously over the time. Incremental mining algorithms are proposed to handle updations of rules when increments to data base occur. It should be done in a manner which is cost-effective, without involving the database already mined and permitting reuse of the knowledge mined earlier. The two major operations involved are (i) Additions: Increase in the support of appropriate itemsets and discovery of new itemsets. (ii) Deletions: Decrease in the support of existing large itemsets leading to the formation of new large itemsets.

1.1.3 Distributed Data Mining

The emergence of network based distributed computing environments such as the internet, private intranet and wireless networks has created a natural demand for scalable techniques for data mining in a distributed manner. Also, the proliferation of data in the recent years has made it impossible to store it in a single global server. Several data mining methods can be applied to find local patterns which can be combined to form global knowledge. Parallel algorithms are designed for very large databases to study the performance implications and trade-off between computation, communication, memory usage, synchronization and the use of problem specific information in parallel data mining.

1.1.4 Sequential Mining

A sequence is an ordered set of item-sets. All transactions of a particular customer made at different times can be taken as a sequence. The term support is used in a different meaning. Here the support is incremented only once, even if a customer has bought the same item several times in different transactions. Usually the Web and scientific data are sequential in nature. Finding patterns from such data helps to predict future activities, interpreting recurring phenomena, extracting outstanding comparisons for close attention, compressing data and detecting intrusion.

The incremental mining of sequential data helps in computing only the difference by accessing the updated part of the database and datastructure. The sequential data are text, music notes, satellite data, stock prices, DNA sequences, weather data, histories of medical records, log files, etc.. The applications of sequential mining are analysis of customer purchase patterns, stock market analysis, DNA sequences, computational biology study, scientific experiments, disease treatments, Web access patterns, telecommunications, biomedical research, prediction of natural disasters and system performance analysis etc..

1.1.5 Clustering

Clustering is the process of grouping the data into classes so that objects within a cluster are similar to one another, but are dissimilar to objects in other clusters. Various distance functions are used to make quantitative determination of similarity and an objective function is defined with respect to this distance function to measure the quality of a partition. Clustering is an example for unsupervised learning. It can be defined as, given n data points in a d-dimensional metric space, partition the data points into k clusters, such that the data points within a cluster are more similar to each other than the data points in different clusters.

Clustering has roots in data mining, biology and machine learning. Once the clusters are decided, the objects are labeled with their corresponding clusters, and common features of the objects in a cluster are summarized to form the class description. For example, a set of new diseases can be grouped into several categories based on the similarities in their symptoms, and the common symptoms of the diseases in a category can be used to describe that group of diseases. Clustering is a useful technique for the discovery of data distribution and patterns in the underlying database. It has been studied in considerable detail by both statistics and database researchers for different domains of data. As huge amounts of data are collected in databases, cluster analysis has recently become a highly active topic in data mining research. Various applications of this method are, data warehousing, market research, seismology, minefield detection, astronomy, customer segmentation, computational biology for analyzing DNA microarray data and World Wide Web.

Some of the requirements of clustering in data mining are scalability, high dimensionality, ability to handle noisy data, ability to handle different types of data etc. Clustering analysis helps to construct meaningful partitioning of a large set of objects based on a divide and conquer methodology. Given a large set of multidimensional data points, the data space is usually not uniformly occupied by the data points, hence clustering identifies the sparse and the crowded areas to discover the overall distribution patterns of the dataset. Numerous applications involving data warehousing, trend analysis, market research, customer segmentation and pattern recognition are high dimensional and dynamic in nature. They provide an opportunity for performing dynamic data mining tasks such as incremental and association rules. It is challenging to cluster high dimensional data objects, when they are skewed and sparse. Updations are quiet common in dynamic databases and usually they are processed in batch mode. In very large databases, it is efficient to incrementally perform cluster analysis only to the updations. There are five methods of clustering; they are (i) Partitioning method (ii) Grid based method (iii) Model based method (iv) Density based method (v) Hierarchical method.

Partition Method: Given a database of N data points, this method tries to form, k clusters, where $k \leq N$. It attempts to improve the quality of clusters or partition by moving the data points from one group to another. Three popular algorithms under this category are k-means, where each cluster is represented by the mean value of the data points in the cluster and k-medoids, where each cluster is represented by one of the objects situated near the center of the cluster, whereas k-modes extends the k-means to categorical attributes. The k-means and the k-modes methods can be combined to cluster data with numerical and categorical values and this method is called k-prototypes method. One of the disadvantage of these methods is that, they are good in creating spherical shaped clusters in small databases.

Grid Based Method: This method treats the database as a finite number of grid cells due to which it becomes very fast. All the operations are performed on this grid structure.

Model Based Method: is a robust clustering method. This method locates clusters by constructing a density function which denotes the spatial distribution of the data points. It finds number of clusters based on standard statistics taking outliers into consideration.

Density Based Method: finds clusters of arbitrary shape. It grows the clusters with as many points as possible till some threshold is met. The e-neighborhood of a point is used to find dense regions in the database.

Hierarchical Method: In this method, the database is decomposed into several levels of partitioning which are represented by a dendrogram. A dendrogram is a tree that iteratively splits the given database into smaller subsets until each subset contains only one object. Here each group of size greater than one is in turn composed of smaller groups. This method is qualitatively effective, but practically infeasible for large databases, since the performance is at least quadratic in the number of database points. Consequently, random sampling is often used in order to reduce the size of the dataset. There are two types in hierarchical clustering algorithms; (i) Divisive methods work by recursively partitioning the set of datapoints S until singleton sets are obtained. (ii) Agglomerative algorithms work by starting with singleton sets and then merging them until S is covered. The agglomerative methods cannot be used directly, as it scales quadratically with the number of data points. Hierarchical methods usually generate spherical clusters and not of arbitrary shapes.

The data points which do not belong to any cluster are called outliers or noise. The detection of outlier is an important datamining issue and is called as outlier mining. The various applications of outlier mining are in fraud detection, medical treatment etc..

1.1.6 Classification

Classification is a process of labeling the data into a set of known classes. A set of training data whose class label is known is given and analyzed, and a classification model is prepared from the training set. A decision tree or a set of classification rules is generated from the clasification model, which can be used for better understanding of each class in the database and for classification of data. For example, classification rules about diseases can be extracted from known cases and used to diagnose new patients based on their symptoms. Classification methods are widely developed in the fields of machine learning, statistics, database, neural network, rough sets and are an important theme in data mining. They are used in customer segmentation, business modeling and credit analysis.

1.1.7 Characterization

Characterization is the summarization of a set of task relevant data into a relation, called generalized relation, which can be used for extraction of characteristic rules.

The characteristic rules present characteristics of the dataset called the target class. They can be at multiple conceptual levels and viewed from different angles. For example, the symptoms of a specific disease can be summarized by a set of characteristic rules. Methods for efficient and flexible generalization of large data sets can be categorized into two approaches: the data cube approach and the attribute-oriented induction approach.

In the data cube approach, a multidimensional database is constructed which consists of a set of dimensions and measures. A dimension is usually defined by a set of attributes which form a hierarchy or a lattice of structure. A data cube can store pre-computed aggregates for all or some of its dimensions. Generalization and specialization can be performed on a multiple dimensional data cube by *roll-up* or *drill-down* operations. A roll-up operation reduces the number of dimensions in a data cube, or generalizes attribute values to higher level concepts. A drill-down operation does the reverse. Since many aggregate values may need to be used repeatedly in data analysis, the storage of precomputed aggregates in a multiple dimensional data cube will ensure fast response time and offer flexible views of data from different angles and at different levels of abstraction. The attribute-oriented induction approach may handle complex types of data and perform attribute relevance analysis in characterization.

1.1.8 Discrimination

Discrimination is the discovery of features or properties that distinguish the class being examined(target class) from other classes(contrasting class). The method for mining discriminant rules is similar to that of mining characteristic rules except that mining should be performed in both target class and contrasting classes synchronously to ensure that the comparison is performed at comparative levels of abstraction. For example, to distinguish one disease from others, a discriminant rule summarizes the symptoms of this disease from others.

1.1.9 Deviation Mining

Deviation mining is the discovery and evaluation of the deviation patterns of the objects in the target data in a time related databases. The expected behavior or norm of the objects is usually given by the user or computed based on some assumptions, such as average, linear growth etc.. For example, one may discover and evaluate set stocks whose behavior deviates from the trend of the majority of stocks during a certain period of time.

1.1.10 Evolution Mining

Evolution mining is the detection and evaluation of data evolution regularities for certain objects whose behavior changes over time. This may include characterization, association, or clustering of time related data. For example, one may find the general characteristics of the companies whose stock price has gone up over 20% last year, or evaluate the trend or particular growth patterns of high-tech stocks.

1.1.11 Prediction

Prediction is the estimation or forecast of the possible values of some missing data or the value distribution of certain attribute in a set of objects. This involves finding the set of attributes of interest and predicting the value distribution based on a set data similar to the selected object. For example, an employee's potential salary can be predicted based on the salary distribution of similar employees in the company.

1.1.12 Web Mining

Web mining is the process of mining massive collection of information, on the world-wide web and has given rise to considerable interest in the research community. The heterogeneous, unstructured and chaotic web is not a database but it is a set of different data sources with unstructured and interconnected artifacts that continuously change. Web is a huge, distributed repository of global information to service the enormous requirements of news, advertisement, consumer information, financial management, education, government, e-commerce etc. The *www* contains huge dynamic collection of hyperlink information and web page access and usage information, providing useful sources for data mining. Web mining is applying data mining techniques to automatically discover and extract useful information from www documents and services. Data mining holds the key to uncover the authoritative links, traversal patterns and semantic structures that brings intelligence and direction to our web interactions. Web mining is a technique that automatically retrieve, extract and evaluate information for knowledge discovery from web documents and services. The web page complexity far exceeds the complexity of any traditional text document collection. Only small portion of web pages contain truly relevant or useful information. The web mining tasks can be classified into following (i) Web structure mining, (ii)Web content mining and (iii) Web usage mining.

The web structure mining generates structural summary about the organization of web sites and web pages. It tries to discover the link structure of the hyperlinks at the inter-document level. Web content mining deals with the discovery of useful information from the web contents, web data, web documents and web services. The contents of the web includes a very broad range of data such as audio, video, symbolic, metadata and hyperlinked data in addition to text. Web content mining focuses on the structure of inner-document, based on the topology of the hyperlinks, while web structure mining categorizes the web pages and generate the information, such as the similarity and relationship between different web sites.

Web usage mining involves data from web server access logs, proxy server logs, browser logs, user profiles, registration files, user sessions or transactions, user queries, book mark folders, mouse clicks and scrolls and any other data generated by the interaction of the users and the web. Web usage mining provides the key to understand web traffic behavior, which can inturn be used for developing policies for web caching, network transmission, load balancing or data distribution. Web content and structure mining utilize the real or primary data on the web, while web usage mining takes secondary data generated by the users interaction with web.

The web usage mining basically extracts useful access information from the web log data. The mined information can be used for analyzing the access patterns and concluding on general trends. An intelligent analysis helps in restructuring the web. Web log analysis can also help to build customized web services for individual users. Since web log data provides information about specific pages popularity and the methods used to access them, this information can be integrated with web content and linkage structure mining to help rank web pages. They can also be used to improve and optimize the structure of a site, to improve the scalability and performance of web based recommender systems and to discover e-business intelligence for the purpose of online marketing.

Web usage mining can also provide patterns which are useful for detecting intrusion, fraud, attempted break-in etc.. It provides detailed feed back on user behavior, providing the web site designer with information to redesign the web organization. Web log data usually consists of URL requested, the IP address from which the request originated and a timestamp. Some of the applications are; improving web site design, building adaptive web sites, analyzing system performance, understanding user reaction and motivation.

1.1.13 Text Mining

Text mining analyzes text document content to investigate syntactical correlation and semantic association between terms. Key concepts and phrases are extracted to represent the document or a section of a document. Text mining includes most of the steps of data mining starting from data cleaning to knowledge visualization. The dominant categories in text mining are text analysis, text interpretation, document categorization, and document visualization.

A text search architecture usually consists of the steps (i) Storage (ii) Indexing (iii) Search Criteria (iv) Index Building (v) Query Optimization. Business applications of text mining products are (i) Drug firms-Biomedical research (ii) Electric utility-Customer opinion survey analysis (iii) Online News paper-searching for a job, car or house. For example, Intelligent Miner for text from IBM has: Feature extraction tool, Clustering tool and Categorization tool.

1.1.14 Data Warehouses

Dramatic advances in data capture, processing power, data transmission, and storage capabilities are enabling organizations to integrate their various databases into data warehouses. A data warehouse is a subject-oriented, integrated, time-variant and non-volatile collection of data in support of management decision making process. They are different from file systems, data repositories and relational data base systems. Usually a data warehouse is built around a subject like a product or sales or customers and so on. It concentrates on the modeling and analysis of data for decision makers. It is constructed by integrating several sources like relational databases, flat files and on-line transaction records. Data cleaning and integration techniques are applied to obtain consistency in encoding structures and delete anamolies. For this purpose update-driven approach can be used.

When a new information source is attached to the warehousing system or when relevant information at a source changes, the new or modified data is propagated to the integrator. The integrator is responsible for installing the information in the warehouse, which may include filtering the information, summarizing it or merging it with information from other sources. The data warehouse is time-variant because it changes over a period of time. The historical data of about 5-10 years old is preserved to study the trends. The data is stored in such a way that the keys always contain the unit of time(day, week etc.) being referred to and non- volatile because a data warehouse is a read only database. Only data needed for decision making is stored. A data warehouse is a dedicated database system and supports decision support system. It helps the knowledge worker to make better and faster decision. A typical data warehouse contains five types of data.

1. *Current detail data*: reflects the most recent events.
2. *Older detail data*: It is moved from disk to a mass-storage medium.
3. *Lightly summarized data*: improves the response and use of data warehouse.
4. *Highly summarized data*: is required by service manager and should be available in compact and easily accessible form. Highly summarized data improves the response time
5. *Metadata*: is the information about the data rather than the information provided by the data warehouse. Administrative metadata includes all the information necessary for setting up and using a warehouse. It usually consists of the warehouse schema, derived data, dimensions, hierarchies, predefined queries and reports. It also contains physical organization such as data partitions, data extraction, cleaning and transformation rules, data refresh and purging policies, user profiles, user authorization and access control policies.

Business metadata includes business terms and definitions, ownership of data and charging policies. Operational metadata includes information that is collected during the operation of the warehouse. A metadata repository is used to store and manage all the metadata associated with the warehouse.

A wide variety and number of data mining algorithms are described in the literature - from the fields of statistics, pattern recognition, machine learning and databases. They represent a long list of seemingly unrelated and often highly specific algorithms. Some of them include; (i) Statistical models, (ii) Probabilistic graphical dependency models, (iv) Decision trees and rules, (v) Inductive logic programming based models, (vi) Example based methods, lazy learning and case based reasoning, (vii) Neural network based models, (viii) Fuzzy set theoretic models, (ix) Rough set theory based models, (x) Genetic algorithm based models, and (xi) Hybrid and soft computing models.

1.2 Soft Computing

Efficient tools and algorithms for knowledge discovery in large data sets have been devised during the recent years. These methods exploit the capability of computers to search huge amount of data in a fast and effective manner. However, the data to be analyzed is imprecise and afflicted with uncertainty. In the case of heterogeneous data sources such as text and video, the data might moreover be ambiguous and partly conflicting. Besides, patterns and relationships of interest are usually vague and approximate. Thus, in order to make the information mining process more robust or say, human-like methods for searching and learning it requires tolerance toward imprecision, uncertainty, and exceptions. Thus, they have approximate reasoning capabilities and are capable of handling partial truth. Properties of the aforementioned kind are typical of soft computing.

Soft computing differs from conventional (hard) computing in that, unlike hard computing, it is tolerant of imprecision, uncertainty, partial truth, and approximation. The guiding principle of soft computing is to exploit the tolerance for imprecision, uncertainty, partial truth, and approximation to achieve tractability, robustness and low solution cost.

The principal constituents of soft computing are fuzzy logic, neural networks, genetic algorithms and probabilistic reasoning. These methodologies of soft computing are complementary rather than competitive and they can be viewed as a foundation component for the emerging field of conceptual intelligence.

1.2.1 Importance of Soft Computing

The complementarity of fuzzy logic, neural networks, genetic algorithms and probabilistic reasoning has an important consequence in many cases. A problem can be solved most effectively by using fuzzy logic, neural networks, genetic algorithms and probabilistic reasoning in combination rather than using them exclusively. A typical example for such combination is *neurofuzzy systems*. Such systems are becoming increasingly visible as consumer products ranging from air conditioners and washing machines to photocopiers and camcorders. Thus the employment of soft computing techniques leads to systems which have high MIQ (Machine Intelligence Quotient).

1.2.2 Genetic Algorithms

Genetic Algorithms have found a wide gamut of applications in data mining, where knowledge is mined from large databases. Genetic algorithms can be used to build effective classifier systems, mining association rules and other such datamining problems. Their robust search technique has given them a central place in the field of data mining and machine learning [6].

GA can be viewed as an evolutionary process where at each generation, from a set of feasible solutions, individuals or solutions are selected such that individuals with higher fitness have greater probability of getting chosen. At each generation, these chosen individuals undergo crossover and mutation to produce a population of the next generation. This concept of survival of the fittest proposed by Darwin is the main cause for the robust performance of GAs. Crossover helps in the exchange of discovered knowledge in the form of genes between individuals and mutation helps in restoring lost or unexplored regions in search space.

1.2.3 Neural Networks

An Artificial Neural Network (ANN) is an information processing paradigm that is inspired by the way biological nervous systems, such as the brain processes information. The key element of this paradigm is the novel structure of the information processing system. It is composed of a large number of highly interconnected processing elements (neurons) working in union to solve specific problems. An ANN is configured for a specific application, such as pattern recognition or data classification, through a learning process. Learning in biological systems involves adjustments to the synaptic connections that exist between the neurons [8]. Neural networks, with their remarkable ability to derive meaning from complicated or imprecise data, can be used to extract patterns and detect trends that are too complex. The neural network can be used to provide projections given new situations of interest and answer *what if* questions. The advantages of neural networks include:

- *Adaptive learning*: An ability to learn how to do tasks based on the data given for training or initial experience.
- *Self-Organization*: An ANN can create its own organization or representation of the information it receives during learning time.
- *Real Time Operation*: ANN computations may be carried out in parallel, and special hardware devices are being designed and manufactured which take advantage of this capability.
- *Fault Tolerance via Redundant Information Coding*: Partial destruction of a network leads to the corresponding degradation of performance. However, some network capabilities may be retained even with major network damage.

An artificial neuron is a device with many inputs and one output. The neuron has two modes of operation; the training mode and the testing mode. In the training mode, the neuron can be trained to fire (or not), for particular input patterns. In the testing mode, when a trained input pattern is detected at the input, its associated output becomes the current output. If the input pattern does not belong in the trained list of input patterns, the firing rule is used to determine whether to fire or not. The firing rule is an important concept in neural networks and accounts for their high flexibility.

1.2.4 Support Vector Machines

Support Vector Machines (SVMs) are a set of related supervised learning methods used for classification and regression. They belong to a family of generalized linear classifiers. They can also be considered a special case of Tikhonov regularization. A special property of SVMs is that they simultaneously minimize the empirical classification error and maximize the geometric margin and hence they are also known as maximum margin classifiers. Two parallel hyperplanes are constructed on each side of the hyperplane that separates the data. The hyperplane is the one that maximizes the distance between the two parallel hyperplanes. An assumption is made that the larger the margin or distance between these parallel hyperplanes, the better the generalization error of the classifier. The SVM builds a model from the training samples which is later used on the test data. This model is built using the training samples that are most difficult to classify (Support Vectors). The SVM is capable of classifying both linearly separable and non-linearly separable data. The nonlinearly separable data can be handled by mapping the input space to a high dimensional feature space. In this high dimensional feature space, linear classification can be performed. SVMs can exhibit good accuracy and speed even with very less training.

1.2.5 Fuzzy Logic

Fuzzy logic is derived from fuzzy set theory dealing with reasoning that is approximate rather than precisely deduced from classical predicate logic. It can be thought of as the application side of fuzzy set. Fuzzy truth represents membership in vaguely defined sets, like likelihood of some event or condition and fuzzy sets are based on vague definitions of sets, not randomness [10].

Since, data mining is the process of extracting nontrivial relationships in the database, the association that are qualitative are very difficult to utilize effectively by applying conventional rule induction algorithms. Since fuzzy logic modeling is a probability based modeling, it has many advantages over the conventional rule induction algorithms. The advantage is that it allows processing of very large data sets which require efficient algorithms. Fuzzy logic-based rule induction can handle noise and uncertainty in data values well. Most of the databases, in general, are not designed or created for data mining. Selecting and extracting useful attributes of target objects becomes hard. Not all of the attributes needed for successful extraction can be contained in the database. In these cases, domain knowledge and user analysis becomes a necessity. The techniques such as neural networks tend to do badly since the domain knowledge cannot be incorporated into the neural networks, therefore Fuzzy logic based models utilize the domain knowledge in coming up with rules of data selection and extraction.

It can be observed that no single technique can be defined as the optimal technique for data mining. The selection of technique used depends highly on the problem and the data set. Hambaba (1996) stressed the need of using hybrid techniques for different problems since each intelligent technique has a particular computational property that suits them appropriately to a particular problem. The real time applications like loan evaluation, fraud detection, financial risk assessment, financial decision making, and credit card application evaluation uses the combination of Neural Networks and Fuzzy Logic Systems.

1.2.6 Rough Sets

The Rough Sets theory was introduced by Zdzislaw Pawlak in the early 1980's, and based on this theory one can propose a formal framework for the automated transformation of data into knowledge. Pawlak has shown that the principles for learning by examples can be formulated on the basis of this theory. It simplifies the search for dominating attributes leading to specific properties, or just rules pending in the data.

The Rough Set theory is mathematically simple and has shown its fruitfulness in a variety of data mining applications. Among these are information retrieval, decision support, machine learning, and knowledge based systems. A wide range of applications utilize the ideas of the theory. Medical data analysis, aircraft pilot performance evaluation, image processing, and voice recognition are few examples. Inevitably the database used for data mining contains imperfection, such as noise, unknown values or errors due to inaccurate measuring equipment. The Rough Set theory comes handy for dealing with these types of problems, as it is a tool for handling vagueness and uncertainty inherent to decision making.

1.3 Data Mining Applications

Data mining is extensively in customer relationship management. Data clustering can also be used to automatically discover the segments or groups within a customer data set. Businesses employing data mining may see a return on investment, but also they recognize that the number of predictive models can quickly become very large. Rather than one model to predict which customers will churn, a business could build a separate model for each region and customer type. Then instead of sending an offer to all people that are likely to churn, it may only want to send offers to customers likely take to the offer. And finally, it may also want to determine which customers are going to be profitable over a window of time and only send the offers to those that are likely to be profitable. In order to maintain this quantity of models, they need to manage model versions and move to automated data mining[7,9].

Data mining can also be helpful to human-resources departments in identifying the characteristics of their most successful employees. Information obtained, such as universities attended by highly successful employees, can help HR focus recruiting efforts accordingly. Additionally, Strategic Enterprise Management applications help a company translate corporate-level goals, such as profit and margin share targets, into operational decisions, such as production plans and workforce levels.

Another example of data mining, often called the market basket analysis, relates to its use in retail sales. If a clothing store records the purchases of customers, a data-mining system could identify those customers who favour silk shirts over cotton ones. Data mining is a highly effective tool in the catalog marketing industry. Catalogers have a rich history of customer transactions on millions of customers dating back several years. Data mining tools can identify patterns among customers and help identify the most likely customers to respond to upcoming mailing campaigns.

In recent years, data mining has been widely used in area of science and engineering, such as bioinformatics, genetics, medicine, education, and electrical power engineering. In the area of study on human genetics, the important goal is to understand the mapping relationship between the inter-individual variation in human DNA sequences and variability in disease susceptibility. In layman terms, it is to find out how the changes in an individual's DNA sequence affect the risk of developing common diseases such as cancer. This is very important to help improve the diagnosis, prevention and treatment of the diseases. The data mining technique that is used to perform this task is known as multifactor dimensionality reduction.

In the area of electrical power engineering, data mining techniques have been widely used for condition monitoring of high voltage electrical equipment. The purpose of condition monitoring is to obtain valuable information on the insulation's health status of the equipment. Data clustering such as self-organizing map (SOM) has been applied on the vibration monitoring and analysis of transformer on-load tap-changers. Using vibration monitoring, it can be observed that each tap change operation generates a signal that contains information about the condition of the tap changer contacts and the drive mechanisms.

Another area of application for data mining is in science/engineering research, where data mining has been used to study the factors leading students to choose to engage in behaviors which reduce their learning and to understand the factors influencing university student retention. Other examples of applying data mining technique applications are biomedical data facilitated by domain ontologies, mining clinical trial data, traffic analysis using SOM, etc..

References

1. Hand, D., Mannila, H., Smyth, P.: Principles of Data Mining. MIT Press, Cambridge (2001)
2. Mehmed, K.: Data Mining: Concepts, Models, Methods, and Algorithms. John Wiley and Sons, Chichester (2003)
3. Han, J., Kamber, M.: Data Mining, Concepts and Techniques. Elsevier, Amsterdam (2007)
4. Ye, N.: The Handbook of Data Mining, Human Factors and Ergonomics (2003)
5. Tan, P.N., Steinbach, M., Kumar, V.: Introduction to Data Mining. Pearson Education, London (2007)
6. Holland, J.H.: Adaptation in Natural and Artificial Systems. University of Michigan Press (2004)
7. Kantardizic, M.M., Zurada, J.: Next Generation of Data Mining Applications. Wiley Interscience, Hoboken (2005)
8. Mitchell, T.M.: Machine Learning. McGraw Hill International Editions, New York (1997)
9. Freitas, A.A.: Data Mining and Knowledge Discovery with Evolutionary Algorithms. Springer, Heidelberg (2005)
10. Jang, J.S.R., Sun, C.T., Mizutani, E.: Neuro-Fuzzy and Soft Computing. Pearson Education, London (2004)

Chapter 2
Self Adaptive Genetic Algorithms

Abstract. Genetic Algorithms(GAs) are efficient and robust searching and optimization methods that are used in data mining. In this chapter, we propose a Self-Adaptive Migration Model GA (SAMGA), where parameters of population size, the number of points of crossover and mutation rate for each population are adaptively fixed. Further, the migration of individuals between populations is decided dynamically. This chapter gives a mathematical schema analysis of the method stating and showing that the algorithm exploits previously discovered knowledge for a more focused and concentrated search of heuristically high yielding regions while simultaneously performing a highly explorative search on the other regions of the search space. The effective performance of the algorithm is then shown using standard testbed functions and a set of actual classification based datamining problems. Michigan style of classifier is used to build the classifier and the system is tested with machine learning databases of Pima Indian Diabetes database, Wisconsin Breast Cancer database and few others.

2.1 Introduction

Data mining is a process of extracting nontrivial, valid, novel and useful information from large databases. Hence data mining can be viewed as a kind of search for meaningful patterns or rules from a large search space, that is the database. In this light, Genetic algorithms are a powerful tool in data mining, as they are robust search techniques. GAs are a set of random, yet directed search techniques. They process a set of solutions simultaneously and hence are parallel in nature. They are inspired by the natural phenomenon of evolution. They are superior to *gradient descent* techniques as they are not biased towards local optima [1, 2, 3, 4]. The steps in genetic algorithm are given in Table 2.1.

In this algorithm, the three basic operators of GA namely, selection, crossover and mutation operators are fixed apriori. The optimum parameters for these operators depend on problem on which the GA is applied and also on the fitness of the current population. A new breed of GA called adaptive GAs [5, 6], fix the parameters for the

K.R. Venugopal, K.G. Srinivasa, L.M. Patnaik: Soft Comput. for Data Min. Appl., SCI 190, pp. 19–50.
springerlink.com

Table 2.1 Genetic Algorithms

```
Genetic Algorithm()
{
Initialize population randomly;
Evaluate fitness of each individual in the population;
While stopping condition not achieved
   {
    Perform selection;
    Perform crossover and mutation;
    Evaluate fitness of each individual in the population;
   }
}
```

GA operators dynamically to adapt to the current problem. Generally, the operators are adapted based on the fitness of individuals in the population. Apart from the operators themselves, even the control parameters can be adapted dynamically. In these adaptive genetic algorithms, the GA parameters are adapted to fit the given problem. The algorithm is shown in Table 2.2.

Table 2.2 Adaptive Genetic Algorithms

```
Adaptive Genetic Algorithm()
{
    Initialize population randomly;
    Evaluate fitness of each individual in the population;
    While stopping condition not achieved
    {
     Perform selection;
     Perform crossover and mutation;
     Evaluate fitness of each individual;
     Change selection, crossover and mutation operators.
    }
}
```

The use of parallel systems in the execution of genetic algorithms has led to parallel genetic algorithms. The Island or Migration model of genetic algorithm [7] is one such genetic algorithm where, instead of one population, a set of populations is evolved. In this method, at each generation all the populations are independently evolved and at some regular interval fixed by migration rate, few of the best individuals are exchanged among populations.

2.2 Related Work

Lobo et al., have proposed an adaptive GA which works even when optimal number of individuals is not known [8]. In this method parallel searches are conducted with different number of individuals, expecting one of the populations to have the appropriate number of individuals that yields good results. However, this method is not truly adaptive in the sense that the appropriate number of individuals is not learnt but is obtained by trial and error. This method is not feasible as it is not realistic to perform large number of blind searches in parallel. Similarly, some of the applications of the parameter-less genetic algorithms [9] and multi-objective rule mining using genetic algorithms are discussed in [10].

An adaptive GA which runs three GAs in parallel is proposed in [11]. Here at each *epoch* fitness values of elite individuals is compared and the number of individuals are changed according to the results. For example, if GA with the largest number of individuals provides best results, then in the next epoch all individuals have large number of individuals. However, the optimum number of individuals required by a population depends on which region of the search space the individuals are in and is not same for all subpopulations.

An adaptive GA where mutation rate for an individual is encoded in the gene of an individual is proposed in [9]. The system is proposed with the hope that finally individuals with good mutation rate survive. However, only individuals with low mutation rate survive in the later phases of the search. An adaptive GA that determines mutation and crossover rate of an individual by its location in a two dimensional lattice plane is proposed in [12]. The algorithm keeps diversity of these parameters by limiting the number of individuals in each lattice.

A meta-GA is a method of using GA to fix the right parameters for the GA. However, in this process the number of evaluations needed is high and the process is expensive. One such meta-GA is proposed in [13]. A GA that adapts mutation and crossover rates in Island model of GA is proposed in [14]. Here adaptation is based on average fitness of the population. Parameters of a population here are updated to those of a neighboring population with high average fitness.

The breeder genetic algorithm BGA depends on a set of control parameters and genetic operators. It is shown that strategy adaptation by competing subpopulations makes the BGA more robust and more efficient in [15]. Each subpopulation uses a different strategy which competes with other subpopulations. Experiments on multi-parent reproduction in an adaptive genetic algorithm framework is performed in [16]. An adaptive mechanism based on competing subpopulations is incorporated into the algorithm in order to detect the best crossovers. A parallel genetic algorithm with dynamic mutation probability is presented in [17]. This algorithm is based on the farming model of parallel computation. The basic idea of the dynamically updating the mutation rate is presented. Similarly, an adaptive parallel genetic algorithm for VLSI Layout Optimization is discussed in [18]. A major problem in the use of genetic algorithms is premature convergence. An approach for dealing with this problem is the distributed genetic algorithm model is addressed in [19]. Its basic idea is to keep, in parallel, several subpopulations that are processed by genetic algorithms, with each one being independent of the others. But all these algorithms either consider mutation rate or crossover rate as a dynamic parameter, but not both at the same time. The application of a breeder genetic algorithm to the problem of parameter identification for an adaptive finite impulse filter is addressed in [20]. A population-based optimization method, called Free Search (FS) is also discussed in [21].

A Self adaptive Genetic Algorithms with Simulated Binary Crossover is discussed in [22]. It talks about a single population GA for starters and it does not have a multipoint crossover let alone adaptive multipoint crossover. The only similarity between [22] and proposed model is that both the methods chooses adaptively the proportion of the population for crossover. Similarly, an adaptive dynamic crossover operators based on fuzzy connectives and adaptive real coded GAs based on fuzzy logic controller is discussed in [23]. But it fails to prove the model analytically and the convergence rate of our proposed model(SAMGA) is better.

The Nix and Vose Punctuated Equillibiria is given in [24]. Markov Modelling for Genetic algorithms and their convergence rate are discussed in [25 - 29]. The average Hamming distance of individuals in the population to derive convergence rate is used in [30]. Genetic Algorithms based on binomially distributed populations is also proposed. In this chapter, the definition of convergence is restated and convergence rate is derived based on expectation value of individuals.

2.3 Overview

Any heuristic search can be characterized by two concepts, exploitation and exploration. These concepts are often conflicting in the sense that if exploitation of a search is increased, then exploration decreases and vice versa. Fortunately, the parallel nature of GA helps in alleviating this problem slightly by simultaneously processing a set of solutions thus providing for exploration, while at the same time survival of the fittest enforces exploitation. However, if a single population processes a set of individuals, then the schemas in the population that give better results only survive as generations proceed and hence search is not very explorative. Further, if the explorative power of search is increased too much using mutation, convergence does not occur. Thus, we need a method wherein, the explorative power of search is increased for individuals in bad region of the state space and exploitation power of search is increased for individuals in the high fitness region of the state space.

In this chapter, the proposed adaptive migration(island) model of GA is built such that, given a search space, the number of individuals in the population that resides in a relatively high fitness region of the search space increases thus improving exploitation. For these high fitness population, the mutation rate and number of points of crossover are decreased thus making the search more focused. On the other hand, for populations in a relatively low fitness zone of search space, the number of individuals is decreased but the mutation rates and number of points of crossover are increased to make the search of these regions more explorative. Thus, the search can be termed as one which exploits the heuristically promising region of search space while exploring other regions. The search is also competitive as an increase in fitness of one population makes the search in other populations with lower fitness more explorative and rigorous.

2.4 Algorithm

The genetic algorithm described here is an adaptive GA where between evaluation of individuals and applying the three operators of GA (selection, crossover and mutation), the parameters used for these operators are updated based on the evaluation of individuals.

2.4.1 *Problem Definition*

Let S be made up of all the 2^k, k bit binary numbers representing the entire search space. Given the objective function f which is a mapping $f: S \rightarrow R$ where R is a set of real numbers, the problem is to find an $x^* \in S$ such that
$f(x^*) \geq f(x)\ \forall\, x \in S$.

2.4.2 *Pseudocode*

Let $E = \{P_1, P_2 ... P_{np}\}$ be an ecosystem with np populations in it. Populations $P_1, P_2 ... P_{np} \subset S$. $P_i[j]$ stands for the jth individual of population P_i, clearly $P_i[j] \in S$. Let n_i be the size of population P_i and nc_i be the number of points of crossover used for reproduction in population P_i. Let $\bar{f}_i$ be the average fitness of a population P_i. Let $f(P_i)$, the fitness of population P_i be the fitness of the best individual of that population. Let $\bar{f}$ be the average fitness of the ecosystem. Let $\bar{n}$ be the average number of individuals per population. Let pm_i be the rate of mutation used for population P_i. The rate of crossover being one. Then, the pseudo code for the Self Adaptive Migration model Genetic Algorithms (SAMGA) can be given as shown in Table 2.3.

The algorithm shown above is adaptive in four respects. The first parameter adaptively changed is the population size of each population. The population size is dynamically varied based on the fitness of the best individual of that population compared to the mean fitness of the population. The number of individuals in population P_i is updated as,

$$n_{i,t+1} = n_{i,t} + \frac{f(P_i)}{\bar{f}} - 1 \tag{2.1}$$

Where t is used to represent time in generations. Using this update, the size of population with fitness greater than the mean population fitness grows while size of population whose fitness is below the mean population fitness shrinks. Thus, it can be visualized that more number of individuals are concentrated in the heuristically better promising region of search space (exploitation).

The second parameter that is dynamically adapted is the number of points of crossover. The update used for number of crossover points is given by,

$$nc_{i,t+1} = nc_{i,t} + \frac{\bar{n}}{n_i} - 1 \tag{2.2}$$

Table 2.3 Self Adaptive Migration Model Genetic Algorithms(SAMGA)

```
begin
var prev = 0;
for i = 1 : np, do
   (i) Set n_i, the number of individuals in the population P_i to some arbitrary value n_0
   (ii) Initialize all individuals of population P_i to some random bit strings
   (iii) Set number of crossover points used for population P_i, nc_i to one
   (iv) Set mutation rate used for population P_i, pm_i to 0.01
next
for gen = 1 : maximum generation limit, do
   var nsum = 0;
   var fsum = 0;
   for i = 1 : np, do
      (a) Evaluate fitness of all the individuals of the population P_i and
      find f(P_i) the best fitness of the population
      (b) nsum = nsum + n_i
      (c) fsum = fsum + f(P_i)
   prev = f̄
   f̄ = fsum / np
   n̄ = nsum / np
   for i = 1 : np, do
      (a) nc_i = nc_i + n̄/n − 1
      (b) pm_i = pm_i + (n̄/n − 1) * 0.0001
      (c) n_i = n_i + f(P_i)/f̄ − 1
      (d) if (n_i == 0)
       Delete population P_i (extinction)
      (e) endif
   next
   for i = 1 : np, do
      Perform elitist selection for population P_i with the modified population size n_i
      Perform nc point non-uniform crossover on selected individuals of population P_i
      Perform mutation on individuals of population P_i with mutation probability pm_i
   next
   if prev == f̄
      Exchange or migrate best individuals between populations.
   endif
next
end
```

Using this update we can see that, if the number of individuals in a population is less compared to the mean number of individuals in a population, then the number of points of crossover is increased. In an indirect way update on the number of points of crossover is linked to fitness of population. From update on population size, it is clear that population size increases with fitness of population but from update of number of points of crossover, we see that for relatively low population sizes the number of points of crossover is increased and hence search is more explorative.

The third parameter considered is the mutation rate whose update is similar to the update on the number of crossover points, given by,

$$pm_{i,t+1} = pm_{i,t} + (\frac{\bar{n}}{n_i} - 1) * 0.0001 \quad (2.3)$$

The significance of this update is similar to the update on the number of points of crossover. Obviously, higher the mutation rate, more explorative is the search. The factor of 0.0001 is chosen arbitrarily as it is a considerably small factor to update probability of mutation.

The final parameter that is adaptive in the algorithm is the rate of migration. Migration refers to copying individuals from one population to another. Migration helps in discovering new schemas got by crossover of schemas of two populations. In the algorithm, there is no exact parameter for migration rate. Migration occurs when the average fitness of the populations remains unchanged between two generations. Thus, when populations have attained a steady state, migration occurs to try and discover a new schema.

The selection scheme used in the genetic algorithm is the elitist scheme. The best individuals of each population are copied unaltered to the next generation population. The remaining individuals are selected based on their fitness. The use of elitist selection guarantees that the best individual of any generation is at least as good as the best individual of the previous generation. It helps in achieving global convergence. Here, 100% of individuals are chosen from the k most fit but k is large to prevent premature convergence.

One interesting phenomenon that can be noticed when this algorithm is used on a complex, search space, is that just like in nature, if a population consistently performs badly, its number finally decreases to zero and the population becomes extinct. This suggests that the total number of individuals in the ecosystem is affected by the problem size and complexity.

2.5 Mathematical Analysis

The Holland's schema theorem for a general case can be given as,

$$M(H,t+1) \geq ((1-p_c)M(H,t)\frac{f(H)}{f_{avg}}$$

$$+p_c[M(H,t)\frac{f(H)}{f_{avg}}(1-losses)+gains])(1-p_m)^{O(H)}$$

Where $M(H,t)$ is the number of individuals in a population with schema H are present in the current generation, f_{avg} is average fitness of population, $O(H)$ is order of schema H and p_c and p_m are rates of crossover and mutation respectively. In our algorithm, we consider p_c to be one. Hence the Schema theorem becomes,

$$M(H,t+1) \geq [M(H,t)\frac{f(H)}{f_{avg}}(1-losses)+gains](1-p_m)^{O(H)}$$

In the proposed algorithm, as the population size varies each generation, the schema lower bound for the varying population size becomes,

$$M(H,t+1) \geq [\frac{M(H,t)}{n_t} n_{t+1} \frac{f(H)}{f_{avg}}(1 - losses)$$

$$+gains](1 - p_m)^{O(H)} \tag{2.4}$$

where n_t is population size at generation t.

If we consider loss to be any crossover that disrupts the schema, then our calculation of gain must account for the preservance of the schema when both parents are of the schema H. Now for an n point crossover to be non-disruptive, even number of crossover points, only can occur between fixed bits of schema [31] as shown in Figure 2.1. The remaining crossover points must be outside defining length. Hence the probability of n point crossover generating only even number of crossovers between fixed bits for a schema of k order hyperplane is $P_{k,even}$. Probability of disruption P_d of n point crossover is bounded by,

$$P_d(n, H_k) \leq 1 - P_{k,even}(n, H_k)$$

Figure 2.1 shows an n point crossover which is non disruptive. d_1, d_2 and d_3 are fixed bits, L_1 and L_2 are distances of d_3 and d_2 from d_1 respectively and L is the string length. As a special case, probability of even number of crossover points falling between fixed bits for a second order hyperplane is given by [32]

$$P_{2,even}(n, L, L_1) = \sum_{i=0}^{\frac{n}{2}} {}^nC_{2i} \frac{L_1}{L}^{2i} \frac{L - L_1}{L}^{n-2i} \tag{2.5}$$

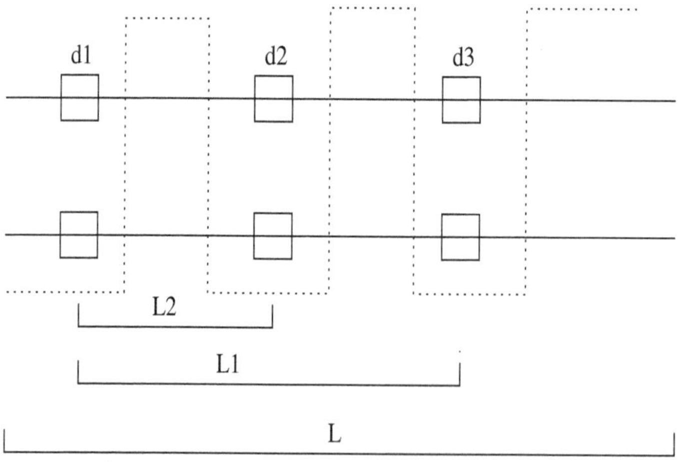

Fig. 2.1 Non-Disruptive n point crossover

That is, the probability is given by the product of number of ways of choosing an even number of points from an n point crossover, the probability of placing an even number of points between the two defining points and the probability of placing the other points outside the defining points. L here is the string length and L_1 is the defining length.

We can extend the probability of disruption for a kth order hyperplane as

$$P_{k,even}(n,L,L_1,...,L_{k-1}) =$$

$$\sum_{i=0}^{\frac{n}{2}} {}^nC_{2i} \frac{L_1}{L}^{2i} \frac{L-L_1}{L}^{n-2i} P_{k-1,even}(n,L_1,L_2,...,L_{k-1}) \tag{2.6}$$

That is, probability that an even number of crossover points fall between k defining bits $P_{k,even}$ is given by the probability that even number of crossover points fall between the first two defining bits and the rest of the points fall outside the defining bits into $P_{k-1,even}$. Hence, taking bound on the probability of disruption

$$P_d(H_k) \leq$$

$$1 - \sum_{i=0}^{\frac{n}{2}} {}^nC_{2i} \frac{L_1}{L}^{2i} \frac{L-L_1}{L}^{n-2i} P_{k-1,even}(n,L_1,...,L_{k-1}) \tag{2.7}$$

Now, as mentioned earlier, the lower bound on the gain is given by the preservance of schema when disruptive crossover occurs between two parents both following the same schema. Hence, gain is given by $gains \geq n * P_d *$, Probability that P_1 and P_2 are in schema H

After selection, the number of individuals in schema H is given by $M(H,t)\frac{f(H)}{f_{avg}}$. Total number of individuals in a population is n. Hence probability that given a parent, it is in schema H is given by $\frac{M(H,t)}{n}\frac{f(H)}{f_{avg}}$. Hence the gain is obtained as,

$$gains \geq n_{t+1} P_d \frac{M(H,t)}{n_t} \frac{f(H)}{f_{avg}} \frac{M(H,t)}{n_t} \frac{f(H)}{f_{avg}} \tag{2.8}$$

Using this lower bound on $gains$ in Equation (2.4) and replacing loss with disruption

$$M(H,t+1) \geq [\frac{M(H,t)}{n_t} n_{t+1} \frac{f(H)}{f_{avg}} (1-P_d)$$

$$+ n_{t+1} P_d (\frac{M(H,t)}{n_t} \frac{f(H)}{f_{avg}})^2](1-p_m)^{O(H)}$$

Simplifying,

$$M(H,t+1) \geq$$

$$\frac{M(H,t)}{n_t} n_{t+1} \frac{f(H)}{f_{avg}} [1 - P_d + P_d (\frac{M(H,t)}{n_t} \frac{f(H)}{f_{avg}})](1-p_m)^{O(H)}$$

But $n_t f_{avg} = \sum f$ for generation t. Therefore, we obtain the schema theorem as

$$M(H,t+1) \geq$$

$$\frac{M(H,t)}{n_t} n_{t+1} \frac{f(H)}{f_{avg}} [1 - P_d(1 - \frac{M(H,t)f(H)}{\sum f})](1-p_m)^{O(H)} \tag{2.9}$$

This is the schema theorem that deals with a single population. An ecosystem of population where populations with better fitness have more number of individuals than those with low fitness population is considered in our algorithm. Consider a low fitness population in the low yielding region of the search space. Consider the case when this population comes across a high fitness point in the search space. Let $\bar{H}$ be the schema with high fitness that comes across the low fitness population. All other individuals in population have a low fitness compared to this high fitness point and hence

$$f(\bar{H}) = \sum f$$

Applying this to Equation (2.9) we get

$$M(\bar{H},t+1) \geq$$

$$\frac{M(\bar{H},t)}{n_t} n_{t+1} \frac{f(\bar{H})}{f_{avg}} [1 - P_d(1 - M(\bar{H},t))](1-p_m)^{O(\bar{H})}$$

Since we have only found a single point with high fitness, $M(\bar{H},t) = 1$. Therefore

$$M(\bar{H},t+1) \geq \frac{M(\bar{H},t)}{n_t} n_{t+1} \frac{f(\bar{H})}{f_{avg}} (1-p_m)^{O(\bar{H})}$$

Thus, there is no disruption. $f_{avg} = \sum f / n_t$ and $f(\bar{H}) = \sum f$ Therefore

$$M(\bar{H},t+1) \geq \frac{M(\bar{H},t)}{n_t} n_t n_{t+1} (1-p_m)^{O(\bar{H})}$$

But $M(\bar{H},t) = 1$. Therefore

$$M(\bar{H},t+1) \geq n_{t+1}(1-p_m)^{O(\bar{H})}$$

Thus if a population P_i in the low fitness region comes across even one high fitness solution, in the very next generation approximately $n_{i,t+1}(1-p_m)^{O(\bar{H})}$ of the $n_{i,t+1}$ individuals of the population in the next generation are in schema $\bar{H}$. This immediate drift towards high fitness region of a low fitness population, suggests the robust performance of the method. The assumption that the fitness of the best solution is approximately equal to sum of fitness of all individuals in the population is used to display the power of the algorithm when the entire population is in a very low fitness area of search space. This analysis is a proof for the performance of the algorithm in the worst case. In general, the drift towards high fitness region is faster when the population lies in a low fitness region. The high fitness populations of the ecosystem drive the low fitness populations to perform a more explorative search while the high fitness populations perform a concentrated exploitational search in the high yielding regions. As soon as the low fitness populations find a better region, the individuals in the population crowd this region of the search space. This competitive nature of populations ensures robust performance. The search conducted by the genetic algorithm can thus be explained as exploitational with respect to high yielding regions of the search space and explorative with respect to other regions. The adaptive updates to the parameters of number of points of crossover and rate of mutation for each population are responsible for the explorative nature of the low fitness population. The adaptive parameter of population size helps in concentrated search in high fitness region of search space. The term explorative indicates that larger area of search space is covered by the search. There is no doubt that as mutation rate increases, the search is more explorative.

Theorem 2.1. *The total number of individuals in the ecosystem is a constant; that is,*

$$\sum_{i=1}^{np} n_{i,t} = \sum_{i=1}^{np} n_{i,t+1} = constant$$

Proof. Let the initial number of individuals in the ecosystem be n_{sum}. Therefore,

$$\sum_{i=1}^{np} n_{i,0} = n_{sum} = constant$$

Now at generation t, let total number of individuals in the ecosystem be n'_{sum}. Therefore,

$$\sum_{i=1}^{np} n_{i,t} = n'_{sum}$$

In the next generation $t+1$, apply the update given by Equation (2.1) to all the populations, then the total number of individuals in the ecosystem in generation $t+1$ is given by,

$$\sum_{i=1}^{np} n_{i,t+1} = \sum_{i=1}^{np} (n_{i,t} + \frac{f(P_i)}{\bar{f}} - 1)$$

$$\sum_{i=1}^{np} n_{i,t+1} = \sum_{i=1}^{np} n_{i,t} + \sum_{i=1}^{np} \frac{f(P_i)}{\bar{f}} - \sum_{i=1}^{np} 1$$

Therefore

$$\sum_{i=1}^{np} n_{i,t+1} = n'_{sum} + \frac{1}{\bar{f}} \sum_{i=1}^{np} f(P_i) - np$$

But

$$\bar{f} = \frac{\sum_{i=1}^{np} f(P_i)}{np}$$

Therefore

$$\sum_{i=1}^{np} n_{i,t+1} = n'_{sum} + np - np = n'_{sum} = \sum_{i=1}^{np} n_{i,t}$$

Substituting $t = 0$ we get

$$\sum_{i=1}^{np} n_{i,0} = \sum_{i=1}^{np} n_{i,t} = n'_{sum} = n_{sum} = constant$$

Thus, the number of individuals in the ecosystem is constant, i.e., the algorithm just groups these individuals into different populations such that, individuals in the same population reproduce and show co-operation while individuals in different populations compete to perform better. Thus the algorithm displays interpopulation competition and intra population co-operation.

2.5.1 Convergence Analysis

The convergence rate of a genetic algorithm is defined in this chapter as the rate at which the hamming distance between the individuals of ecosystem becomes zero. In other words, convergence occurs when all individuals of population have the same genotype. Therefore, the convergence rate of the genetic algorithm is the rate at which the number of copies of this individuals increases till all the individuals of the ecosystem have same genotype (zero hamming distance). Hence we can derive a bound on the rate of convergence by fixing rate of increase of expectation value of individual shown in Equation 2.9.

Before deriving rate of change of expectation value consider the update to number of individuals in a population. From Equation 2.1 we have,

$$n_{i,t} = n_{i,t-1} + \frac{f(P_{i,t-1})}{\bar{f}_{t-1}} - 1$$

Expanding the recursive equation for t generations we get,

$$n_{i,t} = n_{i,0} + \sum_{j=0}^{t-1} \frac{f(P_{i,j})}{\bar{f}_j} - t \tag{2.10}$$

Now, the use of Elitist selection guarantees the convergence of the genetic algorithm. During the process of convergence under steady state, a population is either increasing or decreasing in population size. This suggests a linear relation of the summation term in Equation 2.10 and t. Let the coefficient of linear term be β. Therefore,

$$n_{i,t} = n_{i,0} + \beta t - t$$

Therefore, the population size is linearly dependent on number of generations. Thus,

$$n_{i,t} = n_{i,0} + (\beta - 1)t \tag{2.11}$$

where β is some constant.

Applying this in Equation 2.9 we get,

$$M^i(H,t) \geq M^i(H,t)(n_{i,0} + (\beta - 1)t)$$

$$\frac{f(H)}{\sum \bar{f} n_{i,t-1}}[1 - P_d(1 - \frac{M^i(H,t)f(H)}{\sum f_{i,t}})](1 - p_m)^{O(H)} \tag{2.12}$$

where $M^i(H,t)$ is the expectation value of schema H in population P_i. Let

$$f(H) = \bar{f} + c\bar{f}$$

where c is some constant. That is the average fitness of schema H is above the average fitness of the population by a factor c. Therefore,

$$\frac{f(H)}{\bar{f}} = constant$$

Now,

$$n_{i,t-1} = n_{i,0} + (\beta - 1)(t - 1)$$

Now to calculate a bound on convergence rate, we use the Θ notation. As $n_{i,t}$ and $n_{i,t-1}$ are linearly dependent on t, in the calculation of order of convergence they cancel each other out. Mutation just induces small randomness into the search and it does not alter the order of convergence but only the actual convergence rate.

From Equation 2.12 the order of convergence $\Theta(M^i(H,t))$ as,

$$\Theta(M^i(H,t)) = \Theta(\alpha M^i(H,t-1)^2)$$

Expanding the recursion of $M^i(H,t-1)$ we get,

$$\Theta(M^i(H,t)) = \Theta(\alpha^3 M^i(H,t-2)^4)$$

Further expanding the recursive relation we get,

$$\Theta(M^i(H,t)) = \Theta(\alpha^7 M^i(H,t-3)^8)$$

Finally, expanding the entire recursive relation we get,

$$\Theta(M^i(H,t)) = \Theta(\alpha^{1+2+4+\ldots+2^{t-1}})$$

As the term in power is the summation of geometric series with factor two, we get,

$$\Theta(M^i(H,t)) = \Theta(\alpha^{2^t-1})$$

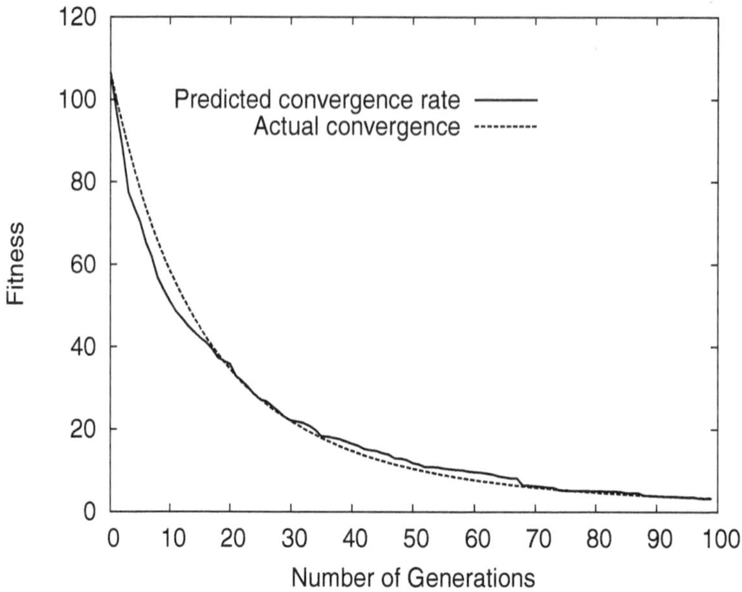

Fig. 2.2 Plot of actual versus predicted convergence

Thus the rate of convergence of the algorithm is of the order,

$$\Theta(M^i(H,t)) = \Theta(\alpha^{2^t})$$

Clearly this is the case of *exponential of exponential* order. This shows that the algorithm has a much better convergence rate as compared to simple GA.

The graph in Figure 2.2 shows a plot of convergence of the proposed Genetic Algorithm for the first De Jongs test bed function (F1) and the convergence rate derived. From the graph, it is clear that the convergence rate of the algorithm is close to predicted convergence rate. The actual convergence rate is compared with the function,

$$C(x) = e^{e^x}$$

Such that,

$$1.54078 \geq x \leq 0.163734$$

This clearly proves that the actual convergence rate of the algorithm is indeed exponential of exponential as predicted from mathematical analysis.

2.6 Experiments

Each population of the ecosystem can be run in parallel on different processors as the algorithm is a parallel genetic algorithm. However, the algorithm is implemented on a single processor system. Figure 2.3 shows a view of the ecosystem

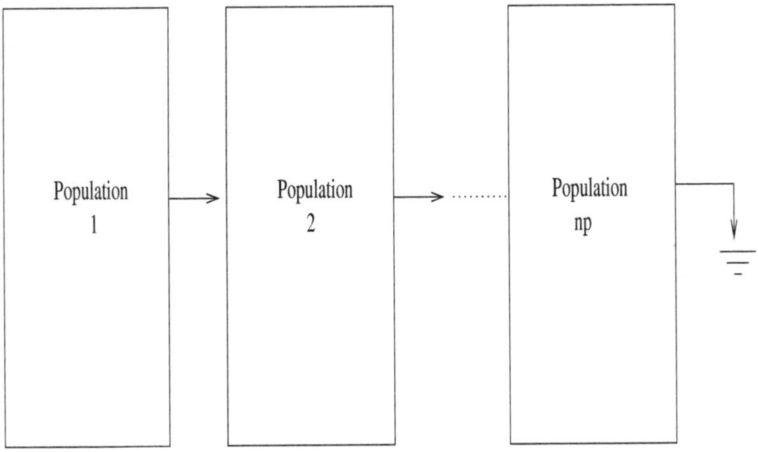

Fig. 2.3 Implementation of Ecosystem

implemented. The ecosystem consists of a linked list of populations and variables to hold average fitness of population and average number of individuals in a population. The linked list structure helps in an easy implementation of the phenomenon of extinction where, when the population becomes extinct that node is simply removed.

Each population which is a node of the linked list is a structure consisting of the population parameters like, number of points of crossover, population size and rate of mutation. Further, the population also contains an array of bit strings which represent the individuals of the population. The genotype to phenotype conversion and fitness evaluation is performed by a fitness function.

Table 2.4 Testbed functions used

F1	$f(x) = \Sigma_1^{10} x_i^2$ $-5.12 \leq x_i \leq 5.12$
F2	$f(x) = 100(x_1^2 - x_2^2) + (1 - x_1)^2$ $-2.048 \leq x_i \leq 2.048$
F3	$f(x) = 200 + \Sigma_{i=1}^{10} x_i^2 - 10cos(2\pi x_i)$ $-5.12 \leq x_i \leq 5.12$
F4	$f(x) = \Sigma_{i=1}^{10} x * sin(x)$ $-5.12 \leq x_i \leq 5.12$

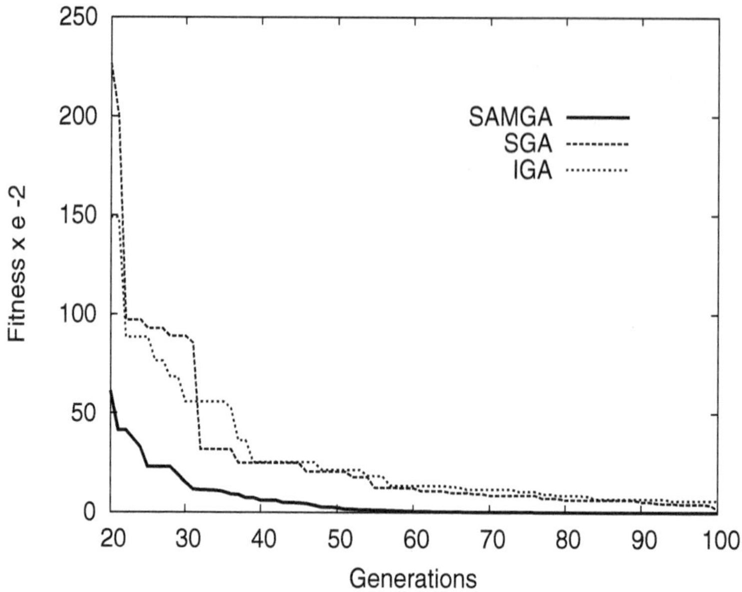

Fig. 2.4 Convergence for function F1

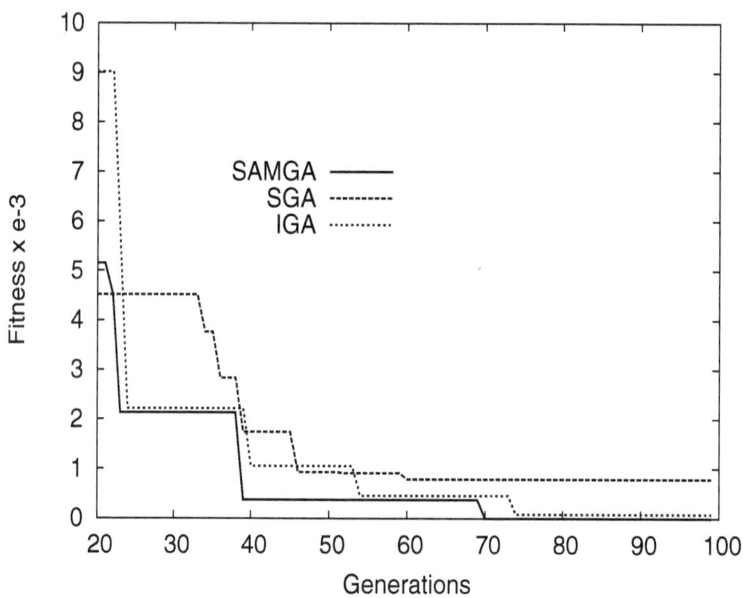

Fig. 2.5 Convergence for function F2

To evaluate the performance of the algorithm, it is used to find the minimal points of complex high dimensional landscapes obtained using the testbed functions shown in Table 2.4. Function F1, which is the De Jong's function 1, is a unimodal function with one distinct peak. Function F2 (Rosenberg function), is a function which is basically an almost flat valley and hence finding the minimal point becomes a difficult problem. Function F3 (Rastragin function) is a multimodal function with many peaks. The function is a good testbed function as any algorithm has a good chance of getting stuck at one of the peaks which is a local minima. Function F4 is again a multimodal function. In the experiments conducted, the IGA and SAMGA both had 10 populations of 60 individuals each. The SGA had 600 individuals in its population. For SGA and IGA the crossover rate chosen is 1 and mutation rate 0.01.

The plot in Figure 2.4 shows the convergence of our algorithm(SAMGA), island model(IGA) and simple genetic algorithm(SGA) for the function F1. It is observed that the Self - Adaptive Migration GA(SAMGA) converges much faster than the other algorithms. The plot in Figure 2.5 shows the convergence of the three algorithms for Rosenberg function. As our function is low dimensional, the fitness is multiplied by 1000 before plotting. Similarly, the plot in Figure 2.6 shows the convergence for Rastragin function. This is a multimodal function with many peaks of almost equal heights. In both the cases, SAMGA outperforms IGA and SGA.

The plot in Figure 2.7 shows the convergence for function F4. Here SAMGA and SGA have similar performance and both outperform IGA. The graphs in Figures 2.8 through 2.11, shows the convergence rate of all the three algorithms for varying generations and functions. In all these cases, the performance of SAMGA is most significant in the later generations nearer to convergence.

To test how the explorative power of the proposed algorithm helps in overcoming GA - deception the algorithm along with simple GA and Island GA is used to solve ugly 3 bit deceptive function. With 250 individuals in four populations, SAMGA is able to overcome deception in 150 generations, whereas IGA overcame deception only after 228 generations. With 1000 individuals SGA is able to overcome deception in 557 generations. This experiment proves that the explorative nature on the lower fitness populations in SAMGA helps in overcoming GA deception better.

Example Search: Consider the landscape given by the equation

$$y = (x-1) * cos(x-1)$$

The proposed GA is used on this landscape with an initialization of three populations of four individuals each. Figure 2.12 shows the function plot along with the initial set of individuals in the ecosystem. Figure 2.13 shows the search result after 10 generations and Figure 2.14 shows the search result after 20 generations. The number of individuals present is not clear from the graph as in the 10th and 20th generations, many individuals have the same bit string. However, from Lemma 1, the total number of individuals is the same in ecosystem. From the result it can be seen that from the initial random distribution, in the 10th generation individuals get settled at the first and second peaks (one individual at the lower peak and eleven

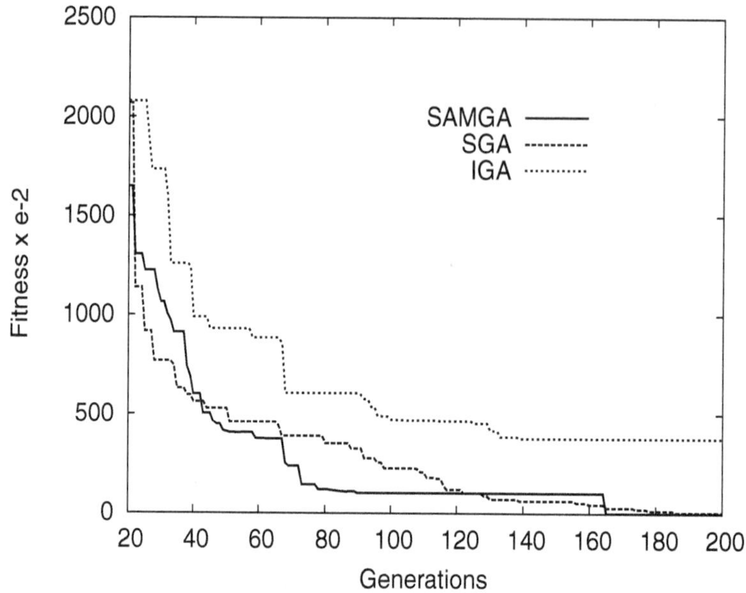

Fig. 2.6 Convergence for function F3

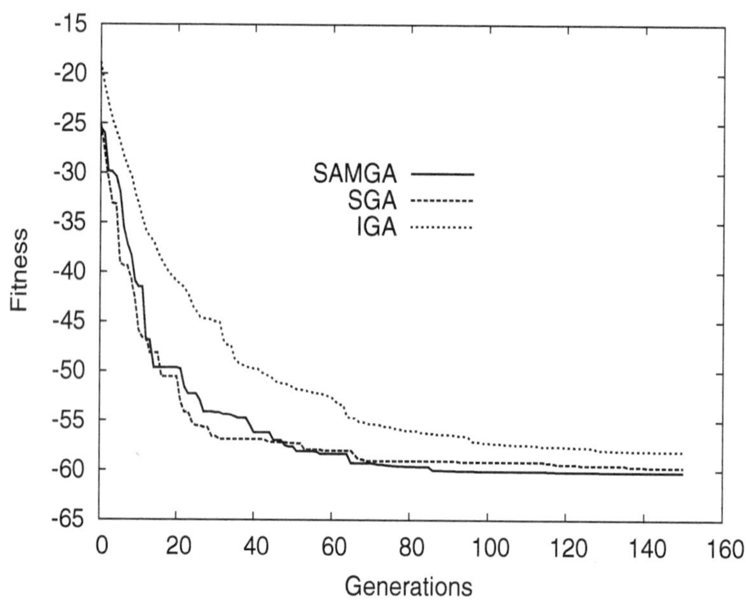

Fig. 2.7 Convergence for function F4

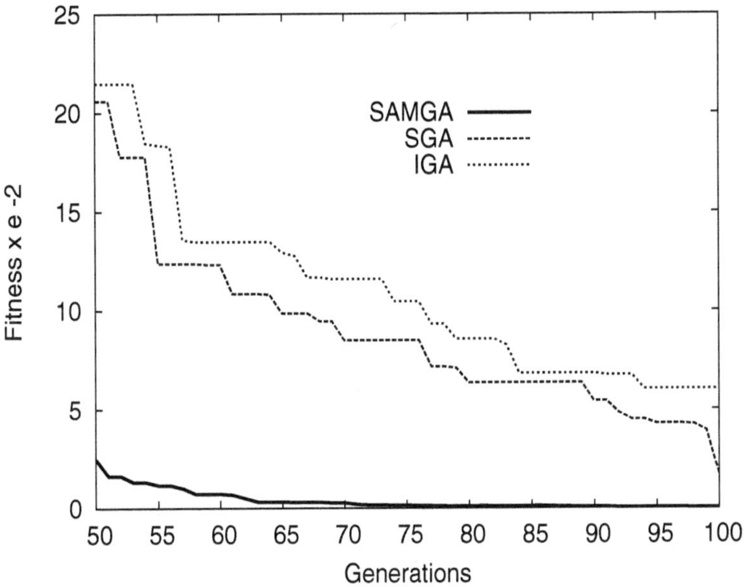

Fig. 2.8 Convergence for function F1

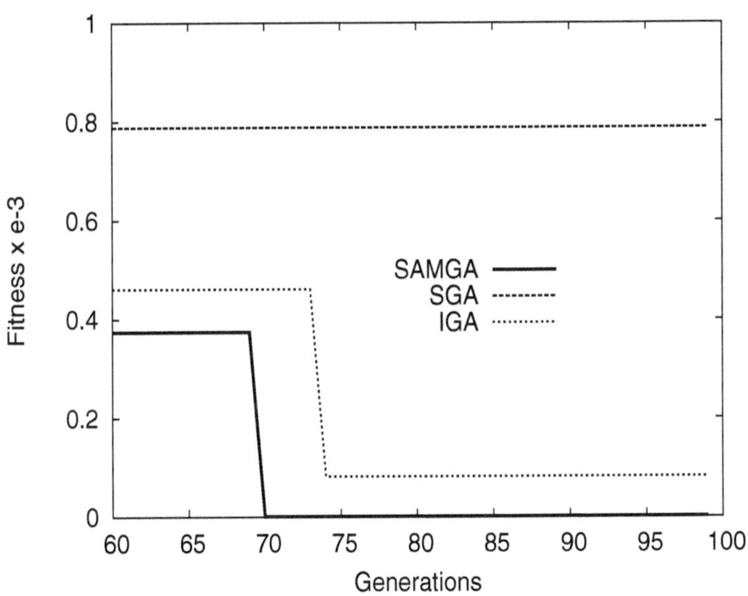

Fig. 2.9 Convergence for function F2

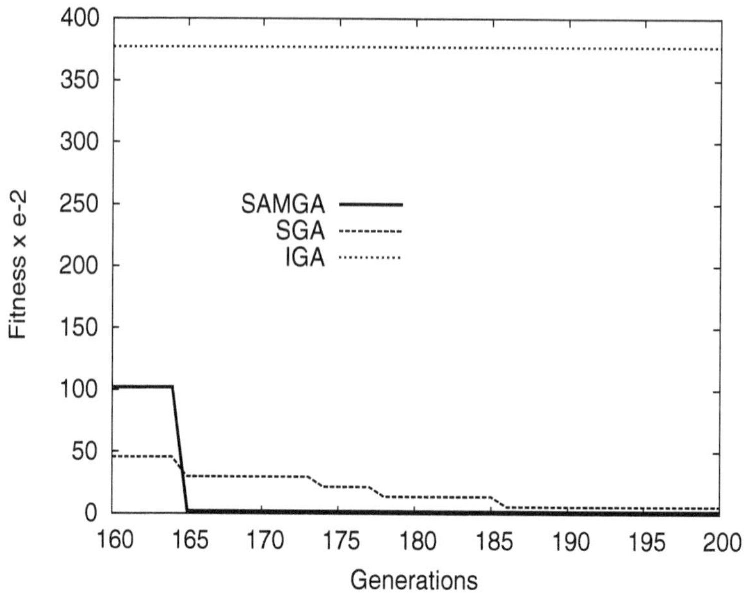

Fig. 2.10 Convergence for function F3

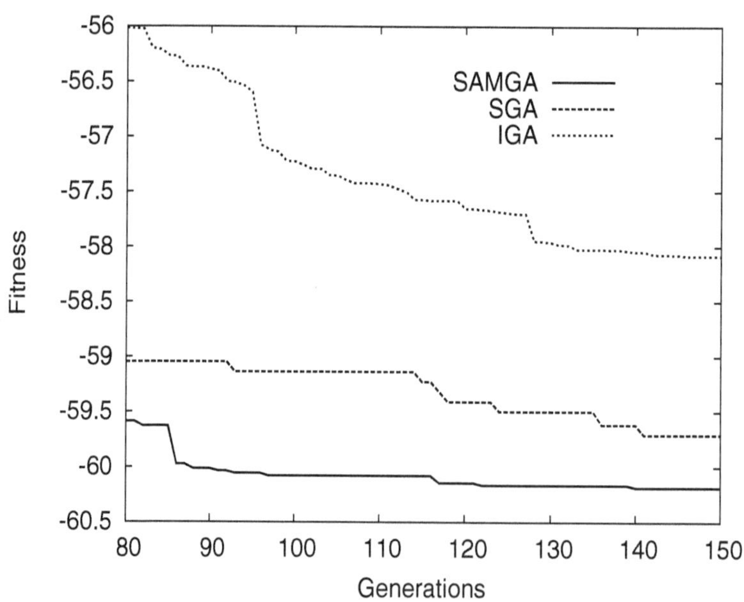

Fig. 2.11 Convergence for function F4

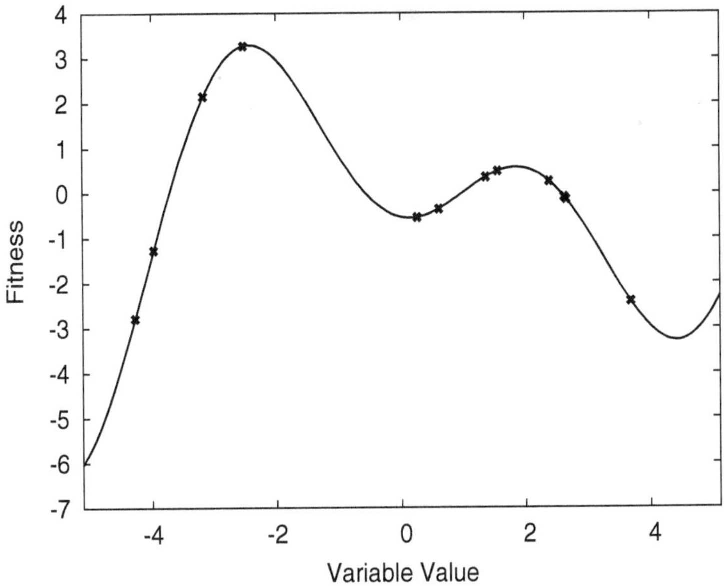

Fig. 2.12 Example : Initial set of Individuals

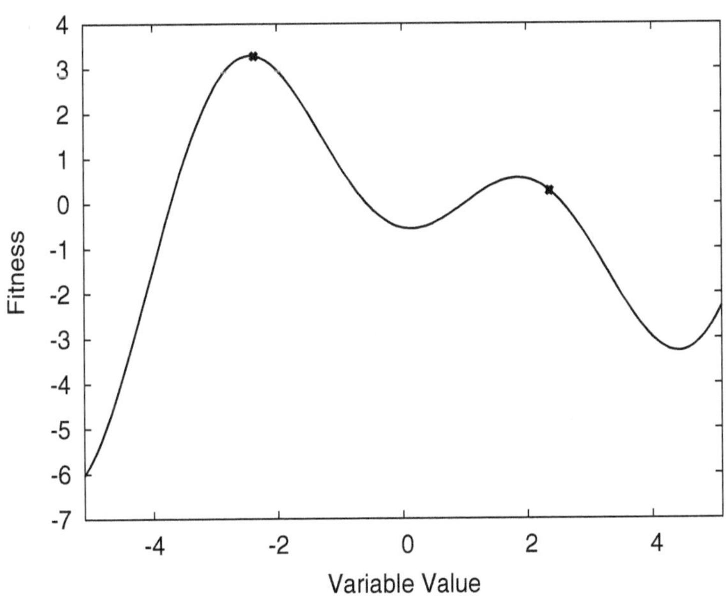

Fig. 2.13 Example : Individuals after 10 generations

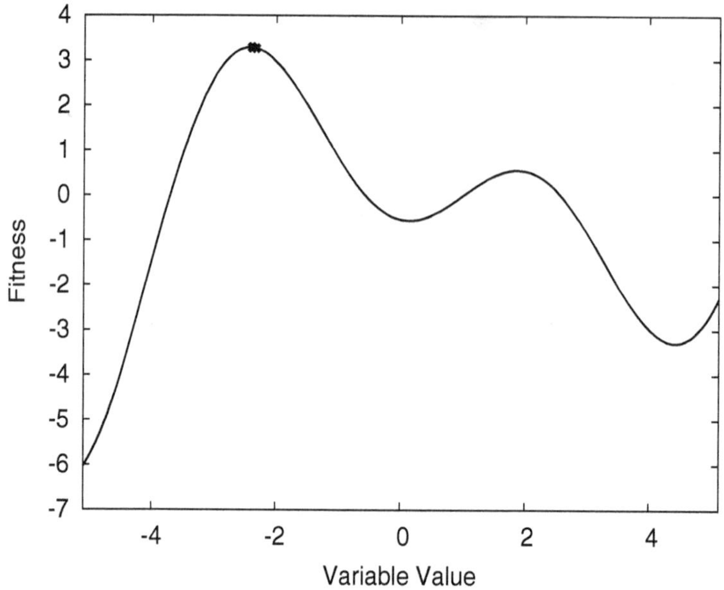

Fig. 2.14 Example : Individuals after 20 generations

at higher peak). Finally, in the 20th generation, all the individuals get settled at the higher peak which is the maxima.

2.7 Performance Analysis

To evaluate the performance of the algorithm on real data mining applications, a Michigan Style classifier is built (see Appendix for details on classifier system) and the classifier is tested on Datamining applications from the UCI Machine learning repository. The training set used is the first 40% of the data. The next 60% is used for testing. The various data mining applications chosen for the test are

- *Pima Indian Diabetes* (PID) Database with 768 cases, 2 classes and 8 attributes
- *Wisconsin Breast Cancer Database* (Wisc) with 699 entries , 2 classes and 9 attributes
- *Hepatitis Database* (Hep) with 155 entries , 2 classes and 19 attributes
- *Ionosphere Database* (Ion) with 351 entries , 2 classes and 34 attributes

The different methods that are compared are given below. For a survey of these methods on machine learning problems refer [33]:

- *Self-Adaptive Migration GA* (SAMGA) - The proposed algorithm.
- *Learning Vector Quantization* (LVQ) - Establishes number of codebooks to approximate various domains of input vector by quantized values.

Data Set \ Methods	SAMGA	ITI	LVQ	LMDT	Bpnn	RBF
Pima	74.6	73.16	71.28	73.51	73.4	71.6
Wisc	96.28	91.14	94.82	95.74	95.8	94.9
Ion	89.4	93.65	88.58	86.89	86.6	87.6
Hep	83.4	75.93	78.3	84.59	75.4	77

Fig. 2.15 Accuracy of the Classification Results

- B*ackpropagation Neural Network* (Bpnn) - A neural network that uses feedback of errors for learning.
- *Radial Basis Function neural network* (RBF) - It is a Neural network
- *Incremental Decision Tree Induction* (ITI) - Builds decision tree incrementally as new training instances are provided.
- *Linear Machine Decision Tree* (LMDT) - Uses multi-class decision trees with multivariate tests at internal decision nodes.

The table in Figure 2.15 summarizes the performance of SAMGA and other machine learning techniques. It can be seen that SAMGA has consistently good performance and relatively high classification rates as compared to other methods. It can be seen that the performance of the methods vary from application to application but SAMGA gives consistent good performance for all the applications.

Similarly, the performance of SAMGA after 10 fold cross validation of the datasets used is given in Figure 2.16. It can be seen that SAMGA shows better performance when compared with the other techniques even after the 10 fold cross validation of the datasets.

Data Set \ Methods	SAMGA	ITI	LVQ	LMDT	Bpnn	RBF
Pima	73	71.6	70	72.51	72.22	70.16
Wisc	94.1	90.40	93.24	93.44	93.80	92.49
Ion	86.2	91.45	85.12	85.90	84.77	84.76
Hep	84.7	72.32	74.31	83.45	74	75.70

Fig. 2.16 Accuracy of the Classification Results after 10 fold cross validation

2.8 A Heuristic Template Based Adaptive Genetic Algorithms

In this section, the concept of heuristic templates for adaptive Genetic Algorithms is presented. The crossover and mutation are totally random processes in the existing genetic algorithms. Here, we use the heuristics to adapt the crossover and mutation for better results. This increases the convergence rate. In our algorithm, as the GA proceeds through the generations, we update a template, which is proportional to the average individual in the population over the generations. Comparing the template and individual to crossover, we derive the point at which, when the crossover is done we get best results. The same is done for mutation. The proposed scheme is useful for applications like pattern recognition and data classification. The genetic algorithm described here is an adaptive GA where crossover and mutation for an individual of a particular population is performed based on heuristic templates saved for that population.

2.8.1 Problem Definition

Let S be made up of all the 2^k, k bit binary numbers representing the entire search space. Given the objective function f which is a mapping $f : S \to R$ where R is a set of real numbers, the problem is to find an $x^* \in S$ such that $f(x^*) \geq f(x)\ \forall\, x \in S$. Let $E = \{P_1, P_2 ... P_{np}\}$ be an ecosystem with np populations in it. Populations $P_1, P_2 ... P_{np} \subset S$. $P_{i,j}$ stands for the jth individual of population P_i, clearly $P_{i,j} \in S$. $P_{i,j}[k]$ is the bit value of the kth bit of individual $P_{i,j}$. Let n be the size of each population P_i. Let p_m be the rate of mutation. Let $T_1, T_2 ... T_{np}$ be the heuristic templates of populations $P_1, P_2 ... P_{np}$ respectively. The heuristic templates is a vector of float values with length equal to the string length of individuals. The pseudocode for the the proposed adaptive GA is given in Table 2.5.

2.9 Example

Consider the problem of minimizing function $f(x) = x^2$. Let us use a single population with 4 individuals in it to solve the problem. Let the initial population consist of individuals

$$P_{1,1} = 5 = 0101_b, P_{1,2} = 11 = 1011_b, P_{1,1} = 6 = 0110_b, and \quad P_{1,2} = 8 = 1000_b$$

Now for first generation the template for the two population is $T_1 = 0,0,0,0$. After proportionate selection using elitist scheme , crossover and mutation we get,

$$P_{1,1} = 5 = 0101_b, P_{1,2} = 13 = 1101_b, P_{1,1} = 3 = 0011_b, and \quad P_{1,2} = 6 = 0110_b$$

Now similarity template becomes $T_1 = -0.5, 0.5, 0, 0.5$. Hence the weights for roulette wheel for crossover becomes $pnts = 1.167, 1, 1.167$. Thus there is a bias towards choosing the point between first two or last two bits for crossover rather than the middle point. Thus after selection, crossover and mutation we have

$$P_{1,1} = 3 = 0011_b, P_{1,2} = 5 = 0101_b, P_{1,1} = 2 = 0010_b, and \quad P_{1,2} = 7 = 0111_b$$

Table 2.5 Heuristic Template based Adaptive Genetic Algorithms(HTAGA)

begin
for i = 1 : np, **do**
(a) Set heuristic template vector T_i to 0 vector or origin
(b) Initialize all individuals of population P_i to some random bit strings
(c) Set mutation rate to some arbitrary small value
(d) Set migration rate to some arbitrary value
next
for gen = 1 : maximum generation limit, **do**
for i = 1 : np, **do**
for j = 1 : n, **do**
for k = 1 : string length of individuals, **do**
if $P_{i,j}[k] == 1$
$T_{i,j}[k] = T_{i,j}[k] + \frac{1}{n}$
else
$T_{i,j}[k] = T_{i,j}[k] - \frac{1}{n}$
endif
next
next
Evaluate fitness of all the individuals of the population P_i and find $f(P_i)$ the best fitness of the population (best solution of that population for the generation).

Perform elitist selection for population P_i.
for j = 1 : string length of individuals - 1, **do**
pnts[j] = 0;
for k = j+1 : string length of individuals, **do**
$pnts[j] = pnts[j] + \frac{abs(T_k)}{(k-j)}$
next
for k = j : 0, **do**
$pnts[j] = pnts[j] + \frac{abs(T_k)}{(k-j)}$
next
next
(i) Perform crossover on selected individuals of population P_i using crossover point selected using roulette wheel selection whose weights for each point is given by the vector pnts.

(ii) Perform mutation on individuals of population P_i with mutation probability pm_i and for mutation, set bit value of the position as 1 if template value is negative at that point and set bit value 0 if template value is positive at that point.
(iii) If template value is 0 invert the bit.
next
if gen == multiple of migration rate
Exchange or migrate best individuals between populations.
endif
next
end

Now similarity template becomes $T_1 = -1.5, 0.5, 0.5, 1$. Hence the weights for rolled wheel for crossover becomes $pnts = 2.583, 2.25, 2.25$. Hence after selection, crossover and mutation we have

$$P_{1,1} = 2 = 0010_b, P_{1,1} = 3 = 0011_b, P_{1,2} = 5 = 0101_b, and \quad P_{1,2} = 4 = 0100_b$$

Now similarity template becomes $T_1 = -2.5, 0.5, 0, 1$. Hence the weights for roulette wheel for crossover becomes $pnts = 3.33, 2.25, 1.83$. Here clearly there is a strong bias towards crossover between first two bits and a slightly lesser bias towards crossover between second and third bit. Hence after selection, crossover and mutation we have

$$P_{1,1} = 2 = 0010_b, P_{1,1} = 3 = 0011_b, P_{1,2} = 5 = 0101_b, and \quad P_{1,2} = 0 = 0000_b$$

Thus we see that we have reached the global minima. For a noisy and peaky search space it can be seen that the heuristic template works even better.

2.10 Performance Analysis of HTAGA

In the experiments conducted, the IGA and HTAGA both have 10 populations of 60 individuals each. The SGA had 600 individuals in its population. For SGA and IGA the crossover rate chosen is 1 and mutation rate 0.01.

The plot in Figure 2.17 shows the convergence of our algorithm HTAGA, IGA and SGA for the function F1. It is observed that the HTAGA converges much faster than the other algorithms. The plot in Figure 2.18 shows the convergence for Rastragin function. Even though this is a multimodal function with many peaks of almost equal heights, HTAGA outperforms IGA and SGA.

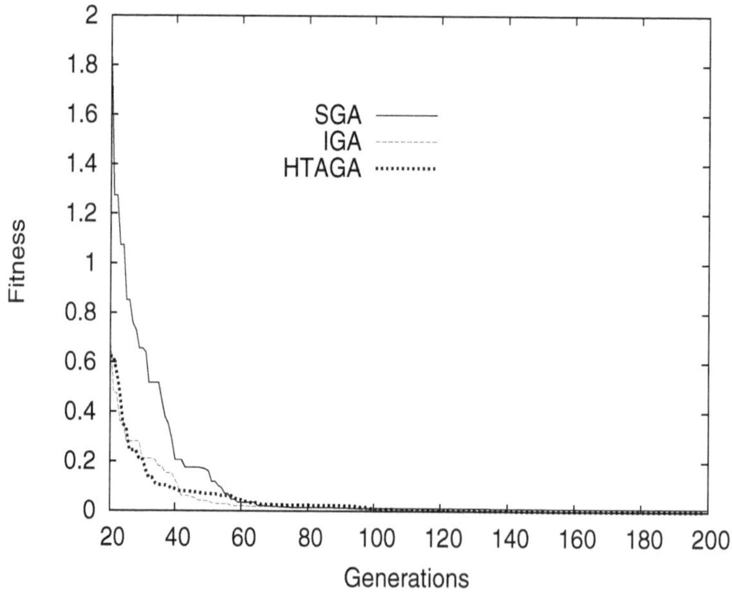

Fig. 2.17 Convergence for function F1

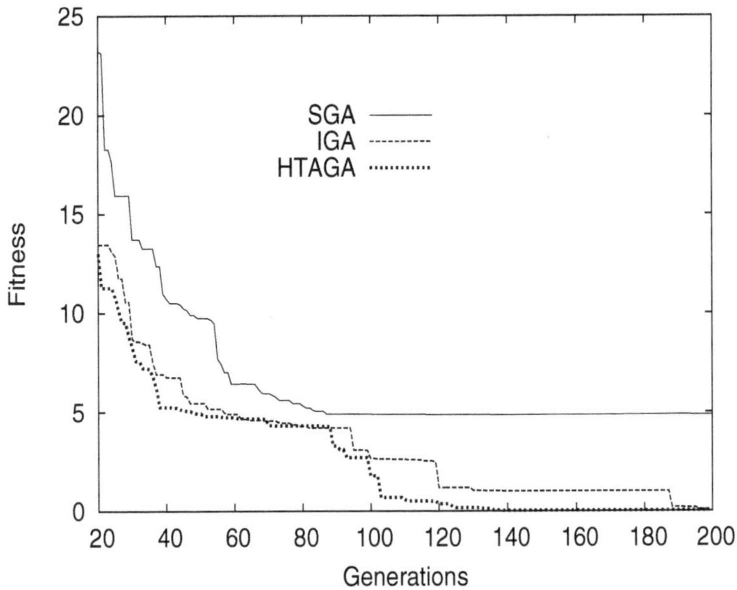

Fig. 2.18 Convergence for function F3

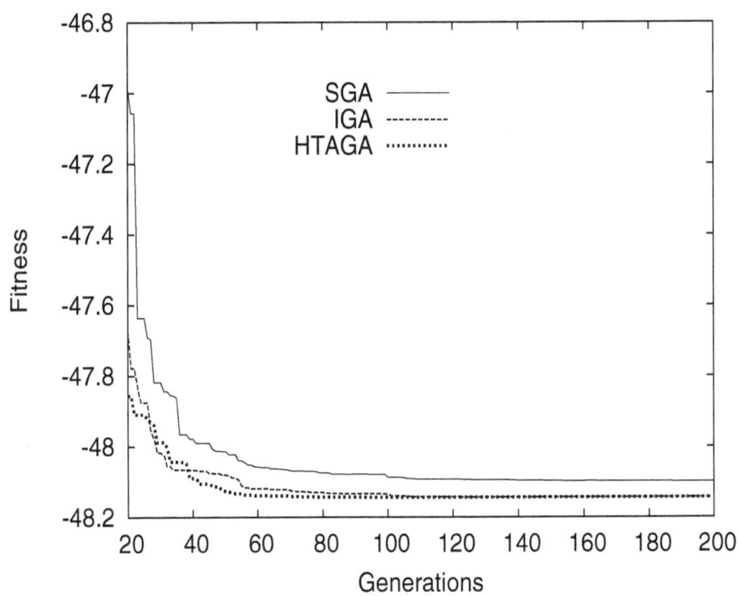

Fig. 2.19 Convergence for function F4

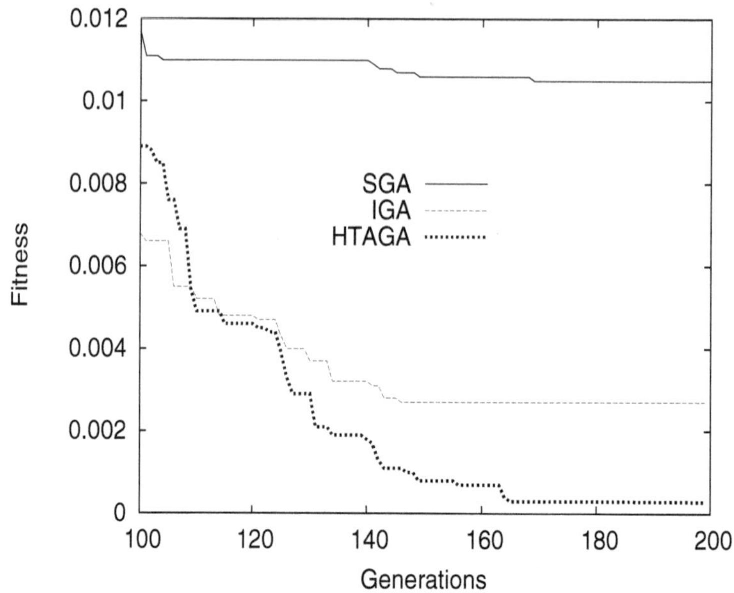

Fig. 2.20 Convergence for function F1

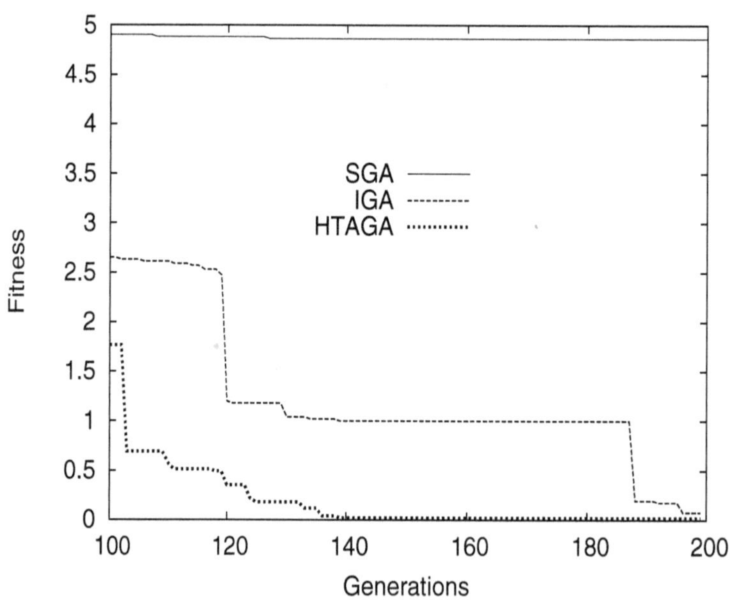

Fig. 2.21 Convergence for function F3

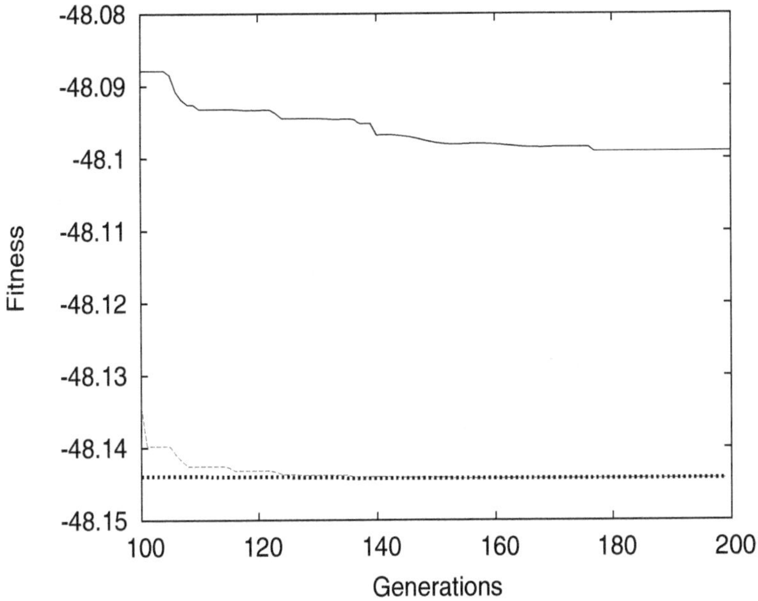

Fig. 2.22 Convergence for function F4

Data Set \ Methods	HTAGA	ITI	LVQ	LMDT	Bpnn	RBF
Pima	73.7	73.16	71.28	73.51	73.4	71.6
Wisc	93.12	91.14	94.82	95.74	95.8	94.9
Ion	87	93.65	88.58	86.89	86.6	87.6
Hep	81.2	75.93	78.3	84.59	75.4	77

Fig. 2.23 Accuracy of the Classification Results for HTAGA

The plot in Figure 2.19 shows the convergence for function F4. Here HTAGA and SGA have similar performance and both outperform IGA. The graphs in Figures 2.20 through 2.22, shows the convergence rate of all the three algorithms for varying generations and functions. In all these cases, the performance of HTAGA is most significant in the later generations nearer to convergence.

The table in Figure 2.23 summarizes the performance of HTAGA and other machine learning techniques. It can be observed that HTAGA has consistently good performance and relatively high classification rates as compared to other methods. It can be seen that the performance of the methods vary from application to application but HTAGA gives consistent good performance for all the applications. From tables in Figure 2.16 through 2.23 it can be observed that SAMGA performs well when compared to HTAGA.

2.11 Summary

Genetic Algorithms have proved to be a robust general purpose search technique. They have a central place in data mining applications due to their ability to search large search spaces efficiently. In this chapter, we have proposed a Self-Adaptive Migration GA search techniques, have two central but competing concepts of exploitation and exploration. The proposed algorithm can be characterized by focused and deeper exploitation of the heuristically promising regions of the search space and wider exploration of other regions of the search space. The algorithm achieves this by varying the number of individuals in a population which helps in better exploitation of high fitness regions of the search space by increasing the number of individuals in the region. The algorithm also helps in better exploration by increasing the number of crossover points and the mutation rates for the low fitness population.

The chapter provides a mathematical analysis of the method using schema theorem which accounts for lower bound on the gain. It has been shown that the method proposed assures that, when a low fitness population of the ecosystem stumbles across a high fitness solution due to its explorative nature there is an immediate drift of that population towards the high fitness region of search space.

The proposed method has been tested on a set of testbed functions and on real data mining problems of classification. The results show that the method achieves faster convergence and provides better accuracies in classification. The convergence of the method has been compared with those using single population genetic algorithm and Island model GA and results prove that the proposed algorithm outperforms both SGA and IGA. The classifier system built using SAMGA is evaluated against other classification systems.

Appendix

Genetic Algorithm classifiers can be divided into two classes, the Michigan Approach classifiers and the Pittsburg approach classifier based on how rules are encoded in the population of individuals. In the Michigan approach, each individual of the population is a single rule of the entire rule base. In the Pittsburg approach, each individual encodes a set of prediction rules. The classifier used in this chapter is a Michigan approach classifier.

In the classifier system, each rule is represented by a bit string. Consider the rule ***if*** *(18≤age≤21) and (50≤weight≤70)* ***then*** *class = normal*. Now if there are only two classes, normal(0) and abnormal(1), then this rule can be represented in the format given by Figure 2.24. Figure 2.25 gives the bit string representation of the same rule. Generally for the fitness evaluation of the rule credit assignments

Age		Weight		Class
18	21	50	70	0

Fig. 2.24 Rule representation

Age		Weight		Class
00010010	00010101	00110010	01000110	0

Fig. 2.25 Bit string representation of the rule

are provided. For the Michigan style classifier, care should be taken to eliminate identical individuals from the population so that the rule base is not made up of the same highly fit rule.

References

1. Holland, J.H.: Escaping Brittleness: The Possibilities of General-Purpose Learning Algorithms Applied to Parallel Rule-based Systems. In: Mitchell, T., et al. (eds.) Machine Learning, vol. 2, pp. 593–623. Morgan Kaufmann, San Francisco (1986)
2. De Jong, K.A., Spears, W.M., Gordon, D.F.: Using Genetic Algorithms for Concept Learning. Machine Learning 13, 161–188 (1993)
3. Mata, J., Alvarez, J.-L., Riquelme, J.-C.: Discovering numeric association rules via evolutionary algorithm. In: Chen, M.-S., Yu, P.S., Liu, B. (eds.) PAKDD 2002. LNCS, vol. 2336, pp. 40–51. Springer, Heidelberg (2002)
4. Shenoy, P.D., Srinivasa, K.G., Venugopal, K.R., Patnaik, L.M.: Evolutionary Approach for Mining Association Rules on Dynamic Databases. In: Proc. of PAKDD. LNCS (LNAI), vol. 2637, pp. 325–336. Springer, Heidelberg (2003)
5. Srinivas, M., Patnaik, L.M.: Adaptive Probabilities of Crossover and Mutation in Genetic Algorithms. IEEE Transactions on Systems Man and Cybernetics 24(4), 17–26 (1994)
6. Back, T.: Self Adaptation in Genetic Algorithms. In: Proceedings of First European Conference on Artificial Life, pp. 263–271 (1992)
7. Martin, J.H., Lienig, J., Cohoon, J.H.: Population Structures: Island(migration) Models: Evolutionary Algorithms Based on Punctuated Equilibiria. In: Handbook of Evolutionary Computation, pp. C6.3:1–C6.3:16. Oxford University Press, Oxford (1997)
8. Lobo, J.H.: The Parameter-less Genetic Algorithm: Rational and Automated Parameter Selection for Simple Genetic Algorithm Operation PhD Thesis, University of Lisbon, Portugal (2000)
9. Lobo, F.G., Goldberg, D.E.: The Parameter-Less Genetic Algorithm in Practice. Information Sciences 167(1-4), 217–232 (2000)
10. Ghosh, A., Nath, B.: Multi-Objective Rule Mining using Genetic Algorithms. Information Sciences 163(1-3), 123–133 (2000)
11. Hinterding, R., Michalewicz, Z., Peachey, T.C.: Self Adaptive Genetic Algorithm for Neumeric Functions. In: Proceedings of the 4th Conference on Parallel Problem Solving from Nature, pp. 420–429 (1996)
12. Krink, T., Ursem, R.K.: Parameter Control Using the Agent Based Patchwork Model. In: Proceedings of The Congress on Evolutionary Computation, pp. 77–83 (2000)

13. Kee, E., Aiery, S., Cye, W.: An Adaptive Genetic Algorithm. In: Proceedings of The Genetic and Evolutionary Computation Conference, pp. 391–397 (2001)
14. Tongchim, S., Chongstitvatan, P.: Parallel Genetic Algorithm with Parameter Adaptation. Information Processing Letters 82(1), 47–54 (2002)
15. Voosen, D.S., Muhlenbein, H.: Strategy Adaptation by Competing Subpopulations. In: Parallel Problem Solving from Nature III, pp. 199–208. Springer, Berlin (1994)
16. Eiben, A.E., Sprinkhuizen-Kuyper, I.G., Thijseen, B.A.: Competing Crossovers in an Adaptive GA Framework. In: Proceedings of the Fifth IEEE Conference on Evolutionary Computation, pp. 787–792. IEEE Press, Los Alamitos (1998)
17. Herrera, F., Lozano, M.: Gradual Distributed Real-Coded Genetic Algorithms. IEEE Transactions on Evolutionary Computation 4(1), 43–62 (2000)
18. Schnecke, V., Vornberger, O.: An Adaptive Parallel Genetic Algorithm for VLSI Layout Optimization. In: Proc. of fourth Intl. Conf. on Parallel Problem Solving from Nature, pp. 859–868 (1996)
19. Herrera, F., Lozano, M.: Gradual Distributed Real-Coded Genetic Algorithms. IEEE Transactions on Evolutionary Computation 4(1), 43–62 (2000)
20. Montiel, O., Castillo, O., Seplveda, R., Melin, P.: Application of a Breeder Genetic Algorithm for Finite Impulse Filter Optimization. Information Sciences 161(3-4), 139–158 (2004)
21. Penev, K., Littlefair, G.: Free Search - A Comparative Analysis. Information Sciences 172(1-2), 173–193 (2005)
22. Deb, K., Beyer, H.G.: Self Adaptive Genetic Algorithms with simulated Binary Crossover. Evolutionary Computation 9(2), 197–221 (2001)
23. Herrera, F., Lozano, M.: Adaptive Genetic Algorithms based on Fuzzy Techniques. In: Proc. of Sixth Intl. Conf. on Information Processing and Management of Uncertainity in Knowledge Based Systems (IPMU 1996), Granada, pp. 775–780 (July 1996)
24. Vose, M.D., Liepins, G.E.: Punctuated Equilibria in Genetic Search. Complex Systems 5(1), 31–44 (1991)
25. Nix, A., Vose, M.D.: Modeling Genetic Algorithms with Markov Chains. Annals of Mathematics and Artificial Intelligence 5, 79–88 (1992)
26. Athreya, K.B., Doss, H., Sethuraman, J.: On the Convergence of the Markov Chain Simulation Method. Annals of Statistics 24, 69–100 (1996)
27. Eiben, A.E., Aarts, E.H.L., Van Hee, K.M.: Global Convergence of Genetic Algorithm: A Markov Chain Analysis. In: Parallel Problem Solving from Nature, pp. 4–12. Springer, Heidelberg (1991)
28. He, J., Kang, L., Chen, Y.: Convergence of Genetic Evolution Algorithms for Optimization. Parallel Algorithms and applications 5, 37–56 (1995)
29. Rudolph, G.: Convergence Analysis of Canonical Genetic Algorithms. IEEE Transactions on Neural Networks 5, 96–101 (1995)
30. Louis, S.J., Rawlins, G.J.E.: Syntactic Analysis of Convergence in Genetic Algorithms. Foundations of Genetic Algorithms, 141–152 (2002)
31. De Jong, K.A.: An Analysis of the Behaviour of A Class of Genetic Adaptive Systems, PhD Thesis, Department of Computer and Communication Sciences, University of Michigan, Ann Arbor (1975)
32. Spear, W.M., De Jong, K.: An Analysis of Multipoint Crossover. In: Foundations of Genetic Algorithms Workshop, Bloomington, pp. 301–315 (1991)
33. Eklund, P.W.: A Performance Survey of Public Domain Supervised Machine Learning Algorithms, Technical Report, The University of Queensland Australia (2002)

Chapter 3
Characteristic Amplification Based Genetic Algorithms

Abstract. This chapter proposes a new approach, wherein multiple populations are evolved on different landscapes. The problem statement is broken down, to describe discrete characteristics. Each landscape, described by its fitness landscape is used to optimize or amplify a certain characteristic or set of characteristics. Individuals from each of these populations are kept geographically isolated from each other. Each population is evolved individually. After a predetermined number of evolutions, the system of populations is analysed against a normalized fitness function. Depending on this score and a predefined merging scheme, the populations are merged, one at a time, while continuing evolution. Merging continues until only one final population remains. This population is then evolved, following which the resulting population contains the optimal solution. The final resulting population contains individuals which have been optimized against all characteristics as desired by the problem statement. Each individual population is optimized for a local maxima. Thus when populations are merged, the effect is to produce a new population which is closer to the global maxima.

3.1 Introduction

A major issue with the genetic algorithm, is that, in a fitness landscape they tend to converge towards the local maxima rather than the global optimum. Once this occurs, it may be difficult for the algorithm to proceed towards the globally optimal solution. This can cause the algorithm to fail when the gap between the local maxima and global optima is large. The approach proposed here attempts to prevent the algorithm from converging at the local maxima and coerces it to proceed till the global maxima [1, 2, 3, 4, 5, 8].

The period between 1990-1992 saw a great deal of work which attempted to find a correct sorting network [1] for $n = 16$. Daniel Hillis attempted to solve the problem, using a genetic algorithm [6, 7]. In order to form pairs of numbers for comparison and exchange, Hillis used two chromosomes. A single pair was formed by the corresponding numbers on each chromosome. His initial attempts gave a result of 65

K.R. Venugopal, K.G. Srinivasa, L.M. Patnaik: Soft Comput. for Data Min. Appl., SCI 190, pp. 51–62.
springerlink.com

comparisons, which was far short of the 60 comparison schemes developed by other methods. The genetic algorithm is suffering from the same problem: It converged at the local maxima, resulting in only a moderately good solution. Hillis' solution, is to evolve the fitness landscape itself. As the algorithm got closer to converging, the landscape would get steeper, forcing the algorithm to evolve further. The landscape is a set of numbers that each candidate sorting network is tested against. Thus, as the algorithm progressed, the "weaker" set of input test cases were eliminated and the stronger ones retained. In biology, this is known as the phenomenon of host-parasite co evolution. To complete the analogy, the host is the sorting network designed while the parasite is the set of input test cases.

One attempt concentrated on optimizing the parameters of the algorithm, such as the probability of mutation and crossover as well as number of individuals in the population. This is done using a neural network, which "learnt" the characteristics of the solution and could thus realize when the algorithm is slowing down. Knowing this, the neural network could then optimally alter the parameters of the GA, causing the fitness landscape to change dramatically. This in turn would cause the algorithm to evolve further, towards the global optima. The overhead of this approach is somewhat large as the background infrastructure needed for a neural network is complex.

Yet another approach towards optimizing the parameters of the algorithm, is done using another genetic algorithm. This approach, aptly named the parameterless GA, required no parameters as input. The parameters were themselves evolved against the solutions provided. This approach is in some ways related to the host-parasite solution provided by Hillis. Instead of evolving the problem set, the parameters of the GA are evolved. This approach has been widely applied, again with varying degrees of success. [6, 7, 9].

3.2 Formalizations

Proposition: Feature set must be disjoint: Let us assume, a genetic algorithm based solution is to be found, for a given problem. Let us further assume that the problem requires a solution with particular characteristics. Mathematically, we define each of these characteristics as feature sets l_1, l_2 ... l_n where

$$L = \{l_1, l_2, l_3 ... l_n\} \tag{3.1}$$

such that

$$\forall l_1, l_2 \in L l_1 \bigcap l_2 = \emptyset \tag{3.2}$$

The design of the fitness functions for each population depends on the feature sets. Thus, if the feature set is not disjoint, i.e., if equation 3.2 is not true for a chosen L more than one fitness function attempts to optimize the same sub-features. This is redundant and causes an excessively larger survey of the search space. This only slows down the entire algorithm and causes unnecessary crossover inputs.

Notations for the genetic machine: We can mathematically define a genetic algorithm for our purposes, as a machine

$$G = (I, F, p_m, p_x) \tag{3.3}$$

where G is the genetic machine itself, I refers to the set of individuals that it comprises of, F is the fitness function that is used to evaluate the fitness function of an individual $i \forall i \in I$ given by $F(i)$ p_m is the probability of mutation used for the machine and p_x is the probability of crossover used. Different machines may be represented as G_1 and G_2, where $G_1 = (I_1, F_1, p_{m1}, p_{x1})$ and $G_2 = (I_2, F_2, p_{m2}, p_{x2})$.

The entire problem is defined as a set of genetic machines P,

$$P = \{G_1, G_2 ... G_n\} \tag{3.4}$$

where each of $G_i \forall i \in [1, n]$ is a genetic machine of the form in equation (3.3). Each of F is designed to converge towards the local maxima. A complementary function, that evaluates towards the global optimum, $\phi(i)$ is also defined, to allow evaluation of the solution to solve the problem, where i is the individual in question.

The genetic merging operator: We now define a new genetic operator, $*$, which denotes the merging of two genetic machines. Consider two genetic machines G_1 and G_2. The merger of the members of G_2 into G_1, to form the new machine G is denoted as:

$$G = G_1 * G_2 \tag{3.5}$$

The effect of the merge is as shown below: $I_G = I_{G_1} + I_{G_2}$
$F_G = F_{G_1}$
$p_{m_G} = p_{m_{G_1}}$
$p_{x_G} = p_{x_{G_1}}$
Thus the merge only causes the individual members of G_2 to be transferred to G_1.

Selection for merging: The evaluation of the population of a machine G, against the global fitness function ϕ, is denoted as $\phi(G)$. We initially let all the machines in P evolve till their local maxima is reached. Then, two machines from P are selected according to the following scheme:

$$N_1(P) = max\{\phi(G_1), \phi(G_2) ... \phi(G_n)\} \tag{3.6}$$

$$N_2(P) = max\{\{\phi(G_1), \phi(G_2) ... \phi(G_n)\} - N_1(P)\} \tag{3.7}$$

After choosing, $N_1(P)$ and $N_2(P)$, we merge them as:

$$G_{new} = N_1(P) * N_2(P) \tag{3.8}$$

Finally, the new machine G_{new} is put back into the population pool P, after removing $N_1(P)$ and $N_2(P)$.

$$P' = P - N_1(P) + N_2(P) + G_{new} \tag{3.9}$$

Thus $|P|$ is reduced by 1. The machines P are then evolved to their local maxima and their population sizes are culled to the initial size. Thus the weakest individuals that result from any merger are directly eliminated.

The algorithm is iterated, till $|P| = 1$. At this stage, only the final machine G remains. This final machine is evolved against its fitness function F and then against ϕ. This ensures that the solutions provided by the machine can solve both the problem at the local maxima as well as the global optimum.

Previous approaches at improving the efficiency of the genetic algorithm have concentrated on optimizing the parameters of the GA during the execution of the algorithm. We propose a method, where the solution assumes that a GA converges at the local maxima. This apriori knowledge is exploited to prepare individuals that are already at the local maxima, to evolve in a population that converges at the global maxima. Thus neither the input, nor the parameters of the GA need to be modified during execution.

Also, the size of the population of a genetic machine is not constant. It increases after a merge occurs, accepting individuals from another machine. This causes a crossover input to the existing machine. Geographic isolation, occurs when individuals are bred separately, without any influence of each other. Since the fitness landscapes for each population are different, each set of individuals is optimized for different characteristics. In nature, when a geographically isolated set of species is allowed to intermingle, the environment poses a different set of challenges to each of the species. Thus the species is forced to evolve, to avoid extinction. In this chapter, the new genetic operator and the merge operator are introduced. These operators are responsible for merging two evolved genetic machines and the individuals from one machine are transferred to another machine.

3.3 Design Issues

We now explore some of the design issues involved in an implementation of our algorithm. These issues concern the design of the genetic machines, their fitness functions, the probabilities of crossover and mutation and the global fitness function.

Feature set optimization: The feature set L as previously mentioned, needs to be chosen with great care. The condition mentioned in equation 3.2 must be true for the chosen set. The problem of finding L, is an optimization problem, that attempts to minimize $|L|$, while ensuring that equation 3.2 is always satisfied.

Mapping from L to P: After the feature set L has been chosen optimally, it must be converted to an appropriate population set of genetic machines. Thus each member of L must be represented by a genetic machine in P. The design of each machine, should be such that the fitness landscape would encourage the population to converge at the local maxima. This is important from the standpoint of the global fitness function and the merging operator. Convergence at the local maxima is a pre-requisite, before the individuals in the population can be considered against the global function. Otherwise, the algorithm does not reach the global optimum.

$\phi(x)$ **: The global fitness function:** The global fitness function $\phi(x)$ is designed with an overall view of the problem. If a single population based classical genetic algorithm must be implemented to solve the problem at hand, this function must be used. When applied, this function provides a landscape suitable to the global optimum. Direct application of the function, causes the GA to get stuck at a local maxima, instead of converging towards the global optimum. Individuals are periodically evaluated against this function to ensure that they do not stray away from the requirements of the final solution.

3.4 Algorithm

The visual depiction of the algorithm is also shown in Figure 3.1. We shall now explore the algorithm in more detail. The given problem statement, must first be carefully analyzed to derive specific characteristics that need to be optimized. Accordingly, once the desired characteristics of the solution have been decided upon,

Algorithm 1. ALGORITHM FOR GI BASED GA

```
n ← Number of features
for i = 0 to n do
    Design fitness[i]
end for
Design normalized_fitness(population)
while num_of_populations > 1 do
    if n > 1 then
        for i = 0 to num_of_populations do
            evolutions ← 0
            while evolutions < saturation[i] do
                evolve(population[i])
            end while
        end for
    end if
end while
for i ← 0 to n do
    Evaluate normalized_fitness(population[i])
end for
while num_of_populations! = 1 do
    Select pop1 and pop2.
    merge_populations(pop1, pop2)
end while
while fitness(final_population) < required_fitness do
    evolve(final_population);
end while
```

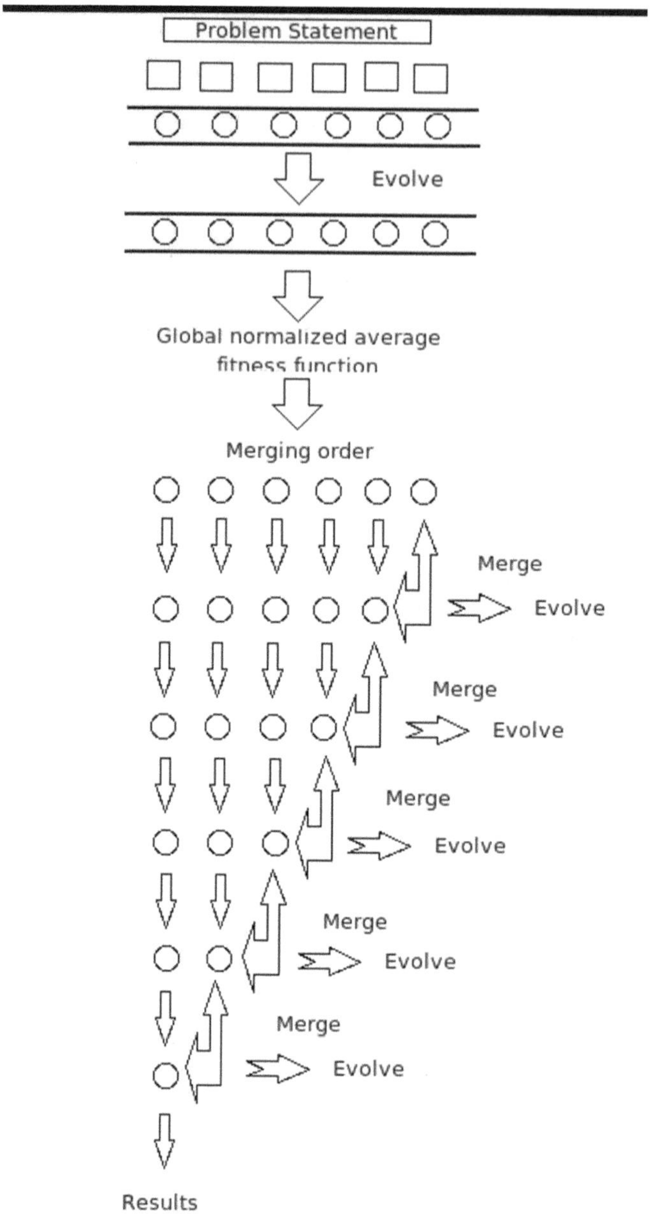

Fig. 3.1 GI based GA: Our proposed approach

the number of populations for the algorithm must be decided. For this purpose, an optimization problem must be solved to find an optimal L. It may be required to use one machine for a set of minor characteristics. In either circumstance, the fitness function for the machine has to be designed appropriately.

The fitness functions, must work on mutually independent characteristics. If there is any overlap of characteristics, thus violating the condition in equation 3.2, the populations would have individuals, with similar characteristics. The overlap, would be at the cost of exploring other candidate solutions. This exploration must be optimized in an exhaustive search, such as ours. More importantly, this overlap causes unnecessary crossover inputs and significantly slows down the algorithm.

Each fitness function, causes the population to saturate at a given point. This saturation is defined at the point, when the population no longer evolves appreciably with any further generations. This point of saturation, can be a fitness value which is reached upon sufficient evolution of the population, or the desired number of times the population has evolved. The point of saturation only serves to define a point at which the algorithm must stop evolving a particular population. This point is not a "hard" point, so to speak, but only serves as a guide as to when a machine has reached its local maxima. At this point, the machine's evolutions must be stopped and its individuals must be evaluated against the global fitness function $\phi(x)$.

The population merging function, defined by $merge\_populations(pop1, pop2)$, serves to move individuals, which have adapted well to one population to evolve in another landscape. This function is a direct implementation of the genetic merge operator $*$ that has previously been introduced. This has the effect of improving the quality of crossover, without changing the crossover rate itself. For implementations, with fewer populations, this merging scheme could be predefined such as a trivial transfer of individuals from one population to another. For more complex implementations, the merging scheme should define the number of individuals to be transferred, as well as the order of merging. The order of merging defines the order of characteristics that are optimized. In general, it would be more useful to transfer individuals from stronger populations to weaker populations, as this would have the effect of optimizing already strong individuals, in a different landscape. In addition, the native population would have a varied mutant quantum.

The global fitness function $\phi(x)$, must to be designed, keeping in mind the overall requirements of the solution. This function is used to evaluate individuals across populations and thus, must cover taking account of all characteristics. This function should be applied when a single population based genetic algorithm is implemented to solve the problem. Once these functions have been setup as desired, the algorithm is executed. Each of the populations are evolved, without any interaction from the other populations, thus realizing the concept of geographical isolation. After all populations have been sufficiently evolved (once their points of saturation have been reached), the global fitness function is called to evaluate the individuals in the machines.. The merging function is then used to merge individuals from different machines, in the order that is decided with the help of the global fitness function. Each single merge is accompanied by iterative evolution, until the point of saturation is reached.

The above process is iterated, until one single population is left, so that $|P| = 1$. This final population is evolved, until a sufficiently fit solution is found. This final solution is optimized against all the characteristics presented by the different machines.

Analysis and Implications: One of the main implications of our algorithm, is that the cardinality of a set of Individuals $|I|$, is not fixed. This size increases after a merging operation. The size stays at twice the original size, until the culling is done. This culling ensures that the number of individuals evaluated against the global function is small. The individuals at the lower end of the fitness curve need not be evaluated, as they do not contribute any more than the ones at the higher end of the curve. Effectively, their existence is unnecessary and thus they can be removed.

The time taken for the complete execution of the algorithm is much larger than a single population genetic algorithm, for shorter problems. This is because the overhead involved due to the use of multiple machines is much larger. For larger and more complex problems, this one-time overhead is much smaller as compared to the evolution time of the machines. The time taken for the algorithm depends on the solution of the optimization problem, that defines L. The fewer the number of machines, the less the time taken to complete execution. However, if too few machines are used, the algorithm continues to totter around the local maxima, taking much longer to complete execution.

3.5 Results and Performance Analysis

In order to validate our algorithm, we have compared it with the output of a standard genetic algorithm. Our test involved the GA to attempt to guess a given ASCII string. The Figures 3.2, 3.3 and 3.4 show the convergence rates for a single GA based

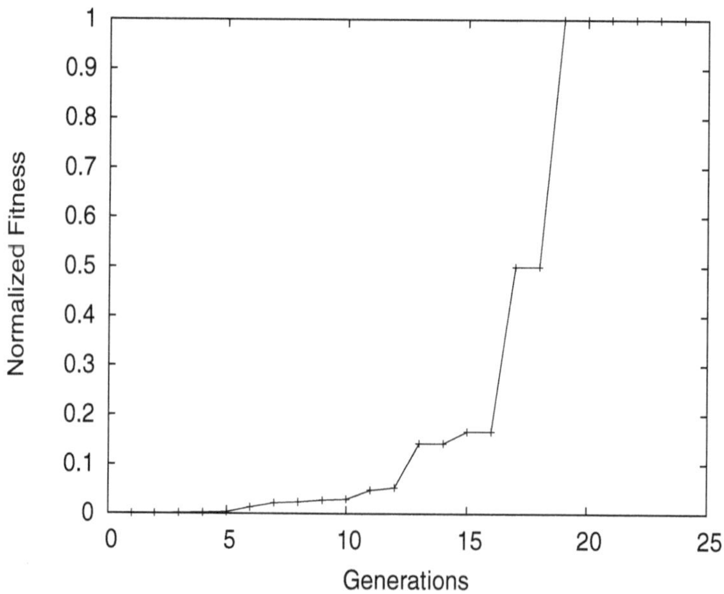

Fig. 3.2 Normal GA: With 4 character string

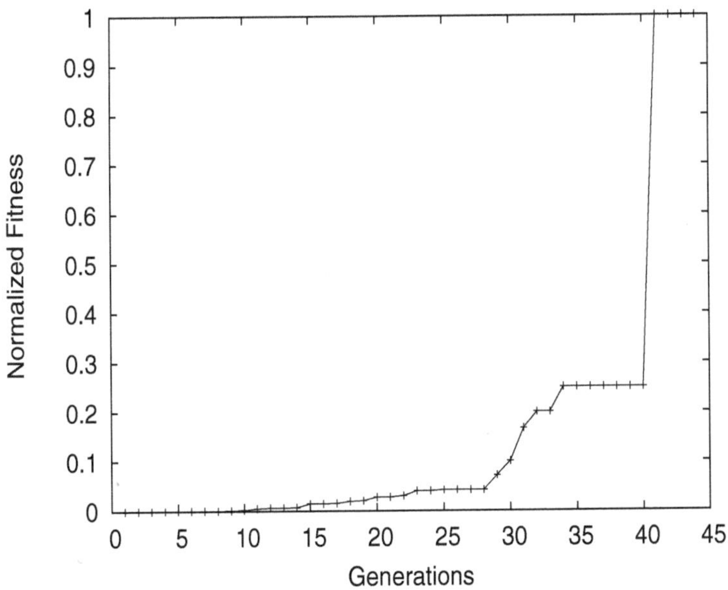

Fig. 3.3 Normal GA: With 8 character string

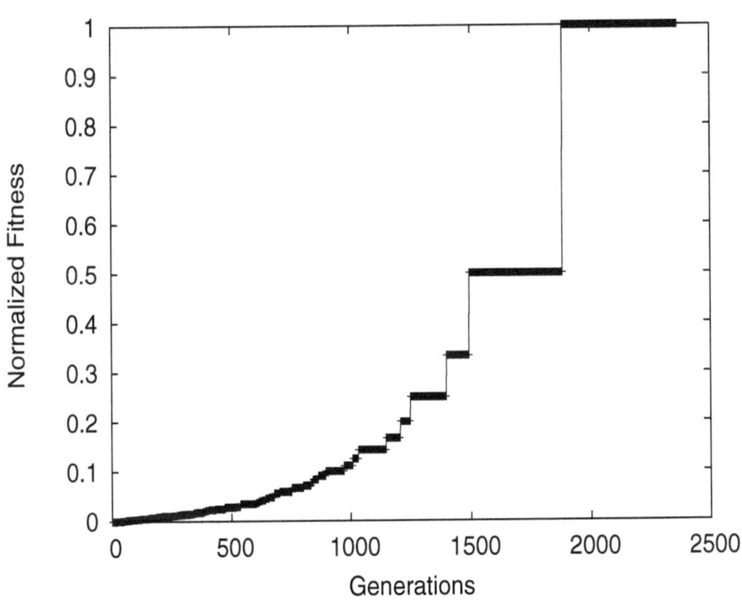

Fig. 3.4 Normal GA: With 32 character string

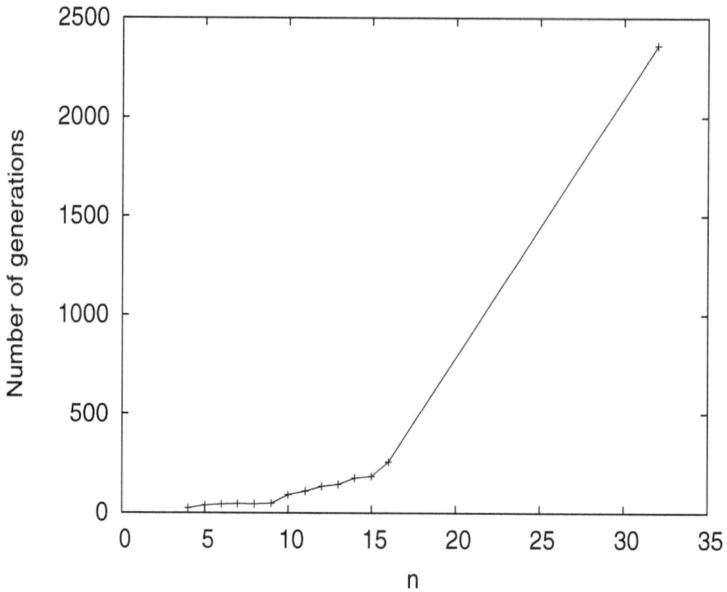

Fig. 3.5 Normal GA: For increasing n

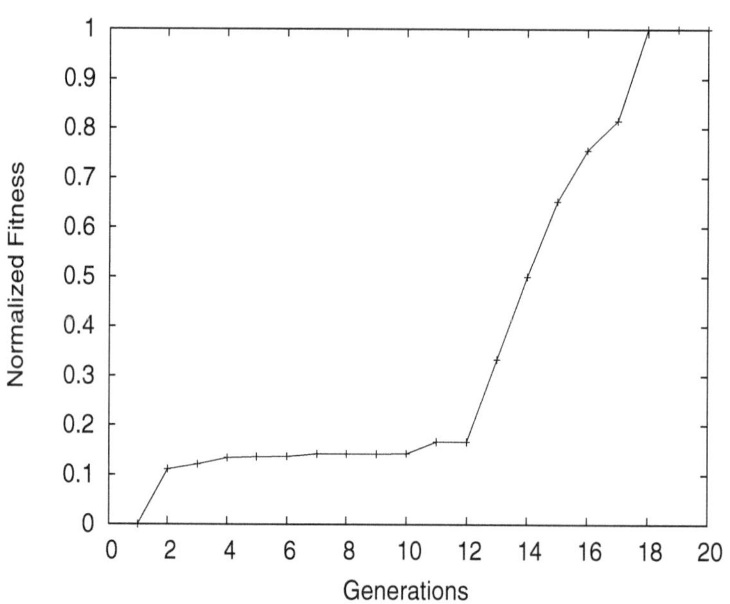

Fig. 3.6 GI based GA: With 8 character string

attempt to guess the string. As n increases, the number of generations considered is shown in Figure 3.5. The convergence rate of our attempt at the same problem, using a geographically isolated genetic algorithm is shown in Figure 3.6.

As it can be seen, as the size of the chromosome increases, the number of generations taken also increases, at a rate that is approximately quadratic. Thus a linear split of the chromosome takes less time to complete execution. We initialize two populations to solve the problem for $n = 8$, one population that solves the first half of the chromosome (the first 4 alleles) and the second population solves the second half of the chromosome (the next 4 alleles). The fitness function used in both GAs is given in Algorithm 2.

Algorithm 2. *Fitness function*

1: $fitness \leftarrow 0$
2: **for** $i \leftarrow 1$ to mw **do**
3: **if** $test\_string[i]$!= $candidate_string[i]$ **then**
3: $fitness = fitness + chr(test\_string[i]) - chr(candidate\_string[i])$
4: **end if**
5: **end for**
6: return $fitness$

For the purpose of plotting the graphs, we have used a normalized value for all fitnesses, normalized over the range $[0, 1]$. It can be clearly seen that the algorithm reaches the local maxima in each of the single population runs. In the geographically isolated populations run, the algorithm directly converges towards the local maxima.

3.6 Summary

The given problem statement is divided into a set of features, which are then optimized. A genetic machine, which has been mathematically defined and formalized, is created for each of the features. These machines are optimized to converge towards their respective local maxima. As each machine converges towards its local maxima, it is merged with another such machine. The genetic merge operator has also been mathematically defined. The net effect, is that before being evaluated against the global function, individuals are optimized against specific characteristics. Thus they need only evolve from the local maxima towards the global optimum, instead of evolving from random points in space. When compared with standard genetic algorithm the proposed algorithm produces better results.

References

1. Mitchell, M.: An Introduction to Genetic Algorithms. MIT Press, Cambridge (1996)
2. Koza, J.R.: Genetic Programming: On the Programming of Computers by Means of Natural Selection. MIT Press, Cambridge (1992)

3. Koza, J.R.: Genetic Programming II: Automatic Discovery of Reusable Programs. MIT Press, Cambridge (1994)
4. Packard, N.H.: A Genetic Learning Algorithm for the Analysis of Complex Data. Complex Systems 4(5), 543–572 (1990)
5. Chughtai, M.: Determining Economic Equilibria using Genetic Algorithms. Imperial College (1995)
6. Hillis, W.D.: Co-evolving Parasites Improve Simulated Evolution as an Optimization Procedure. Physica D-42, 228–234 (1990)
7. Hillis, W.D.: Co-evolving Parasites Improve Simulated Evolution as an Optimization Procedure. In: Artificial Life II. Addison-Wesley, Reading (1992)
8. Forest, S., Mithell, M.: What makes a Problem Hard for a Genetic Algorithm? Some Anamalous Results and their Explanation. Machine Learning 13, 285–319 (1993)
9. Deb, K., Beyer, H.-G.: Self-Adaptive Genetic Algorithms with Simulated Binary Crossover. IEEE, Los Alamitos (2004)
10. Harik, G.: Finding Multiple Solutions in Problems of Bounded Difficulty, Information Sciences (1994)

Chapter 4
Dynamic Association Rule Mining Using Genetic Algorithms

Abstract. A large volume of transaction data is generated everyday in a number of applications. These dynamic data sets have immense potential for reflecting changes in customer behaviour patterns. One of the strategies of data mining is association rule discovery which correlates the occurrence of certain attributes in the database leading to the identification of large data itemsets. This chapter seeks to generate large itemsets in a dynamic transaction database using the principles of Genetic Algorithms. Intra Transactions, Inter Transactions and Distributed Transactions are considered for mining Association Rules. Further, we analyze the time complexities of single scan technique DMARG(Dynamic Mining of Association Rules using Genetic Algorithms), with Fast UPdate (FUP) algorithm for intra transactions and E-Apriori for inter transactions. Our study shows that the algorithm DMARG outperforms both FUP and E-Apriori in terms of execution time and scalability, without compromising the quality or completeness of rules generated.

4.1 Introduction

Mining association rules is a process of discovering expressions of the form $x \Rightarrow y$. For example, customers buy bread (x) along with butter (y). These rules provide valuable insights to customer buying behaviour, vital to business analysis. Let $I = \{i_1, i_2 ... i_m\}$ be a set of literals called items. Let DB denote a set of transactions where each transaction T is a set of items such that T is a subset of I. Associated with each transaction is a unique identifier, called Transaction IDentifier (TID). A transaction T contains x, a set of some items in I, if $x \subseteq T$. An association rule is an implication of the form $x \Rightarrow y$ where $x \subseteq I$, $y \subseteq I$ and $x \cap y = 0$. The rule $x \Rightarrow y$ holds in a transaction set DB with confidence c, if $c\%$ of transactions in DB that contain x also contain y. The rule $x \Rightarrow y$ has support s in the transaction set DB if $s\%$ of transactions contain $x \cup y$. The process of discovering association rules can be split into two domains, (i) finding all itemsets with appreciable support(large itemsets). (ii) generate the desired rules. Finally, rule $x \Rightarrow y$ is generated if, support $(x \cup y)$/support $(x) \geq$ minimum confidence.

K.R. Venugopal, K.G. Srinivasa, L.M. Patnaik: Soft Comput. for Data Min. Appl., SCI 190, pp. 63–80.
springerlink.com

Consider the example of a departmental store. Let *DB* consist of 5,00,000 transactions, of which 20,000 transactions contain bread, 30,000 transactions contain chips and 10,000 transactions contain both bread and chips. The frequency of occurence of bread and chips together as a percentage of the total transactions is called support, i.e., 2% in this case$(10,000/5,00,000)$. The measure of dependence of a particular item on another is called confidence. 20,000 transactions contain bread of which 10,000 also contain chips. Hence, when people buy bread, there is a 50% probability of them buying chips. This is called as intra transaction association rule mining [1].

4.1.1 Inter Transaction Association Rule Mining

Conventional association rule mining has focussed primarily on intra transaction association rules that are limited in scope. This has led to the development of more powerful inter transaction association rules. Inter transaction association rule offers insight into correlations among items belonging to different transaction records. The classical association rules express associations among items purchased by a customer in a transaction i.e., associations among the items within the same transaction record. Inter transaction rules express the relations between items from different transactional records. A typical example would be that when the prices of *Wipro* shares go up on first day and *Infosys* share prices go up on second day, then *TCS* share prices go up on fourth day with 80% probability. Intra transaction associations can be treated as a special case of inter transaction association.

Mining inter transaction associations presents more challenges than intra transaction associations as the number of potential association rules becomes extremely large. A frequent inter transaction itemsets must be made up of frequent intra-transaction itemsets. It has wide applications in predicting stock market price movements, traffic jams by discovering traffic jam association patterns among different highways, flood and drought for a particular period from weather database, etc.. We consider a sliding window of some specific size (size is the number of transactions to be handled at a time) to generate inter-transaction association rules. Mining intra transaction association rules, is the special case of inter transaction association rule mining when the window size is one.

Let $I = \{i_1, i_2 ... i_s\}$ be a set of literals called items. A transaction database *DB* is a set of transactions $\{t_1, t_2 ... t_m\}$, where t_i $(1 \leq i \leq m)$ is a subset of I. Let $x_i = < v >$ and $x_j = < u >$ be two points in one-dimensional space, then the relative distance between x_i and x_j is defined as $\Delta(x_i, x_j) = < u - v >$. A single dimensional space is represented by one dimensional attribute, whose domain is a finite subset of non-negative integers. A single dimensional inter transaction association rule is an implication of the form $x \Rightarrow y$, where $x \subset I_e$, $y \subset I_e$ and $x \cap y = 0$. A single dimensional inter-transaction association rule can be expressed as $\Delta_0(a)$, $\Delta_1(c) \Rightarrow \Delta_3(e)$, which means if the price of stock a goes up today, and price of stock b increases next day, then most probably the price of stock e will increase on the fourth day.

Dynamic Data Mining: New association rules, which reflect the changes in the customer buying pattern, are generated by mining the updations in the database and is called Dynamic mining [2]. Dynamic mining algorithms are proposed to handle updation of rules when increments or decrements occur in the database. It should be done in a manner which is cost effective, without involving the database already mined and permitting reuse of the knowledge mined earlier. The two major operations involved are (i) Additions: Increase in the support of appropriate itemsets and discovery of new itemsets. (ii) Deletions: Decrease in the support of existing large itemsets leading to the formation of new large itemsets. The process of addition and deletion may result in the invalidation of certain existing rules.

Distributed Data Mining: Distributed data mining is a result of further evolution of data mining technology. It embraces the growing trend of merging communication and computation. It accepts the fact that data may be inherently distributed among loosely coupled sites connected by a network and these sites may have heterogenous data. It offers techniques to discover new knowledge through distributed data analysis and modeling using minimal communication of data. Consider the large amount of data present on the net. There exist sites for weather, city demography etc.. If the database is large it becomes practically impossible to download everything and build a global data model. Instead partial data models would correctly lead to the computation of global data model.

4.1.2 Genetic Algorithms

Recent trends indicate the application of GA in diverse fields ranging from character script recognition to cockpit ambience control [3]. The objective of this chapter is to generate association rules by applying genetic algorithms on dynamic databases. The resulting single-scan technique, Dynamic Mining of Association Rules using Genetic algorithms (DMARG) is presented. It is an iterative technique which works with a solution set, called population, directly. The population of solutions undergo selection using operators of reproduction, crossover and mutation. The fitness function, a property that we want to maximise, is used to evaluate quality of solutions. The genetic algorithm used in this chapter is summarized below.

1. Create an initial population of schema ($M(0)$), where each member contains a single one in its pattern. For example, for transactions of length four, the initial population would contain 1*** , *1** , **1* , ***1.
2. Evaluate the fitness of each individual with respect to the schema. Fitness of each individual, m, in the current population $M(t)$ is denoted by $u(m)$ where $u(m) = \frac{number\,of\,matching\,bits - number\,of\,non\_matching\,bits}{Total\,number\,of\,fields}$
3. Enter the schema into the Roulette wheel. The selection probabilities $p(m)$ for each m in M(t) determines the size of each slot in the Roulette wheel.
4. Generate $M(t+1)$ by probabilistically selecting individuals from $M(t)$ to produce offspring *via* crossover. The offspring is a part of the next generation.

5. Repeat step (2) until no new schema are generated.
6. Steps (2), (3) and (4) create a generation which constitutes a solution set.

Every transaction comprising of a set of items is represented by a fixed number of bits forming a string, where 0 and 1 represent the absence or presence of an item respectively. For example, in the string 1001, the bit positions one and four have been set to one while the bit positions two and three have been set to zero. Hence, the items corresponding to bit positions one and four viz., A and D have been purchased, while those corresponding to bit positions two and three viz., B and C have not been purchased.

Reproduction is a process in which fit strings are passed to the next generation where they have a higher probability of contributing one or more offspring. Once a string has been selected for reproduction, an exact replica of it is made which enters the mating pool - a tentative new population for further genetic action.

Random pairs of these strings are chosen for crossover from the mating pool. An integer position k is then selected along the string at random between 1 and l - 1, l being the length of the string. Two new strings are created by swapping all characters between positions k+1 and l of the parents. For example, consider two strings X = 0101 and Y = 1010. Choose a random number between 1 and 3 , say 2. The resulting crossover yields two new strings X*= 0110 and Y*= 1001 . These are the members of the new generation.

Flip Mutation is the operation of changing a 1 to 0 and vice-versa. For example, 0000 becomes 0010 after mutation. Schemata are representative of a set of strings with similarities at certain positions. Schemata is a pattern of $\{1, *\}$, where 1 represents the presence of an item in a transaction and * is a don't-care-condition, which may correspond to either the presence or absence of an item.

4.2 Related Work

The incremental updating technique proposed in [4] is similar to Apriori [5] and operates iteratively on the increment database and in each iteration makes a complete scan of the current database. At the end of the kth pass all the frequent itemsets of size k are derived. The performance evaluation of the Fast Updation Technique(FUP) shows that it is much better than the direct application of the Apriori algorithm on the entire database and needs $O(n)$ passes over the database where n is the size of the maximal large itemset. FUP is 2 to 16 times faster than rerunning Apriori or DHP(Dyanamic Hashing and Pruning) [6] and number of candidate sets generated are about 2-5% of DHP and overhead of running FUP is only 5-20%. This algorithm is extended to handle general updations like additions, deletions and other modifications in FUP2 [7].

DEMON [8], Mining and Monitoring Evolving data handles updation to databases in a systematic manner. It takes care of the warehousing environment by adding or deleting data in blocks. Here the concepts of Block Sequence Selection and Data Span Windows are used to select the appropriate blocks and to control the fine-grained block selection respectively. The algorithm ECUT mines the data on TIDlist which helps in reducing the candidate itemsets. The use of negative border (Negative border is the set of minimal itemsets that do not satisfy the constraints) concept as a constraint relaxation technique to generalize incremental mining of association rules with various types of constraints is handled in [9].

Efficient one-pass incremental algorithms based on negative-border concept are proposed in [10, 11]. The algorithms can be used in conjunction with algorithms like Apriori or Partition based algorithms. The advantage of this algorithm is that, it needs the full scan of the entire database only when the database update causes the negative border of the set of large itemsets to expand. They performed better than the Apriori algorithm as it takes fewer scans of the database, but the drawback of the above methods are that too many candidates are generated in the negative border closure resulting in the increase of the overall mining time. The negative borders are computed by applying the Apriori repeatedly. The Delta proposed in [12] takes care of multi support, local count, identical and skewed databases.

Mining inter-transaction association rules was first introduced in [13], in which extended Aprori and extended hash apriori are proposed to deal with single and multidimensional inter transaction association rules. The algorithm FITI proposed in [14] is much faster than EH-Apriori.

R Agrawal and J C Shafer [15] introduced count distribution, data distribution and candidate distribution algorithm, which effectively uses aggregate main memory to reduce synchronization between the processors and has load balancing built into it. FDM [16] tries to generate small number of candidate sets and substantially reduces the number of messages passed at mining association rules. The intelligent data distribution algorithm and hybrid distribution proposed in [17], efficiently use aggregate memory of the parallel computer by employing intelligent candidate partitioning scheme and uses efficient communication mechanism to move data among the processors. Parallel algorithms for generalized association rules with classification hierarchy was introduced in [18].

4.3 Algorithms

Consider a large database *DB* with horizontal layout consisting of a finite set of transactions $\{t_1, t_2, ... t_m\}$. Each transaction is represented by a binary string of fixed length n, where n is the total number of items present. Let an incremental set of new transactions *db* be added to *DB*. The objective of this paper is to generate all the large intra/inter transaction association rules in $DB \cup db$ without involving *DB* using scan only once technique through the algorithm DMARG.

Assumptions

- A transaction is represented by a binary coded string.
- The items in a transaction are ordered.
- The transactions are generated by synthetic data generator [19].

Algorithm: DMARG

1. Initialize Database *DB*.
2. Read_Window_Size(Window_Size)
3. Prepare_Database(*DB*, Window_Size)
4. Initialize itemsets database *I*.
5. [Incremental operation causes increase in the support of the relevant itemsets]
   Read an itemset *i* from the incremental database *db*.
   Update those Itemsets in *I* which share one or more common transactions with *i*.
6. [Discover new itemsets]
   Repeat step 4.

   a. Apply Genetic Algorithms to members of *I*.

      i. Crossover parents with high fitness values.
      ii. Mutate certain itemsets to obtain a new itemset, if complete set not found.

   b. [Add new combinations of itemsets discovered to the existing itemset database]
      Update itemsets until complete solution is obtained.

7. [Repeat process for the entire incremental transaction database]
   If end of database *db* has not been reached, goto Step 3.
8. [Identify all large itemsets]
   Identify large itemsets above minimum support.
9. [Generate rules]
   Generate all rules above minimum confidence.
10. [Repeat if more increments]
    If the incremental operation is to be repeated, goto Step 1.

The algorithm is inherently incremental in nature. The database initialization is done by converting the given database into an initial population of itemsets in the form of schemata ({*, 1}*), with their support zero. This information is stored in an itemset file, whose contents are updated after each iteration. Since the algorithm works both for intra transaction and inter transaction association rule mining, the function Read_Window_Size() takes the input from the user about the type of association rules to be mined. If the window_size is one, the algorithm mines the intra transaction association rules, if the window_size > 1, then the algorithm mines the inter transaction association rules. Depending upon the window_size the function Prepare_Database(), does the preprocessing of the initial data. If the window_size is greater than one, then the function Prepare_Database() concatenates the transactions upto the window_size and hence prepares the database for mining inter transaction association rules.

Once the database is represented in the form of a schemata, the itemsets database *I* is created. Initially, the itemsets database is a list consisting of all the distinct items in the database with their support as zero. As the algorithm reads the transactions, the itemsets database is updated with the new itemsets if their support is greater than the minimum support. The Genetic Algorithm is applied on the members of I to discover the new itemsets. Finally, the algorithm generates the association rules with the minimum support and confidence. The working principle of the algorithm is explained through an example in Section 4.4.

Given n hosts $\{h_1, h_2, ...h_n\}$ where each h_i $(1 \leq i \leq n)$, contains a local database db_i, we mine db_i to discover large itemsets local to h_i and generate local association rules. The local large itemsets are combined to generate global rules. The local databases are horizontally partitioned. Databases are uniformly distributed across all the nodes of the homogeneous network.

Algorithm: DDMARG

1. The server waits for the request from the clients.
2. Clients register themselves within the allotted time frame.
3. Server registers the clients which participate in mining process.
4. Server requests the clients to check for the integrity of the executables required for distributed mining.
5. Clients return the status of the association rules mined from local database db_i.
6. Server waits for clients to perform local mining on the database (db_i) and places it in WAIT state.
7. Clients interrupt server when the local mining on db_i is completed and generates the local association rules using the algorithm DMARG.
8. Server requests the clients to send the entire rule chromosome mined.
9. Server builds the universal database of rule chromosomes after receiving the complete set of association rules mined in all the registered clients.
10. Server generates global association rules from the result obtained from step 9.
11. Server closes connections with all the clients.
12. End.

4.4 Example

Intra Transaction Association Rule Mining: Consider an example transaction database given in Table 4.1. The first column represents the transaction identifier TID. The various items purchased in a particular transaction is given in column 2. The third column consists of the encoded binary string. As the number of items considered in the database is five(n = 5), the length of the encoded transaction is also five.

Table 4.1 A Simple Transaction Database

TID	Transaction	Encoded Trans.
1	AD	10010
2	ABE	11001

Table 4.2 Strings of Initial Population and their Support

String	Itemset	Support
1****	A	0
1*	B	0
***1**	C	0
****1*	D	0
*****1	E	0

The algorithm scans the database one transaction at a time and tries to approximate or classify the data by the nearest itemset if not already present and updates its support. Assuming, this is the first run of the code and the mode of operation is incremental, the initial population contains the following strings with individual support as zero. The initial population, the itemsets and corresponding supports are shown in Table 4.2.

When the first transaction record is read, it is compared with the existing strings in the itemset list and the fitness value of the string is computed with respect to the encoded transaction TID 1 as follows,

$$fitness = \frac{No.of\,matching\,bits - No.of\,non\_matching\,bits}{No.of\,items\,in\,the\,transaction(n)}$$

If the fitness is one, then the support of the string in question is incremented. In Table 4.3, the fitness of the item A is one, therefore its support is incremented to 1. All strings with fitness 1, are placed in a dynamically created array which serves as a Roulette wheel.

In the case of TID 1, the Roulette wheel would contain strings 1**** and ***1*(as shown in Table 4.3). A random crossover point (position 3) is chosen and crossover operation is applied on the two parents strings 1**** and ***1* to obtain chidren 1**1* and *****. The null string is discarded as it serves no purpose.

Table 4.3 Updations and Reproduction by the Transaction TID 1

String	Itemset	Support	Fitness
1****	A	1	1
1*	B	0	0.6
***1**	C	0	0.6
****1*	D	1	1
*****1	E	0	0.6

No. of Matching Bits = 5, No. of Non Matching Bits = 0.
Fitness(10010,1****) = $(5-0)/5 = 1$.
Fitness(10010,*1***) = $(4-1)/5 = 0.6$.
Roulette Wheel(10010); 1****, ***1*, Crossover at Position 3.
Parent1 = $1**||**$ Child1 = 1**1*(Itemset AD).
Parent2 = $***||1*$ Child2 = *****(Invalid).

Table 4.4 Itemset List after the First Iteration

String	Itemset	Support	Fitness
1****	A	1	1
1*	B	0	0.6
***1**	C	0	0.6
****1*	D	1	1
*****1	E	0	0.6
1**1*	A,D	1	1

Roulette Wheel; 1****, *1***, ****1.
Parent1 = 1 * *|| * * Child1 = 1***1(Itemset AE).
Parent2 = * * *|| * 1 Child2 = *****(Invalid).

Table 4.5 Itemset List after the Second Iteration

String	Itemset	Support	Fitness
1****	A	2	1
1*	B	1	1
***1**	C	0	0.6
****1*	D	1	0.6
*****1	E	1	1
1**1*	A,D	1	0.6
1***1	A,E	1	1

Roulette Wheel; 1****, *1***, ****1, 1***1.
Parent1 = 1 * || * *1 Child1 = 1****(Itemset A).
Parent2 = *1|| * ** Child2 = *1**1(Itemset BE).

Table 4.6 Itemset List after the Final Iteration

String	Itemset	Support	Fitness
1****	A	2	1
1*	B	1	1
***1**	C	0	0.6
****1*	D	1	0.6
*****1	E	1	1
1**1*	A,D	1	0.6
1***1	A,E	1	1
11	B,E	1	1
11***	A,B	1	1
11**1	A,B,E	1	1

Roulette Wheel; 1****, *1***, ****1, 1***1, *1**1, 11***, 11**1.
Reproduction: Perfect match between 11**1 and 11001.
Processing stops for this transaction.

Since a perfect match(number of ones in the encoded transaction 10010 matching the number ones in the generated string 1**1*) has been reached, the process stops and the updated table with the string 1**1* is as shown in the Table 4.4.

The next transaction data is read(TID 2) from the database and genetic operators are applied on the updated database as shown in the Table 4.5. There is a possibility that a perfect match may not be obtained in the first trial, as in the case of trial 1 of TID 1. However, new itemsets are created which are added to the list of itemsets strings and also to the Roulette wheel. The process is repeated for a fixed number of trials and if a perfect match is not found, mutation operator is applied. The iteration is repeated and the final itemset list is as shown in the Table 4.6. A perfect match for the itemset ABE is generated after ten iterations and hence the process stops.

Inter Transaction Association Rule Mining: When the window size is greater than one, then the algorithm generates inter transaction association rules. The database is prepared for mining inter transaction association rules. The major task in this phase is to organize the transactions. We consider equal length transactions and the window size is taken as three. The output of the Prepare_Database() is the augmented transaction T'. Now we run the algorithm DMARG on this augmented transaction T'. This data preparation step is not necessary in the case of intra transaction association rule mining as the window size is one.

Since the rules generated on the augmented transaction database T' is explosive, only few rules of interest are considered in the Table 4.8. As the string length is three and the window size considered is also three, the first three bits in the augmented

Table 4.7 Transaction Database and Corresponding Augmented Transactions

TID	Transaction	Binary Encoded String	Augmented Trans T'
T_0	AB	110	110@101@111$(T_0 \cup T_1 \cup T_2)$
T_1	AC	101	101@111@110$(T_1 \cup T_2 \cup T_3)$
T_2	ABC	111	111@110@011$(T_2 \cup T_3 \cup T_4)$
T_3	AB	110	
T_4	BC	011	

Table 4.8 Inter Transaction Association Rules

Example Rules	Support	Rules generated
1** *** ***	3	A[0]
1* * ***	2	B[0]
**** 1** ***	3	A[1]
11* *** ***	2	A[0]B[0]
1** 1** ***	3	A[0]A[1]
1* 1 ***	2	B[0]A[1]
11* 1** ***	2	A[0]B[0] A[1]

transaction T' belong to TID T_0, similarly second three bits belong to TID T_1 and the last three bits belong to TID T_2 in the first row of Table 4.7. Considering the itemsets, A[0]B[0]A[1] and A[0]B[0] with support 2 in Table 4.8., we can conclude a valid inter transaction association rule i.e., $A[0]B[0] \Rightarrow A[1]$.

Distributed Mining of Association Rules: Consider the example of a transaction database present in two client nodes H_1 and H_2 as shown in Table 4.9. The genetic operators are applied on the database of each node to generate local large itemsets as shown in the Table 4.10.

Given the local itemsets of H_1 and H_2 in Table 4.10, the next step is to obtain a global list which is the aggregate of rule strings produced in all the hosts. This is the responsibility of the server. The server polls each of the clients on a round robin basis. Node H_1 is first prompted to send its rule string to the server. In reply, H_1 sends the first string in its list (i.e., 1****). The server now sends a copy of this rule string to all clients. These nodes are instructed to send the local count or support of the string, if it is present in their database. Thus, node H_1 sends the value 2 and node H_2, 1. The process is repeated for all the strings in the database of node H_1. Once node H_1 exhausts its list of strings, the server moves to the next node, i.e., node H_2. Node H_2 then searches for strings from its database that have not been considered so far, for example, *1**1. It then sends the string to the server. The server gathers the global support of this string by surveying the other nodes. The above algorithm is repeated until every node in the network has been scanned for local rules. The result is shown in Table 4.11.

Table 4.9 Transaction Database in Two Nodes H_1 and H_2

TID	Node H_1 Transaction	Encoded Trans.	TID	Node H_2 Transaction	Encoded Trans.
100	AD	10010	103	BE	01001
101	ABE	11001	104	ABD	11010

Table 4.10 Local Rules Generated at Nodes H_1 and H_2

List	Node H_1 Rule String	Sup	List	Node H_2 Rule String	Sup
1	1****	2	1	1****	1
2	*1***	1	2	*1***	2
3	***1*	1	3	***1*	1
4	****1	1	4	****1	1
5	1**1*	1	5	*1**1	1
6	1***1	1	6	11***	1
7	11***	1	7	1**1*	1
8	11**1	1	8	*1*1*	1
9	*1**1	1	9	11*1*	1

Table 4.11 Global Rules Generated at Server

List	Rule String	Global Count	List	Rule String	Global Count
1	1****	3	7	1**1*	2
2	*1***	3	8	*1*1*	1
3	***1*	2	9	*1**1	2
4	****1	2	10	11**1	1
5	1***1	1	11	11*1*	1
6	11***	2			

4.5 Performance Analysis

Transaction data, in the form of binary strings of desired length, was generated using a Synthetic Data generator[19]. The number of items in a transaction was 1000, the average size of each, 10, the maximum number of items, 25 and the average size of maximal potentially large itemsets was 4. DMARG is compared with a standard algorithm, FUP (Fast UPdate), which works on the principles of Apriori.

Figure 4.1, shows the plot of the execution time of DMARG versus the number of transactions. The transaction databases are varied in the range of [100k-500k]. The support considered is 1. This low value has been explicitly chosen to exhaustively generate all the possible combinations of itemsets. The execution time varies from 2 minutes to 25 minutes. To avoid losing any small itemsets which may become large

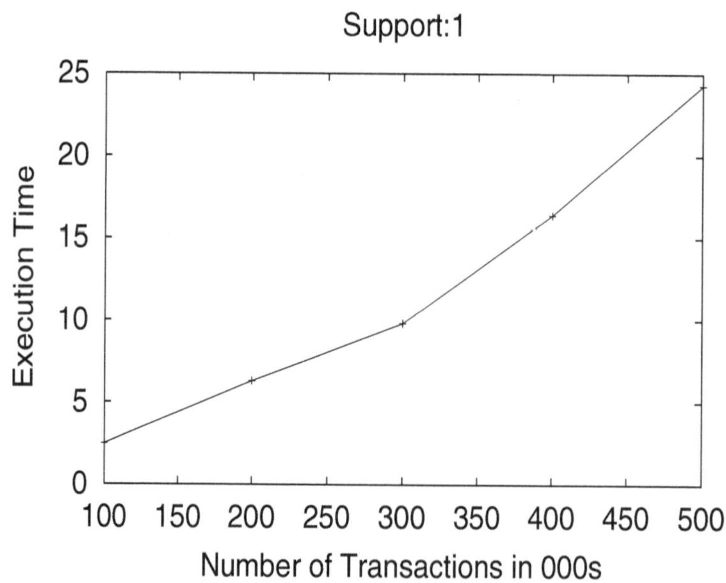

Fig. 4.1 Execution Time of DMARG vs Number of Transactioons (Intra Transaction)

in future, pruning is not employed. It is evident from Figure 4.1, that DMARG can handle large databases without compromising performance and has excellent scalability. The experimental analysis shows that the time complexity of the algorithm DMARG is linear with respect to the number of transactions in the database.

Figure 4.2, compares the execution time of DMARG with the standard algorithm for incremental mining, FUP. The initial database DB was of size 100k. The incremental database *db* varies from 1% to 100% of DB. The execution time for DMARG vary from 2.6 seconds to 269.5 seconds. Under identical conditions, the execution time of FUP varied from 20.7 seconds to 861.4 seconds. It is obvious from Figure 4.2, DMARG is faster than FUP. This difference in the time complexity between DMARG and FUP is due to the single scan feature of DMARG. Figure 4.3, plots the variation of the execution time of DMARG for various decrements of DB. The initial database DB considered is of size 100k. The decrements are varied from 1% to 80% of DB. DMARG is versatile in handling both increments and decrements to DB. For inter transaction association rules the database of size of 1,00,000 transactions with 300 items and the window size of three is taken. The execution time of DMARG is compared with Extended-Apriori. Figure 4.4, shows the graph of comparision of execution times for E-Apriori and DMARG and DMARG is faster than E-Apriori.

In case of distributed mining, simulations are performed on an Ethernet Local Area Network. Data is distributed uniformly amongst all the processors. Figure 4.5, shows the variation of execution time in minutes versus the number of processors. The execution time decreases as the increase in the number of processors. The time

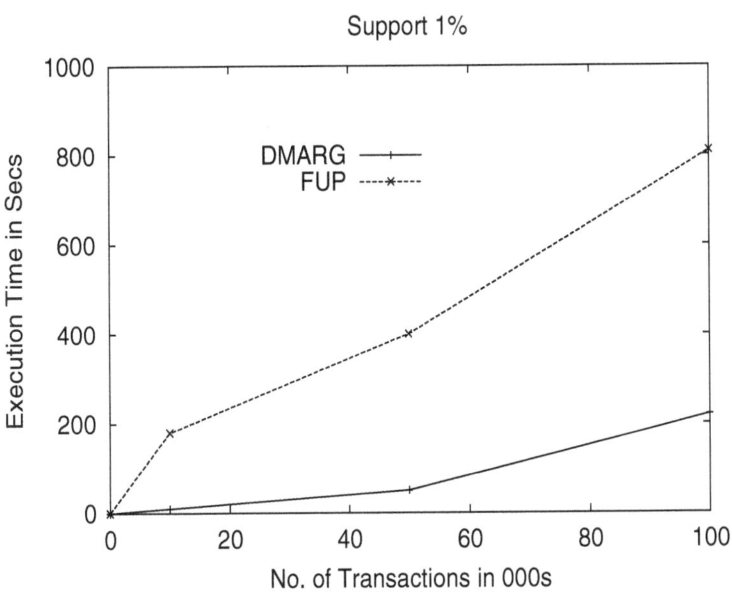

Fig. 4.2 Execution Time vs Increment Database (Intra Transaction)

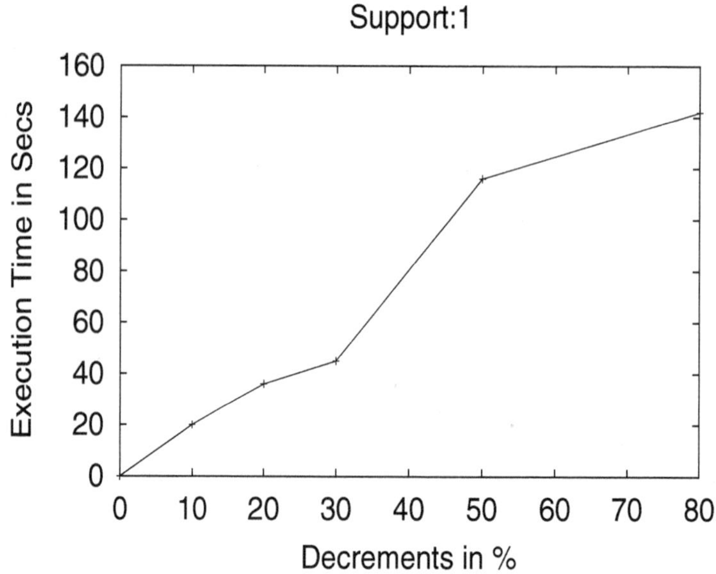

Fig. 4.3 Execution Time vs Decrement Database (Intra Transaction)

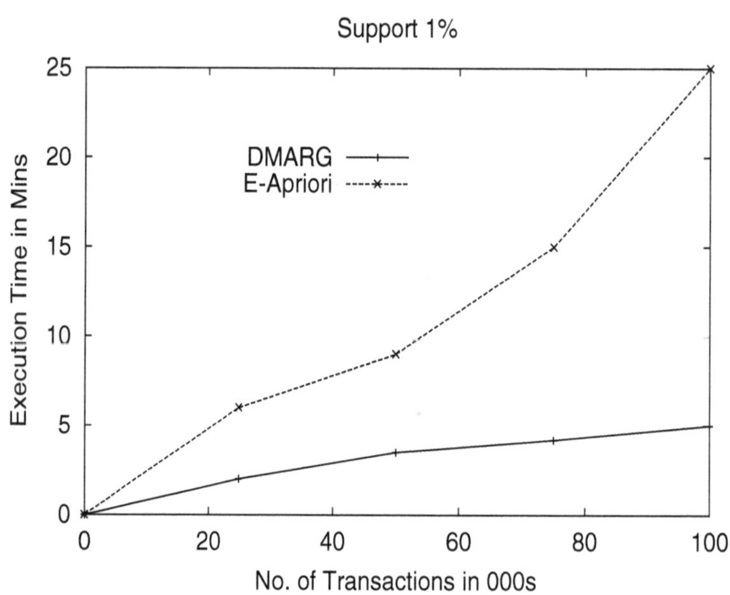

Fig. 4.4 Execution Time vs Number of Transactions (Inter Transactions with Window Size 3)

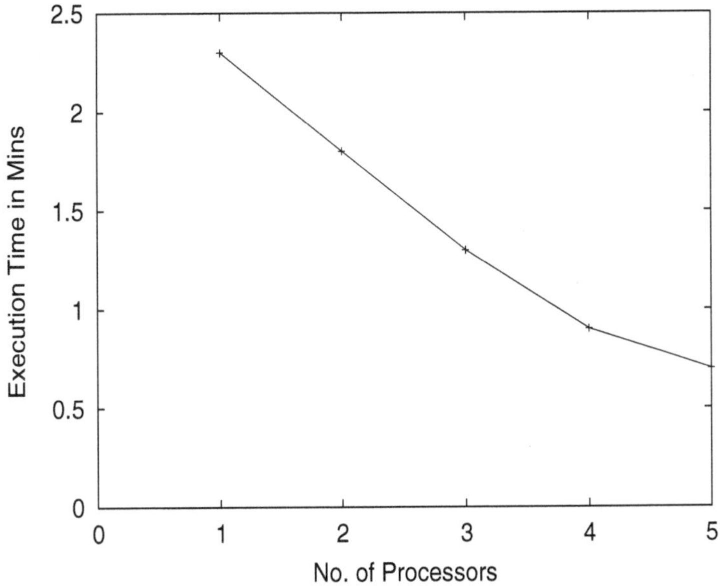

Fig. 4.5 Execution Time vs Number of Transactions (Distributed Mining)

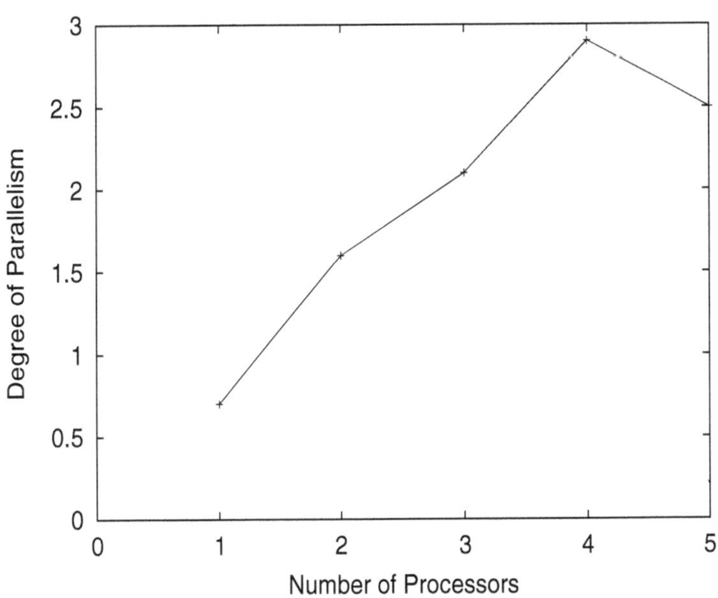

Fig. 4.6 Degree of Parallelism vs Number of Processors (Distributed Mining)

taken for the synchronization is negligible. Instead of synchronizing at every iteration, the synchronization is done at the final stage without pruning away redundant rules, hence the DDMARG consumes minimal time for synchronization. In our experiments, the database is kept constant and the number of processors are varied. We define the degree of parallelism as ratio of the execution time for running DMARG on the entire database versus the execution time for running DDMARG on n processors, where the database is distributed equally amongst all the processors. The Figure 4.6, shows the variation of degree of parallelism versus the number of processors (n). The ratio varies from 1 to 2.5. The curve does not follow the linearity because of communication and synchronization overheads.

4.5.1 Experiments on Real Data

The experiments are conducted on BMS-POS dataset [20]. The IBM-POS dataset contains several years worth of point-of-sale data from a large electronics retailer. The product categories in this dataset are used as items. The transaction in this dataset is a customer's purchase of all the product categories at a time. Our objective is to find associations among the different product categories purchased by the customers. The dataset contains 5,15,597 transactions, 1,657 distinct items and 164 being the maximum transaction size, where as the average transaction size is 6.5. Figure 4.7 shows the number of association rules generated with respect to minimum support. Table 4.12 gives the summary of the number of association rules generated for a given minimum support and the corresponding running time requirements in seconds.

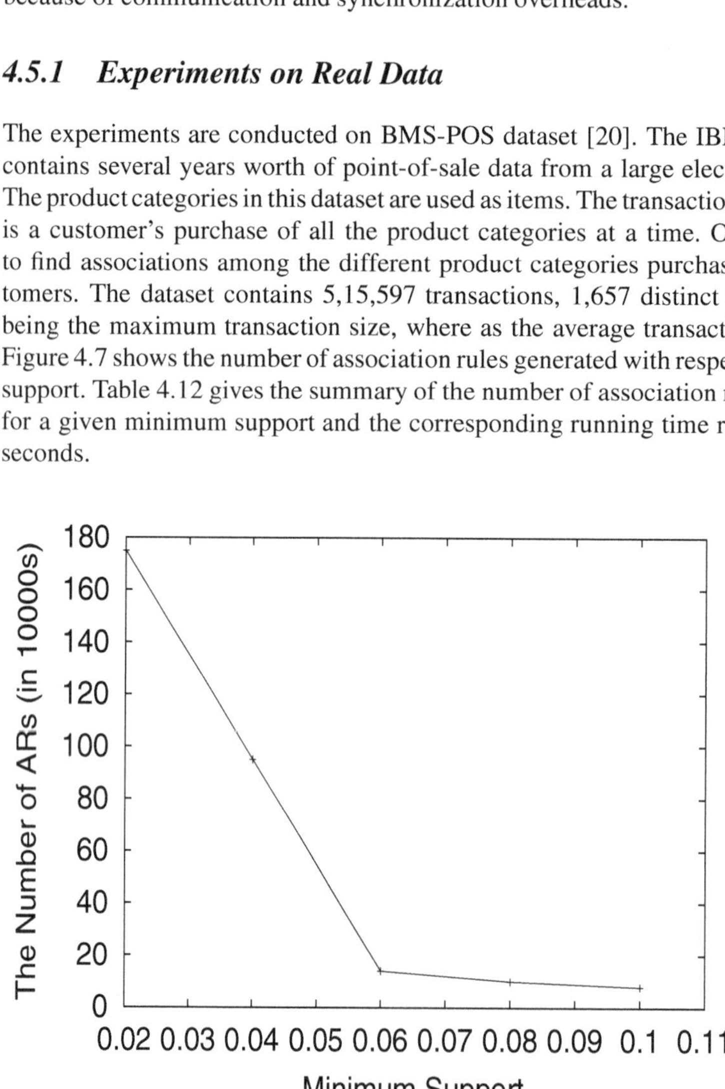

Fig. 4.7 Total Number of Association Rules versus Minimum Support(%) on IBM–POS Dataset

Table 4.12 Number of Association Rules and Running Time Requirement for a given Minimum Support for IBM-POS Dataset

Min. Support	No. of ARs	Running Time
0.02	17, 564, 123	23.2
0.04	1,189, 422	18.4
0.06	2, 14, 799	14.2
0.08	1, 43, 298	13.8
0.10	1, 26, 226	11.6

4.6 Summary

We examine in this chapter, the issue of mining intra and inter transaction association rules for dynamic databases. The algorithm proposed is effective in generating large itemsets and in addressing updations in the database. The rules generated using DMARG, are qualitatively sound. The single scan feature reduces computational costs significantly. The algorithm generates inter transaction association rules using the augmented transactions. The performance of DMARG is studied and compared with that of FUP(in case of intra transactions) and E-Apriori(in case of inter transactions), the results show that DMARG is faster and more efficient than FUP and E-Apriori. DMARG also takes care of decremental transactions. DMARG handles identical and skewed databases. The algorithms DMARG and DDMARG are efficient and effective; Efficient, because of the inherent speed and scalability of genetic algorithms in performing data mining and a low communication overhead achieved when building global knowledge. It is effective because, the quality of the rule chromosomes are not compromised in the process. Further, pruning is not done on the local string chromosomes set. This ensures that no rule is lost until the final filtering process on the server side. In DMARG, a single scan technique, the support is required only before the association rule generation, to build stronger and relevant rules.

References

1. Pujari, A.: Data Mining Techniques. University Press (2000)
2. Shenoy, P.D.: Evolutionary Approach for Mining Association Rules on Dynamic Databases. In: Whang, K.-Y., Jeon, J., Shim, K., Srivastava, J. (eds.) PAKDD 2003. LNCS, vol. 2637, pp. 325–336. Springer, Heidelberg (2003)
3. Srinivas, M., Patnaik, L.M.: Genetic Algorithms: A Survey. IEEE Computers 27(6), 17–26 (1994)
4. Cheung, D.W., Han, J., Vincent, T., Wong, C.Y.: Maintenance of Discovered Association Rules in Large Databases: An Incremental Updating Technique. In: Proc. of Intl. Conf. on Data Engineering, pp. 106–114 (Feburary 1996)
5. Agrawal, R., Srikant, R.: Fast Algorithms for Mining Association Rules. In: Proc. of Intl. Conf. on Very Large Data Bases, Santiago, pp. 478–499 (September 1994)

6. Park, J.S., Chen, M.S., Yu, P.: An Effective Hash based Method for Mining Association Rules. In: Proc. of ACM-SIGMOD Intl. Conf. on Management of Data, pp. 175–186 (May 1995)
7. David, W., Cheung, L., Lee, S.D., Kao, B.: A General Incremental Technique for Maintaining Discovered Association Rules. In: Proc. of 5th Intl. Conf. on Database Systems for Advanced Applications, Melbourne (1997)
8. Ganthi, V., Gehrke, J., Ramakrishnan, R.: DEMON: Mining and Monitoring Evolving Data. IEEE Trans. on Knowledge and Data Engineering 13(1), 50–62 (2001)
9. Thomas, S., Chakravarthy, S.: Incremental Mining of Constrained Associations. In: Prasanna, V.K., Vajapeyam, S., Valero, M. (eds.) HiPC 2000. LNCS, vol. 1970, pp. 547–558. Springer, Heidelberg (2000)
10. Thomas, S., Bodagal, S., Alsabti, K., Ranka, S.: An Efficient Algorirhm for the Incremental Updation of Association Rules in Large Databases. In: Proc. of KDD 1997, pp. 263–266 (1997)
11. Fildman, R., Aumann, Y., Amir, A., Manila, H.: Efficient Algorithms for Discovering Frequent Sets in Incremental Databases. In: Proc. of SIGMOD Workshop on Research Issues in DMKD, pp. 59–66 (1997)
12. Harishta, J.: Technical Report on Data Mining, Dept. of CSA, IISc (1997)
13. Lu, H., Feng, L., Han, J.: Beyond Intra-Transaction Association Analysis: Mining Multi-Dimensional Inter-Transaction Association Rules. ACM Transactions on Information Systems 18(4), 423–454 (2000)
14. Anthony, K., Tung, H., Lu, H., Han, J., Feng, L.: Breaking the Barrier of Transactions: Mining Inter-Transaction Association Rules. In: Proc. of Intl. Conf. on Knowledge Discovery and Data Mining, pp. 297–300 (August 1999)
15. Agrawal, R., Shafer, J.: Parallel Mining of Association Rules. IEEE Tran. on Knowlegde and Data Engineering 8(6), 962–969 (1996)
16. Cheung, D.W., Han, J., Ng, V.T., Fu, A.W., Fu, Y.: A Fast Distributed Algorithm for Mining Association Rules. In: Proc. of IEEE Intl. Conf. on Parallel and Distributed Information Systems, pp. 31–42 (December 1996)
17. Han, E.-H., Karypis, G., Kumar, V.: Scalable Parallel Data Mining for Association Rules. In: Proc. of ACM-SIGMOD Intl. Conf. on Management of Data, pp. 277–288 (1997)
18. Shintani, T., Kitsuregawa, M.: Parallel Mining Algorithms for Generalized Association Rules with Classification Hierarchy. In: Proc. of ACM SIGMOD, Intl. Conf. on Management of Data, pp. 25–36 (1998)
19. Synthetic Data Generator, `http://www.almaden.ibm.com/cs/quest/syndata.html`
20. Real Data Set from KDD CUP, `http://www.ecn.purdue.edu/KDDCUP/data/BMS-WebView-1.dat.gz`

Chapter 5
Evolutionary Approach for XML Data Mining

Abstract. Extensible Markup Language(XML) has emerged as a medium for interoperability over the Internet. The XML technology, with its self-describing and extensible tags, is significantly contributing to the next generation semantic web. The present search techniques used for HTML and text documents are not efficient to retrieve relevant XML documents. In chapter four, Self Adaptive Genetic Algorithms for XML Search(SAGAXSearch) is presented to learn about the tags, which are useful in indexing. The indices and relationship strength metrics are used to extract fast and accurate semantically related elements in the XML documents. Experiments are conducted on the DataBase systems and Logic Programming (DBLP) XML corpus and are evaluated for precision and recall. The proposed SAGAXSearch outperforms the existing algorithms XSEarch and XRank with respect to accuracy and query execution time.

As the number of documents published in the form of XML is increasing, there is a need for selective dissemination of XML documents based on user interests. In the proposed technique, a combination of Self Adaptive Genetic Algorithms and multi class Support Vector Machine is used to learn a user model. Based on the feedback from the users, the system automatically adapts to the users preference and interests. The user model and a similarity metric are used for selective dissemination of a continuous stream of XML documents. Experimental evaluations performed over a wide range of XML documents, indicate that the proposed approach significantly improves the performance of the selective dissemination task, with respect to accuracy and efficiency.

On similar grounds, there is a need for categorization of XML documents into specific user interest categories. However, manually performing the categorization task is not feasible due to the sheer amount of XML documents available on the Internet. A machine learning approach to topic categorization which makes use of a multi class SVM for exploiting the semantic content of XML documents is also presented. The SVM is supplemented by a feature selection technique which is used to extract the useful features. Experimental evaluations performed over a wide range of XML documents indicate that the proposed approach significantly improves the performance of the topic categorization task, with respect to accuracy and efficiency.

K.R. Venugopal, K.G. Srinivasa, L.M. Patnaik: Soft Comput. for Data Min. Appl., SCI 190, pp. 81–118.
springerlink.com

5.1 Semantic Search over XML Corpus

Extensible Markup Language (XML) has been recognized as a standard for describing the data format and its meaning. The user defined tags associate semantics with the contents of XML documents. Hence XML is a medium for interoperability over the Internet. With these advantages, the amount of data that is being published on the Web in the form of XML is growing enormously and many naive users find the need to search over large XML document collections. The keyword query is a search technique that does not require the users to know the structure of the underlying data. There is no need to learn complex query languages to discover knowledge. Thus, the keyword search over XML documents in the present context is of significant importance.

Keyword search over large document collections has been extensively used for text and HTML documents [1] and it has two main drawbacks. First, Search engines are not as intelligent as their users. For example, a keyword search *Kevin Database Technology* retrieves documents in which *Kevin* is the author and also documents in which *Kevin* is mentioned in the references with equal priority, though the former is more semantically relevant to the user. The second drawback is that keyword queries are inherently flexible in nature and can produce large number of results. The results are of varying relevance to the user and they need to be ranked. The time taken to rank the results should be a small portion of the total query execution time. In contrast, a structured query language retrieves only the most relevant results, but the complex query syntax makes it unsuitable for naive users. Thus an approach which has the flexibility of keyword queries that still retains the accuracy of a query language would be most suitable. The keyword search over XML documents presents many new challenges [2].

Search engines for XML can be classified into two general categories: database-oriented and information retrieval-oriented. In the database approach [3], the XML documents are decomposed and stored in relational database. However, query processing becomes expensive since, in many cases, an excessive number of joins is required to recover information from the fragmented data. Object-oriented databases have been associated with XML document collections [4]. In this case, retrieval of information from XML documents is considered as an object view problem.

Extensive research has been done on structured declarative queries over XML documents. A structured declarative query is supported by XQuery [5], which is analogous to SQL queries over relational databases. Though XQuery can achieve perfect precision and recall, they require user to learn query semantics and in cases where the user is unaware of the document structure, a search cannot be performed. An improvement over XQuery that has elegant syntax and semantics is developed in [6].

Information retrieval techniques can consider XML documents as normal text documents, with additional markup overhead. There are several ways of handling the tags. For simplicity the tags can simply be ignored but the document loses its semantics, leading to lower retrieval performance. When tags are taken into consideration, search can retrieve documents containing certain tags, or certain words. Keyword search over XML documents falls under this category.

In Information Retrieval, Genetic Algorithms have been used in several ways [7] but in a different context. Genetic Algorithms have been used to modify user queries [8, 9] and for automatic retrieval of keywords from documents. In [10], GA is applied to adapt multiple matching functions obtained from the combination of scores using individual matching functions. This is used to rank and retrieve documents. In [11] GA has been used for mining of HTML structures. The algorithm learns the important factors of HTML tags through a series of queries.

Keyword search over XML documents is supported by XKeyword [12], XRANK [13] and XSEarch [14]. All these keyword search techniques have elaborate ranking schemes. The simplicity of the search queries i.e., keywords make these techniques suitable for naive users. But, precision and recall values tend to suffer and the extensive ranking function employed acts as an overhead during query execution.

In XRANK, the hierarchical and hyperlinked structure of XML documents are taken into account while computing the ranks for the search results. A ranking technique at the granularity of XML elements is considered here. XRANK can query over a mix of XML and HTML documents.

XSEarch introduces a concept known as interconnected relationship. However, checking for the interconnected relationship is a huge overhead during runtime. Moreover, XSEarch suffers from drawbacks similar to other keyword search engines: unimpressive precision and recall values. In the proposed SAGAXsearch algorithm, the association of Self Adaptive Genetic Algorithms with keyword queries ensures high accuracy i.e., very few non-relevant fragments (high precision) and most of the relevant fragments (high recall) will be selected as results.

5.2 The Existing Problem

Consider a keyword search query *operating system* over the XML document in Table 5.1. The keyword is associated with the <inproceedings> tag (line 19) and is a relevant result, but returning only the elements related to the <inproceedings> tag (line 19) would be more intuitive than returning the whole document. Thus, granularity of the search terms must be refined when searching over XML document corpus [14]. Second, the result of a keyword search over XML documents must be semantically interconnected document fragments. Consider the Keyword search *Ananth Synchronization mechanism* over the XML document shown in Table 5.1. Though, the keywords exist independently in the XML document (line 4, 15 in Table 5.1), they belong to different <inproceedings> tags. The author *Ananth* is not semantically interconnected to the title *synchronization mechanism*. Thus, only semantically interconnected document fragments should contribute to the search results.

Finally, XML documents include large amounts of textual information and part of this is rarely searched. Building a single index for the whole document makes the index bulky and difficult to manage. Thus, there is a prospect of improving the search accuracy and query execution time by separating the frequently searched tags from the occasionally searched ones and building separate indices for both, and this

Table 5.1 Example of a XML Document

```
<dblp>
<inproceedings>
<author>I S Vipin</author>
<author>C G Ananth</author>
<author>G Sarah</author>
<title>Land Use: Problems and Experience</title>
<pages>135-172</pages>
<year>1979</year>
<crossref>conf/ibm/1979</crossref>
<booktitle>Data Base Techniques</booktitle>
<url>db/conf/ibm/db79.htm#zaGM79</url>
</inproceedings>
<inproceedings>
<author>A N Ravi</author>
<title>Synchronization Mechanisms</title>
<pages>2-22</pages>
<year>1980</year>
<crossref>conf/ibm/1980</crossref>
<booktitle>Operating Systems</booktitle>
<url>db/conf/ibm/db80.htm#saito80</url> </inproceedings>
</dblp>
```

is explored by the use of SAMGA(Refer to Chapter 2). A search over the XML documents in the decreasing order of the importance of the tags is accurate and efficient.

The human search strategy which is efficient for small documents, is not viable when performing search over enormous amounts of data. Hence, making search engines cognizant of the search strategy using GA, can help in fast and accurate search over large document collections. We have explored the possibility of Retrieval and ranking of XML fragments based on keyword queries. Self adaptive and real coded Genetic Algorithms are used for learning tag information. A measure of distance metric between the keywords among the XML documents is proposed. Genetically learned tag information is used to retrieve semantically interconnected document fragments.

5.2.1 *Motivation*

Consider the XML document fragments, an excerpt from a health-care record. Consider a keyword search *Vinu salbutamol* over the XML document in Table 5.2. A standard HTML search engine would consider the whole document in Table 5.2 as a suitable response, due to the presence of both the terms in the search query. However, in XML environment the two search terms occur as totally unrelated elements in the document as they belong to the medical records of different patients. In the XML document of Table 5.2 the keyword *penicillin* appears in two different contexts; first it is associated with the <administer> tag and then with the <drug_allergy> tag and the tag name precisely categorizes between the two

Table 5.2 Example of Health Care Record

```
<medical_records>
<patient>
<name>Vinu Krishnan</name>
<record_id>4312</record_id>
<administer>Penicillin</administer>
<drug_allergy>None</drug_allergy>
</patient>
<patient>
<name>Victor James</name>
<record_id>4313</record_id>
<administer>Salbutamol</administer>
<drug_allergy>Penicillin</drug_allergy>
</patient>
</medical_records>
```

occurrences. Additional information like name, record identifiers are also explicitly captured using application specific self explanatory tags. This is useful in keyword search over XML documents. Thus exploiting the tagged and nested structure of XML can help in effective knowledge discovery.

5.3 XML Data Model and Query Semantics

In this section, we briefly describe the XML data model and the keyword query semantics for search over XML documents.

Data Model: The Extensible Markup Language (XML) is a human readable, machine understandable, general syntax for describing hierarchical data, applicable to a wide range of applications. XML allows users to bring multiple files together to form a compound document. The XML document consists of nested elements starting from the root and corresponding associated values. The XML document can be considered as a directed, node-labeled *data graph* $G = (X,E)$. Each node in X corresponds to an XML element in the document and is characterized by a unique *object identifier*, and a *label* that captures the semantics of the element. Leaf nodes are also associated with a sequence of keywords. E is the set of edges which define the relationships between nodes in X. The edge $(l,k) \in E$, if there exists a directed edge from node l to node k in G. The edge $(l,k) \in E$ also denotes that node l is the parent of node k in G. Node l is also the ancestor of node k if a sequence of directed edges from node l leads to node k. An example XML document tree is shown in Figure 5.1.

Query Semantics and Results: Let the XML document tree be called τ. Let x be an interior node in this tree. We say that x directly satisfies a search term k if x has a leaf child that contains the keyword k and x indirectly satisfies a keyword k if some descendant of x directly satisfies the search term k. A search query $q = \{k_1, k_2, ...k_m\}$ is satisfied by a node x iff x satisfies each of $k_1, k_2, ...k_m$ either directly or indirectly.

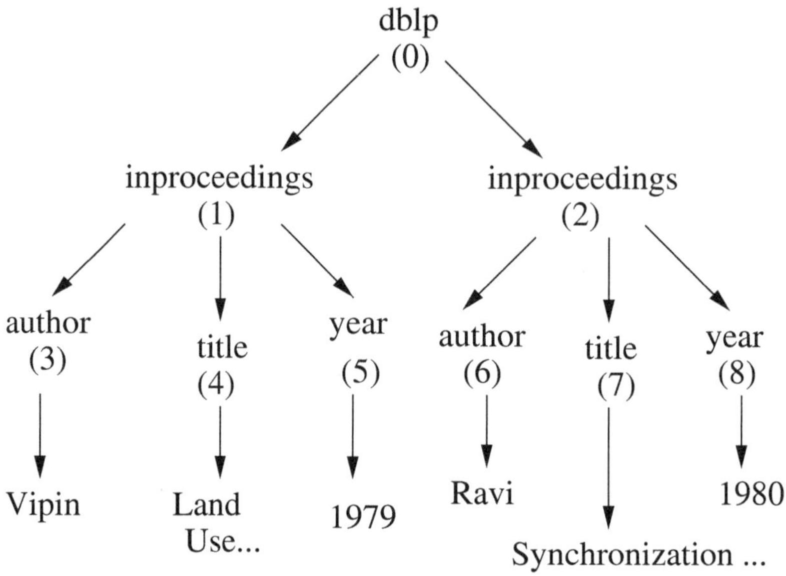

Fig. 5.1 Example XML Document Tree

For example, in the XML tree shown in Figure 5.1, *inproceedings(1)* satisfies the search term *Vipin* and the search term *Vipin 1979* but not the term *Vipin 1980*.

The nodes obtained as result should also be semantically related. Semantically related nodes are nodes that appear in the same context. The various steps in the working of SAGAXSearch are enlisted below.

1. A representative training set is chosen to assist the genetic learning of tags.
2. The keyword queries and the relevant search results are collected from the user.
3. The GA retrieves tag combination which can answer a maximum training queries.
4. Separate indices are built for the frequently used and occasionally used tag combinations.
5. A search over the XML documents in the decreasing order of importance of tags is performed.
6. The search produces only semantically related results.

5.4 Genetic Learning of Tags

XML documents include extensible tags for formatting the data. These tags are self-describing and thus represent the semantics of the contents associated with them. For example, consider the keyword *Widows Location*. Due to the ambiguity in the keywords, it is not possible to determine context and the exact meaning of the keywords. But in the case of XML, using self-describing tags such as <Operating Systems> or <Building Plan> the context of the keywords can be precisely highlighted. The combination of tags in the XML documents also helps in revealing more information about the contents of the documents. For example, the tag combination <author>, <age> <date of birth> describes the personal details of the author. Whereas, the tags <author>, <books>, <publication year> are more concerned about the work of the author rather than his personal details. An XML document usually includes a large number of tags and only a small number of these may be of interest to a user. Hence, a user profile that stores only the tag combinations interesting to a user is more accurate. Using genetic algorithms to learn user profiles has two advantages. First, the tag combinations which are interesting to a user can be extracted. This task can be automatically done using the search terms and the relevant feedback given by the users. The tags which are not interesting to a user can be omitted from the user profile. Second, the context of the search terms given by the users can be adjudged and a profile can be constructed accordingly.

Consider an XML document collection with n documents. Each of the distinct tags in this document collection is stored in a tag pool. Let $T = \{t_1, t_2, ...t_m\}$ be the tag pool with m tags and t_i represents i^{th} tag. Usually, for document collection the tag pool is huge. The purpose of SAMGA is to select from the tag pool, the tag combinations which are interesting to a user. For the system to learn the user interests, the user has to first issue a set of search queries $q = \{k_1, k_2, ...k_m\}$. The documents satisfying the search terms are retrieved as results. The user has to classify the results relevant to him. This is the feedback given to the system in order to learn the user interest. The fitness function used in the GA is given by,

$$fitness = \alpha * (\sum_{i=1}^{N} \frac{freq(i, S_{tag})}{rank(i)}) + (1 - \alpha)N \tag{5.1}$$

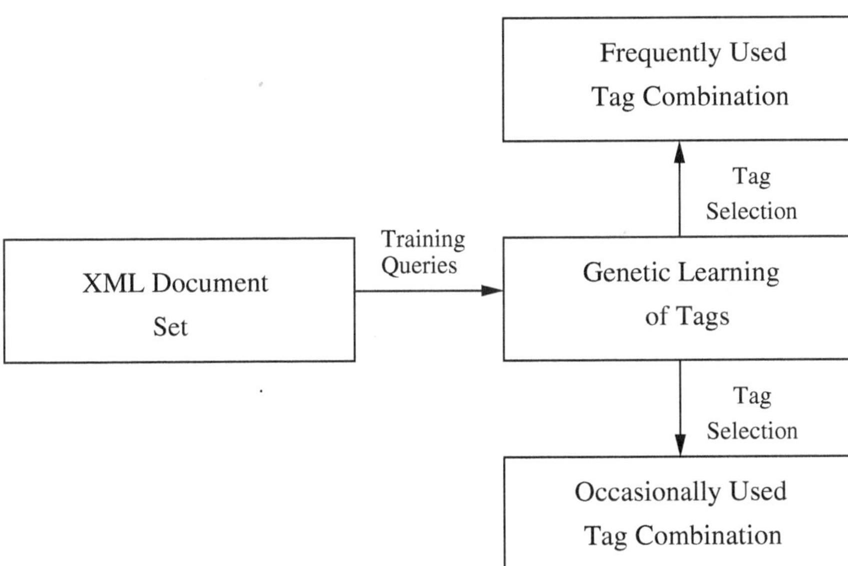

Fig. 5.2 Genetic Learning of Tags

Table 5.3 Genetic Algorithms for Learning Tag Information

Begin
Initialize the population P_i by assigning random tag weights to $j = \{j_1, j_2, ...j_l\}$.
for gen = 1 : maximum generation limit, *do*
(a) Order the tags by their decreasing weights and select the top k tags. Let $S_{tag} = \{t_1, t_2, ...t_k\}$ represent the selected tags.
(b) Evaluate fitness using Equation 5.1.
(c) For the population P_i perform a selection with *stochastic universal sampling* as the selection operator.
(d) Perform *discrete recombination* on the selected individuals of the population P_i.
(e) Perform mutation on the individuals of the population P_i.
Next.
End

where N is the number of documents retrieved with a specific tag configuration, S_{tag} is the set of top k tags with highest tag weights. $freq(i, S_{tag})$ is the frequency of occurrence of the terms of the query $q = \{k_1, k_2, ...k_m\}$ within the tags in S_{tag} in the ith retrieved document. The retrieved documents are ranked according to the frequency of occurrence of the terms. The $rank(i)$ denotes the rank of the ith retrieved document provided the document is also classified as relevant by the user. α is a parameter that is used to express the degree of user preference for accuracy of the search results or the total number of documents that are retrieved. The architecture of the genetic learning system is illustrated in Figure 5.2. A real coded GA is used for learning the tag information in GaXsearch, and is explained in Table 5.3 and the application of SAMGA for XML search(SAGAXSearch) is given Table 5.4.

Consider a training set with n documents. Let $q = \{q_1, q_2, ...q_m\}$ be a collection of typical user queries where q_i represents the ith query and m is the total number of queries. The chromosome is represented as $j = \{j_1, j_2, ...j_l\}$ where j_i denotes the

Table 5.4 Self Adaptive Migration Model Genetic Algorithms(SAMGA) for Learning Tag Information

Initialize the population size and the mutation rate for each population.
Associate random tag weights with tags in the tag pool τ. This represents individuals of the initial population.
for generation = 1: maximum generation limit
for each population
for each individual select top k tags with highest tag weights.
Let $S_{tag} = \{t_1, t_2, ...t_k\}$ represent the selected tags.
Evaluate the fitness function using Equation 5.1.
Modify mutation rate using Equation 2.3.
Modify population size according to Equation 2.1.
Select individuals and perform the recombination operation.
If the average fitness of the ecosystem fails to change over two successive generations, migrate best individuals between populations.

weight of the tag i, l is the total number of distinct tags appearing in the document corpus which can be determined from the Document Type Definition (DTD) of the corpus. Thus, a tag weight is associated with each of the distinct tags appearing in the document collection.

The selection operator used in the algorithm is *stochastic universal sampling*. Here individuals of the population are assigned contiguous segments on a straight line based on their fitness values. Let b be the total number of individuals selected, which are placed on equidistant points over a line. The distance between the points is given by $\frac{1}{b}$. Such a selection scheme has a zero bias and minimum spread, and is found suitable for our algorithm.

The recombination operator used is intermediate recombination, where the variable values of the offspring are around and between the variable values of the parents. Offspring are generated according to the rule $O = a_1 * \rho(a_2 - a_1)$ where ρ is a scaling factor chosen over some interval and a_1, a_2 are the parents in the current generation. Geometrically intermediate recombination produces variables with a slightly larger hypercube than that defined by the parents but constrained by the values of ρ. A real valued mutation operation is also applied in the algorithm to explore new regions and make sure that good genetic material is never lost.

The result of the GA is the classification of tags as either frequently used or occasionally used. This precise categorization helps in maintaining separate indices for the information within the tags. The information within the frequently used tags is stored in an index called Most Frequently used Index (MFI) and the information within the occasionally used tags is stored in an index called Less Frequently used Index (LFI).

5.5 Search Algorithm

The response to a search over an XML document is not the document in its entirety but only semantically relevant document fragments. In this section we discuss identification schemes and semantic relationship between nodes in XML tree.

5.5.1 Identification Scheme

The granularity of search over XML documents is not at the document level, but at the node level in the XML document tree. Hence, an identification scheme for the nodes in the document tree is required. This is accomplished by encoding the position of each node in the tree as a data value before storing it in an index. Given the identification values of the nodes, the scheme must also be able to reconstruct the original XML tree. An identification scheme called Hierarchical Vector for Identification (*hvi*) is derived.

Let x be a node in the XML document tree τ. Then the Hierarchical Vector for Identification of x is given by, $hvi(x) = [\tau_{id(x)} p(x) s_j(p)]$, Here, τ_{id} is the unique identification number assigned to the XML document tree τ, and $p(x)$ is a vector which is recursively defined as, $p(x) = [p(parent(x)) s_j(parent(p(x)))]$ and $s_j(p)$ denotes

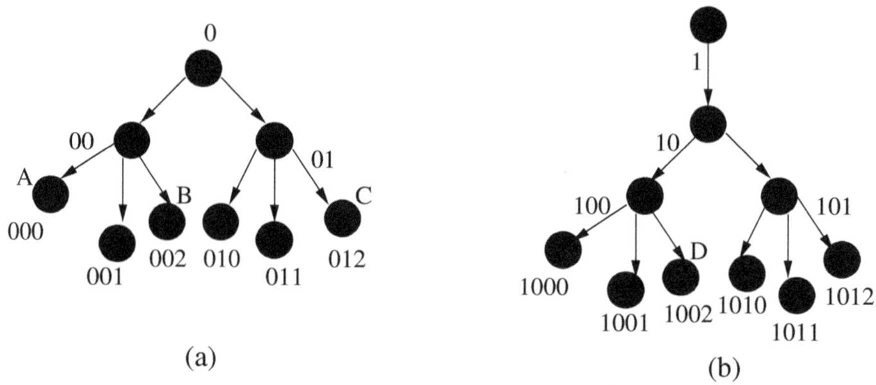

Fig. 5.3 Semantic Interconnection

the jth sibling of the parent p. With this identification scheme, each node captures its absolute position within the whole document. The *hvi* of a node identifies itself and all its ancestors. The *hvi* of various nodes in two XML documents are shown in Figure 5.3(a) and 5.3(b).

Theorem 1: Let $\tau_1, \tau_2, ...\tau_n$ represent the XML document trees of the documents with identification numbers (1, 2, 3,...*n*), where *n* is the number of documents. Then, $\{\exists x_i \in \{\tau_1, \tau_2, ...\tau n\} \wedge \exists x_j \in \{\tau_1, \tau_2, ...\tau n\}, such \quad that \quad x_i \neq x_j, hvi(x_i) \neq hvi(x_j)\}$, i.e., there exist no two distinct nodes among all the XML documents in the collection, such that they have the same *hvi*.

Proof
Case 1: Consider the two nodes x_i and x_j are present in different documents. i.e., $x_i \in \tau_i$ and $x_j \in \tau_j$ such that $\tau_i \neq \tau_j$. Since $\tau_i \neq \tau_j$, $\tau_{id}(x_i) \neq \tau_{id}(x_j), hvi(x_i) \neq hvi(x_j)$.

Case 2: Consider the two nodes x_i and x_j that are present in the same document. i.e., $\{x_i, x_j\} \in \tau$. Since both the nodes are in the same XML document $\tau_{id}(x_i) = \tau_{id}(x_j)$. But, since $x_i \neq x_j$ (from the statement of the theorem) and $\{x_i, x_j\} \in \tau$ there exist two possibilities $p(x_i) = p(x_j)$ or $p(x_i) \neq p(x_j)$. If $p(x_l) \neq p(x_j)$ then $hvi(x_l) \neq hvi(x_j)$. If $p(x_i) = p(x_j)$ then x_i and x_j represent different siblings of the same parent; therefore $hvi(x_i) \neq hvi(x_j)$. Thus, each element of all the XML documents in the document collection is assigned a unique identifier. From Figure 5.3(a) and 5.3(b), it can be observed that there are no two nodes with the same *hvi* values. The same is true for a collection of *n* documents.

5.5.2 Relationship Strength

Let $hvi(x_i)$ and $hvi(x_j)$ represent the *hvi* of two distinct nodes x_i and x_j, existing in the XML document tree τ. The length of the longest common prefix(*lcp*) for both the *hvi* is denoted as $lcp(x_i, x_j)$. Consider two keywords k_1, k_2. The relationship strength between these two keywords, denoted as $RS(k_1, k_2)$ is defined as,

$RS(k_1,k_2) = lcp(x_i,x_j)$, such that x_i directly satisfies k_1 and x_j directly satisfies k_2. The condition that the node should *directly satisfy* the keyword ensures that only those nodes satisfying the keyword and also having the longest length of their identification vectors (*hvi*), are selected while evaluating the Relationship Strength(RS). This is important because a node and all its ancestors satisfy a keyword, but a true measure of RS is represented only by the node which directly satisfies the keyword.

Consider two keywords k_1 and k_2 such that x_i directly satisfies k_1, x_j directly satisfies k_2. If $x_i \in \tau_i$ and $x_j \in \tau_j$ such that $\tau_i \neq \tau_j$, then $RS(k_1,k_2) = lcp(x_i,x_j) = 0$, since they do not share a common prefix. Thus a Relationship Strength value of zero indicates unrelated keywords (keywords in different documents). If $\tau_i = \tau_j$ and both the keywords are directly satisfied by the same node i.e., $x_i = x_j$, then $RS(k_1,k_2) = lcp(x_i,x_j) = length(hvi(x_i)) = RSmax$. Thus Relationship Strength values of two keywords can take integer values in the range of [0: RSmax] based on their occurrence and proximity in the XML document. The concept of Relationship Strength can be extended to a query $q = \{k_1,k_2,...k_m\}$ consisting of m terms. For example, in the document trees in Figure 5.3(a) and 5.3(b), the nodes A and B have a common prefix of length two. Thus, they have a RS value of two; similarly nodes A and C have an RS value of one. Whereas, nodes A and D have an RS value zero since they belong to different document trees.

5.5.3 Semantic Interconnection

In terms of the XML document tree, two nodes are semantically interconnected if they share a common ancestor and this ancestor is not the root of the document tree. As an illustration, consider the XML document tree in Table 5.1. The keywords *Vipin* and 1979 have a common ancestor, *inproceedings*(1). Thus, they are semantically interconnected. Whereas the keywords *Vipin* and 1980 have a common ancestor, *dblp*(0), which is the root of the document tree. Hence, the two keywords are not semantically connected.

Theorem 2: Two keywords k_1 and k_2 are semantically interconnected iff, $RS(k_1,k_2) > level_i + 1$, where $level_i$ is the first such level in the document tree where the degree of the node is greater than one.

Proof: Consider $level_i = 0$. The document tree is as shown in Figure 5.3(a). Since $level_i = 0$, $RS(k_1,k_2) > 1$. If $RS(k_1,k_2) > 1$, then there exist two nodes x_i,x_j that directly satisfy k_1,k_2 and with Hierarchical Vectors for Identification $hvi(x_i)$ and $hvi(x_j)$, such that $lcp(x_i,x_j) \geq 2$. Thus the two keywords have at least two common ancestors and of these only one can be the root of the document tree. The two keywords k_1,k_2 share at least one common ancestor apart from the root, hence they are Semantically Interconnected. For $level_i > 0$, the document tree is as shown in Figure 5.3(b). For example, in the XML document tree in Figure 5.3(a), since $level_i = 0$, RS must be greater than one for the nodes to be semantically relevant. The nodes A and B have an RS value of two and are semantically relevant. Whereas, nodes A and C have an RS value of one, and hence are not semantically relevant.

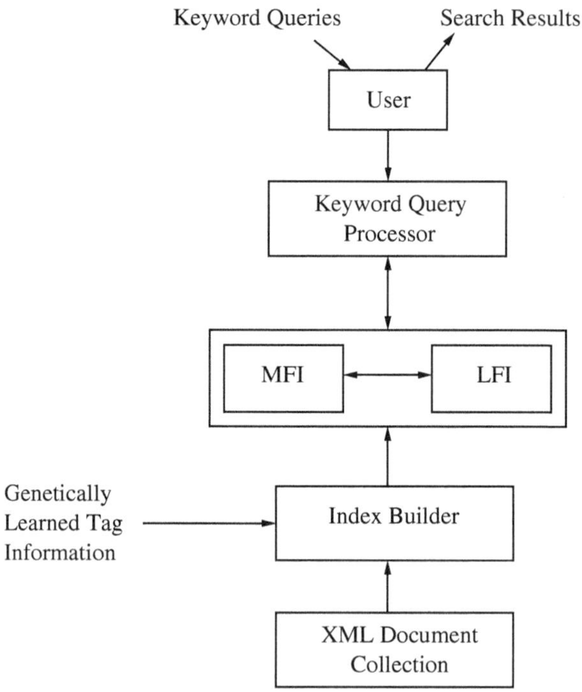

Fig. 5.4 Architecture of SAGAXSearch

The RS can also be used to rank the semantically interconnected keywords. The semantically interconnected keywords with higher RS values are the more relevant results and hence are better ranked than those having low RS values. The architecture to compute semantically related results from the Most Frequently used Index (MFI) and Less Frequently used Index (LFI) is shown in Figure 5.4.

The algorithm SAGAXSearch works as follows. Let $q = \{k_1, k_2, ...k_m\}$ be the search query where k_i represents the ith term in the query q and m represents the total number of terms in the query. The search algorithm first checks the length of the keyword query. If the query consists of a single term, a search over the MFI is performed. If the search is not successful, the algorithm continues search over the LFI. A failure to retrieve results from both MFI and LFI implies that the term is not found. The same technique is extended when searching with queries having more than one term. The only change is that, at each stage the semantic interconnection of the results is checked. Only semantically interconnected nodes are considered as the search results.

For example, in the XML document tree in Figure 5.3(a), since $level_i = 0$, RS must be greater than one for the nodes to be semantically relevant. The nodes A and B have an RS value of two and are semantically relevant. Whereas, nodes A and C have an RS value of one, and hence are not semantically relevant.

Table 5.5 Partitioned Index Search

```
if (m = 1) { s = search the MFI with query q }.
   if (s = NULL) { s = search LFI with query q; search_result = s;}.
elseif (m > 1)
   if search with q in MFI is successful
      s = semantically interconnected nodes in the search results.
      if (s = NULL) No semantically related nodes.
      else search_result = s.
   else
      continue search with q in LFI.
      s = semantically related nodes in the search results.
      if (s = NULL) No semantically related nodes.
      else search_result = s.
```

Let $q = \{k_1, k_2, ...k_m\}$ be the search query where k_i represents the ith term in the query q and m represents the total number of terms in the query. The algorithm to find the semantically interconnected elements is given in Table 5.5.

The search algorithm first checks the length of the keyword query. If the query consists of a single term, a search over the MFI is performed. If the search is not successful, the algorithm continues search over the LFI. A failure to retrieve results from both MFI and LFI implies that the term is not found. The same technique is extended when searching with queries having more than one term. The only change is that, at each stage the semantic interconnection of the results is checked. Only semantically interconnected nodes are considered as the search results.

5.6 Performance Studies

In this section, we analyze the efficiency, and accuracy of SAGAXSearch, which is implemented in Java, via experiments on real data. The real data are XML files from the DBLP database [15].

Test Set: The test set used in SAGAXSearch is the DBLP XML corpus. The DBLP XML corpus is a collection of 200,000 XML documents and is repository for the details of a wide range of scientific articles. The document in the DBLP corpus can be classified into two major categories: journal articles and conference papers. The structure and the elements (nodes) used to represent the two types of documents are different. The corpus makes use of 36 different elements. The elements can be unique to a document category or they might be shared by document categories.

The GA used in SAGAXSearch takes a small number of user queries and the documents adjudged as relevant by the user as inputs. The input queries are sampled randomly from a large collection of user queries. Weights are associated with tags and the GA tries to explore all possible tag combinations and finds the best tag combination which satisfies the maximum number of queries. This is used to build the MFI and LFI. As the generation continues, the weights of the tags in the XML document are adjusted. The adjustments are such that maximum number of

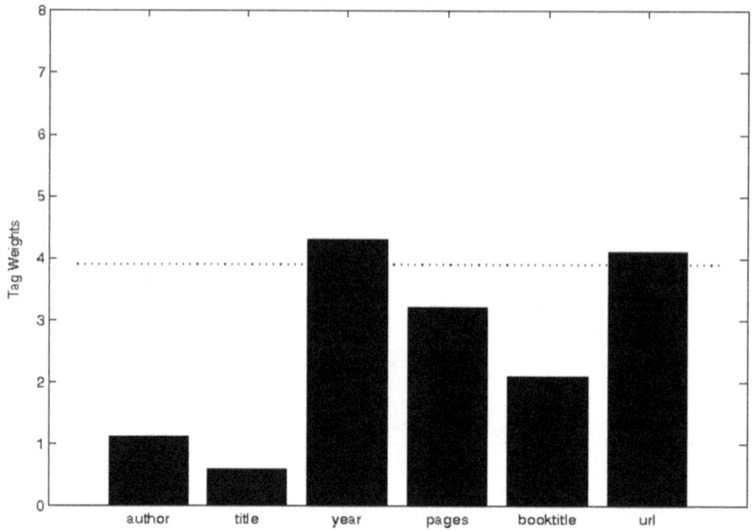

Fig. 5.5 Tag weights at the start of GA

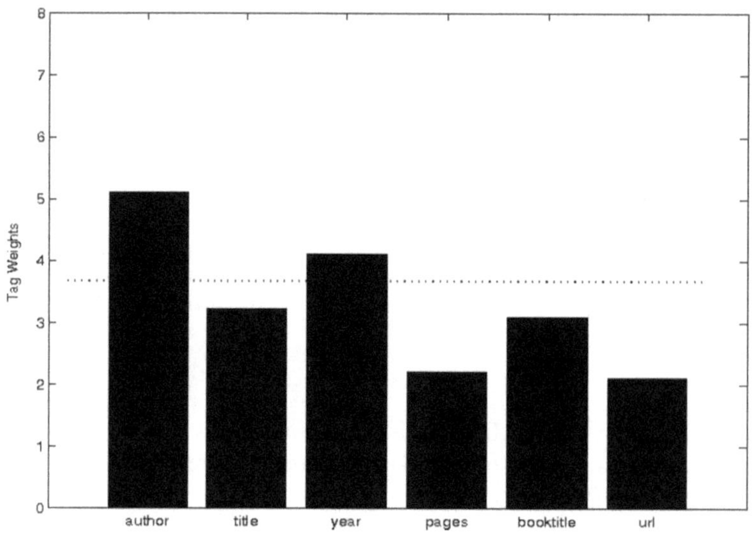

Fig. 5.6 Tag weights at termination

relevant results are retrieved. The DBLP XML database has a large number of distinct tags (>30). The evolution of these tag weights with the generations of the GA is shown in Figures 5.5 and 5.6. Due to space constraints the evolution of all the tags cannot be represented; so we illustrate the evolution in the weights of only six tags: < *author* >, < *title* >, < *year* >, < *pages* >, < *booktitle* >, and < *url* >. The average tag weight represents the average of the weights assigned to all the tags in the document.

The weight of a tag, when compared to the average weight of all tags in the document, is a measure of the importance of the tag within the document. Figure 5.5 shows the random tag weights assigned to the tags at the start of GA. As the generations continue, the user queries are evaluated and the tag weights are adjusted so as to obtain maximum fitness values. The tag weights after the termination of GA are the real measure of the importance of tags, and are as shown in Figure 5.6. For the DBLP dataset, based on randomly sampled user queries, tags like $< author >$, $< title >$, $< year >$, and $< booktitle >$ are classified as important. The tags like $< pages >$, $< url >$, $< cite >$, and $< ee >$ failed to classify as important.

Query Performance: We now evaluate the performance of keyword queries over a subset of the DBLP XML corpus, using SAGAXSearch. Here, we compare the performances of a search using a normal index and a search using MFI and LFI. A normal index is one which stores all the XML tag information within a single flat index structure. For large XML collections, such an index becomes huge and is difficult to manage. In contrast, the MFI has an index size which is much smaller, but still is capable of satisfying a majority of the user queries. During experimentation, the normal index which we built from the subset of the DBLP XML document had a size of 51.8 MB. The same document was indexed into two separate partitions by making use of the knowledge learnt from GA. The two partitions, MFI and LFI, had sizes of 20.4 MB and 31.6 MB respectively. In addition to this, MFI was capable of satisfying about 70% of the user keyword queries and for the remaining 30% of the queries, search had to be continued with LFI.

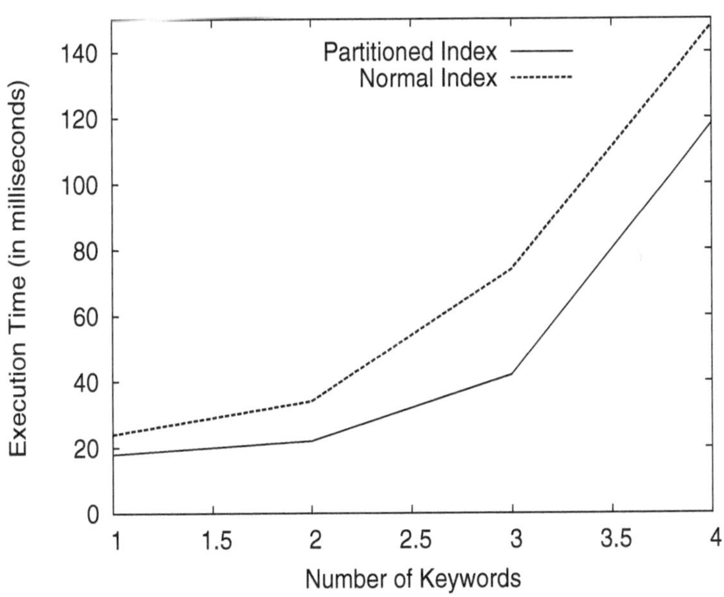

Fig. 5.7 Low frequency of occurrence of keywords

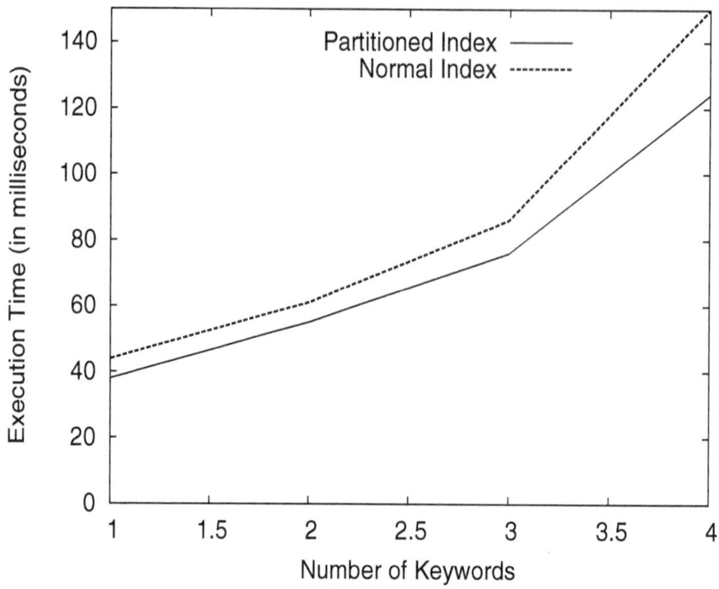

Fig. 5.8 High frequency of occurrence of keywords

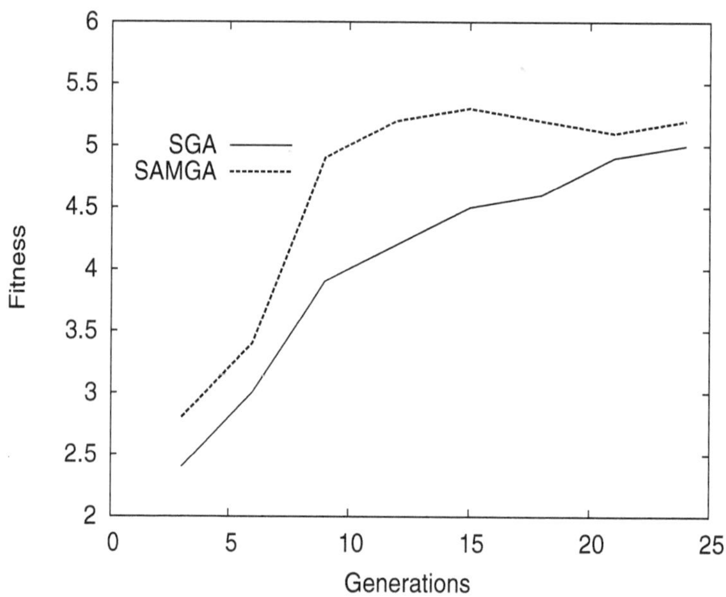

Fig. 5.9 Average Fitness of the Populations

The query execution time is shown in Figure 5.7 and Figure 5.8. The query execution time depends upon several factors like the number of terms in the search query, the desired number of search results and the frequency of occurrence of the

Table 5.6 Top Four Tags and their corresponding Weights

Generation No.	Tag1 : Weight	Tag2 : Weight	Tag3 : Weight	Tag4 : Weight
1	Month : 7.71	Author : 6.67	ee : 6.16	url : 4.26
5	Author : 7.63	Pages : 7.15	School : 6.74	Cite : 5.23
10	Title : 7.72	Year : 6.35	Cite : 5.92	Book Title : 5.87
15	Author : 8.1	Year : 7.63	Title : 7.19	Pages : 6.53
20	Author : 8.03	Year : 7.51	Title : 6.83	Pages : 6.60

keywords. We experimented with variations in all these factors and found that the frequency of occurrence of keywords was the main factor which decided the query execution time. Terms like *database*, *conference*, *technique* had very high frequency of occurrence in the DBLP document, and search queries involving these terms took longer to execute.

The Self Adaptive GA used in SAGAXSearch takes a small number of user queries (10-20 queries) and the documents adjudged as relevant by the user as inputs. The input documents to the GA are XML fragments from the DBLP XML database [16]. The GA tries to explore all possible tag combinations from the DBLP database and tries to find the best tag combination which satisfies the maximum number of queries. The experimental result in Figure 5.9 shows the average fitness for the generations of population. Note that the fluctuations in the curve representing Self Adaptive Migration model GA (SAMGA) is because of the adaptiveness introduced in the migration rate and population size. For SAMGA the average fitness steadily raises until about the fifteenth generation and then the fitness increases slowly. As the generation progresses further, the increment of fitness falls, as most of the individuals have already converged to their best fitness values. In contrast, a Simple GA (SGA) fails to converge even after 20 generations. Thus, the application of SAMGA helps in faster convergence when compared to SGA.

Table 5.6, shows the tag weights of the top four tags with the largest tag weights at the end of every five generations. It can be observed that the tags like $< author >$, $< title >$, $< year >$, $< booktitle >$ are given higher tag weights when compared to the other tags.

Precision and Recall: Precision of the search results is the proportion of the retrieved document fragments that are relevant. Relevance is the proportion of relevant document fragments that are retrieved. For precision and recall, we compare SAGaXsearch with XSEarch [14] and the naive results. Naive results are those which satisfy the search query, but are not semantically interconnected. All these techniques yield perfect recall i.e., all relevant documents are retrieved and the precision values vary. This is because of the factor that apart from the relevant results, some irrelevant results are also retrieved. The precision values of SAGaXsearch are found to be higher than those of XSEarch and naive approaches, when compared with the DBLP XML dataset. The small loss in precision occurs when the same keywords are present in both the MFI and LFI and the intention of the user is to retrieve information from the LFI. In such cases, the algorithm has already retrieved

Precision			
Naive Approach	XSEarch Tag	XSEarch Tag+KW	SAGAXSearch
0.01	**0.02**	**0.09**	**0.85**

Fig. 5.10 Comparison of Precision Values

the results from the MFI, it will not continue search over the LFI. The possibility of such an event is quite rare, and hence SAGaXsearch manages to exhibit high precision values. The comparison of precision values of these techniques is shown in Figure 5.10.

Example: We elaborate the working of SAGAXSearch using examples. The scalability of SAGAXSearch is confirmed by the performance measures on real data as explained earlier. For simplicity and better understanding, we consider small example documents. All the example documents are that of a shopping portal and have a common structure, which is shown in Table 5.7.

Various XML documents conforming to this structure are considered as training and test sets. The training set also consists of keyword queries and the results classified as relevant by the users. During experimentation, the keyword queries are randomly sampled, from the feedback of relevance given by a large number of simulated users. Thus, the training set is not biased towards the preference of any particular user. Training queries like CD pack India, IBM Notebook Rs 50000 Credit card, cartridge for inkjet printer, used car Ford Fusion, etc., are sampled along with the relevance feedbacks of the results. The relevance feedback is a set of XML documents which the users consider as relevant. The genetic algorithm is used to find the tag combinations which can answer the maximum number of sampled queries. The initial random weights associated with tags in the XML document are shown in Table 5.8.

Table 5.7 Structure of example documents

```
<item>
<location> </location>
<price_range> </price_range>
<manufacturer> </manufacturer>
<name> </name>
<description> </description>
<item_id> </item_id>
<payment> </payment>
<condition> </condition>
<seller> </seller>
</item>
```

Table 5.8 Initial weights of tags in the example XML documents

location	price_range	manufacturer	name	
7.3	5.6	1.2	7.3	
description	item_id	payment	condition	seller
2.5	4.1	4.5	6.1	1.2

Table 5.9 Tag weights after termination of GA

location	price_range	manufacturer	name	
1.1	6.1	7.2	7.3	
description	item_id	payment	condition	seller
6.5	2.1	3.6	2.1	0.8

The GA starts with this tag configuration and terminates with the tag combination satisfying the maximum number of queries. The genetically learned tag configuration which can satisfy the maximum number of queries is found to be <price_range>,<manufacturer>, <name>, <description>. Tags like <item_id>, <payment>, <condition>, <location>, <seller> are less frequently used during search. The tag weights after the termination of GA are shown in Table 5.9. With this information, two separate indices are built. MFI contains the information in the tags <price_range>,<manufacturer>, <name>, <description>. LFI contains the information in the remaining tags.

Consider a keyword search query like Motorola mobile GSM on the test set XML documents. The intention of the user is to find Motorola mobile phones with GSM technology. The indexing of the XML test set documents is performed offline and the *hvi* values for the various elements is stored in the index. When a search is performed, all elements satisfying the search terms are retrieved as results. Consider the result elements with *hvi* values [1 2 4], [1 2 7]. They have two prefix $<1,2>$ in common and hence have RS values greater than one. i.e., they share a common parent apart from the root. Thus, they are semantically interconnected. The search results with *hvi* values [1 2 4], [1 3 5] have a common prefix 1. They have an RS value of one. Such results have a single ancestor in common and hence are semantically unrelated elements. Note that the query Motorola mobile GSM can be answered by the MFI alone without the help of LFI and thus the query results are obtained faster. A modified query like Motorola mobile Bangalore credit card payment after searching over the MFI continues search with LFI, but the system has learnt that such specific queries are rarely encountered. Hence, even when such specific queries are encountered there is no loss in precision.

5.7 Selective Dissemination of XML Documents

The ability of individuals to be cognizant of information interesting to them is hampered by the abundant availability of knowledge. One method of addressing this problem is Selective Dissemination of Information (SDI).

Information overload is a major problem on the Internet. The deployment of Internet as a medium for news, electronic mail, digital libraries and e-commerce activities has led to an explosion in the amount of information published on it. The ability of individuals to be cognizant of information interesting to them is hampered by the abundant availability of knowledge. One method of addressing this problem is Selective Dissemination of Information (SDI). An SDI system helps users to cope with the large amount of information by automatically disseminating the knowledge to the users in need of it. Such systems maintain user profiles to judge the interests of the users and their information needs. The new documents are filtered against the user profiles, and the relevant information is delivered to the corresponding users. XML being a markup language, associates structure to the contents of the document using user defined tags. Both the structure and contents of the document can help in effective filtering. The self describing tags and the hierarchical structure of the knowledge can help in efficient selective dissemination of XML documents. Selective dissemination of XML documents presents several new challenges. The utilization of user defined tags is of great importance to improve the effectiveness of the dissemination task. The hierarchical structural information of the XML documents can help in efficient filtering. Thus, a system that can handle the structural information of XML documents can help in accurate and fast dissemination.

The growing amount of information on the internet has led to increased interest regarding selective dissemination systems. Majority of the research on the selective dissemination systems till now has focused on text documents. The SIFT SDI [17] makes use of an inverted list for indexing the user profiles. The incoming documents are matched with the user profiles to select the target users to whom the information has to be sent. The SIFT SDI can be extended to handle XML documents, but this setup cannot handle the tagged and the nested structure of XML documents. Thus, the system fails to produce accuracy values which are expected from XML applications.

The SDI systems for XML can be classified into two types: Boolean or Similarity based. A Boolean approach finds exact match between the incoming documents and the user profiles. The Boolean approach is supported by XFilter [18] which allows the users to define their interests using the XPath query language. The structural information available in the XML documents is used to construct highly expressive user profiles. A Finite State Machine is also used for matching incoming documents with the user profiles. However, such a system has limited scope as it can only find documents with exact match for the dissemination task. In [19], the similarity based approach is satisfied, user profiles are constructed and stored in a multi level indexing structure. The items previously classified as valuable by the users are clustered together using a similarity measure. This approach can perform a fast dissemination of XML documents even in the presence of large number of user profiles.

The Support Vector Machine (SVM) is an efficient method used for classification and regression purposes. The SVM builds a model from the training samples which is later used on the test data. This model is built using the training samples that are most difficult to classify (Support Vectors). SVMs can exhibit good accuracy and speed even with very less training. In this section, we propose a framework for selective dissemination of XML documents based on user profiles learnt using SAMGA and SVM.

Consider the XML documents which share a common set of tags. Sharing a common set of tags implies that the XML documents have the same topic and thus, are similar to each other. However, for text documents the similarity is measured by making use of the words that the documents share, and the accuracy of such a similarity assessment is often low. Thus, in the XML environment, similarity refers to topic similarity rather than the similarity of words. Topic similarity has a close resemblance to the human perception of similarity.

The XML document is structured collection of self describing tags. The tags associate semantics to the contents of the XML documents. For example, the <item> tag is used to specify that the item *Ford* is a commodity for sale. The parent child relationship among the elements can also be used to impart special meanings to plain text contents. Thus, exploiting the tagged and nested structure of XML can help in effective SDI. We have explored the possibility of selective dissemination of XML documents based on a user model. Self adaptive and real coded genetic algorithms are used to learn the user interests. SVM is used to build a user model based on user interests. A similarity metric between the continuously streaming XML documents and the profiles stored in the user model is proposed.

5.8 Genetic Learning of User Interests

The SAMGA of Chapter two used in this approach extracts from the tag pool, the combination of tags which are interesting to a user. It builds the profile of a user. Such a profile can be generated using a single query and relevance feedback session with the system or by using multiple sessions to generate aggregate profile. The accuracy and the utility of the selective dissemination tags improve with the number

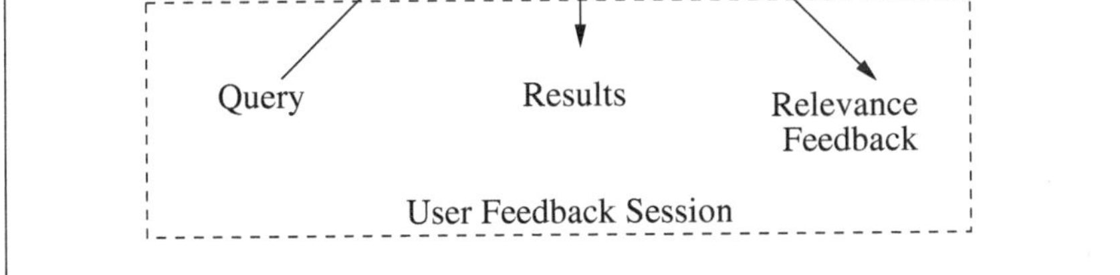

Fig. 5.11 User Profile Construction

of sessions of interaction between the user and the system. The SAMGA used in our approach extracts from the tag pool, the combination of tags which are interesting to a user as shown in Figure 5.11.

5.9 User Model Construction

The Self Adaptive Genetic Algorithms constructs a user profile, which is a set of tag combinations interesting to a user. A user model which can judge the interest category of a user from his profile is to be constructed. We consider this as a supervised learning problem. From the collection of user profiles, random profiles are sampled and a decision on the category to which they belong is made. A feature extractor and a SVM are used for the user model construction.

Feature Extraction: The user profiles constructed using SAMGA can have tags which are rarely used and also tags spread over all the categories. Such tags can deteriorate the performance of SVM. Hence a feature extractor is used to eliminate such tags. The feature extraction is performed using the measure of expected entropy loss. The entropy loss ranks the features high that are discriminators among the categories. It also assigns low ranks to features that cannot act as discriminators.

Let x be an event that a user profile belongs to a specific category and y be an event that the profile contains the specified tag. Let P( ) represent their probability. Let $\overline{x}$ and $\overline{y}$ be the negations of x and y respectively. All the profiles which belong to a category are considered as positive for the category and the remaining profiles are considered as negative. The following probabilities are computed.

$$P(x) = \frac{no.\ of\ positive\ profiles}{no.\ of\ profiles}$$
$$P(\overline{x}) = 1 - P(\overline{x})$$
$$P(y) = \frac{no.\ of\ profiles\ with\ tag\ y}{no.\ of\ profiles}$$
$$P(\overline{y}) = 1 - P(\overline{y})$$
$$P(x/y) = \frac{no.\ of\ positive\ profiles\ with\ tag\ y}{no.\ of\ profiles\ with\ tag\ y}$$
$$P(\overline{x}/y) = 1 - P(x/y)$$
$$P(x/\overline{y}) = \frac{no.\ of\ positive\ profiles\ without\ tag\ y}{no.\ of\ profiles\ without\ tag\ y}$$
$$P(\overline{x}/\overline{y}) = 1 - P(x/\overline{y})$$

The initial entropy of a class distribution is given by,

$$e = -P(x)lg(P(x))P(\overline{x})(lgP(\overline{x}))$$

The posterior entropy for a class when the tag is present is given by,

$$e_p = -P(x/y)(lgP(x/y))P(\overline{x}/y)(lgP(\overline{x}/y))$$

Similarly, the posterior entropy when the tag is absent is given by,

$$e_{\overline{p}} = -P(x/\overline{y})lgP(x/\overline{y}) - P(\overline{x}/\overline{y})lgP(\overline{x}/\overline{y})$$

The expected entropy loss is given by,

$$e - (e_p * P(y) + e_{\overline{p}} * P(\overline{y}))$$

and the entropy loss ratio for a tag is given by,

$$E = e - (e_p * P(y) + e_{\overline{p}} * P(\overline{y}))/e$$

E takes values in the range of (0:1). Higher values of E indicate more discriminatory tags. All the tags with entropy loss ratio values greater than a predefined variable are chosen as features for the SVM.

5.9.1 SVM for User Model Construction

SVM is a machine learning approach for the task of classification which is based on structural risk minimization [20]. Here, the decision surface chosen must minimize the test error on unseen samples. The binary SVM can be extended to support multiclass classification using the *one against one* approach. Here $k(k-1)/2$ SVMs are used where, k is the number of classes. Each SVM trains data from two different classes. A voting vector with a dimension for each class is also used for classification. There are as many votes as the number of SVMs and the class having the maximum number of votes is the result. The result of application of SVM is the user model. The user model has two functions. First, it classifies the various profiles into user interest category. Second, the same model can assign a user interest category to an incoming XML document from among the various prespecified categories.

5.10 Selective Dissemination

The selective dissemination is the task of disseminating the documents to the users, based on their profiles to whom the incoming documents would be most relevant. The first step in this task is determining the user interest category of an incoming XML document. Next, the similarity between the incoming XML document and the user profiles belonging to the same user interest category are determined. A high similarity value indicates that the document is relevant to the corresponding user.

The similarity between the XML document and the user profile is determined by modeling the XML document as a directed tree $G = (V_g, E_g)$. Each node in V_g corresponds to an XML element and E_g is a set of edges which defines the relationships between the nodes in V_g. A vertex in V_g is said to be at a level lev_i if it is at distance of lev_i from the root. Let $level_i(Dx)$ represent the set of all tags of an XML document D_x at a level lev_i. Let $userp_j$ represent the jth user profile and $userp_j = \{tag_1, tag_2, ...tag_l\}$, where l is the total number of tags in the jth user profile. The similarity between a user profile $userp_j$ and the incoming XML document D_x is given by

$$S(D_x, userp_j) = \frac{\sum_{i=1}^{d} \frac{|userp_j \cap level_i(D_x)|}{i * |level_i(D_x)|}}{userp_j \cup D_x} \tag{5.2}$$

where d is the depth of the XML document tree. The following observations can be made about the similarity metric.

- $0 \leq S(D_x, userp_j) < 1$; Since the XML document tree and the user profiles are structures of different kinds, a perfect similarity between the two can never be achieved.
- $S(D_x, userp_j) = 0$, iff there exists no common tags between the XML documents and the user profile.
- $S(D_x, userp_j) = S(userp_j, D_x)$
- Let D_{x_1} and D_{x_2} be two XML documents so that $|userp_j, D_{x_1}| > |userp_j, D_{x_2}|$ i.e., the number of tags shared between $userp_j$ and D_{x_1} is greater than the number of tags shared between $userp_j$ and D_{x_2}. However, this does not imply that $S(D_{x_1}, userp_j) > S(D_{x_2}, userp_j)$ i.e., the number of tags shared between the incoming XML document and the user profiles is not the only factor which decides their similarity.

Definition: The similarity between the user profile and the XML document depends upon two factors:

- The level in the document tree where a dissimilarity occurs. A dissimilarity is said to occur in $level_j$ iff $|level_i(Dx) - userp_j| \geq 1$.
- The Degree of Congruity (dc) in dissimilar levels also effects the similarity. The degree of congruity between the user profile and a level in a XML tree Dx_1 is given by,

$$dc(userp_j, level_m(Dx_1)) = \frac{|userp_j \cap level_m(Dx_1)|}{|level_m(Dx_1)|}$$

Proof: Let $level_m(Dx_1)$ and $level_n(Dx_2)$ represent the tags in the mth and n^{th} levels of two XML documents Dx_1 and Dx_2 respectively. Assume that the mth level in Dx_1 and nth level in Dx_2 are the only levels which have dissimilarity with the user profile $userp_j$.

Case 1: Let. $dc(userp_j, level_m(Dx_1)) = dc(userp_j, level_n(Dx_2))$ From Equation 5.5, it is clear that the similarity depends upon the values of m and n. Thus, $S(Dx_1, userp_j) > S(Dx_2, userp_j)$ iff $m < n$. The similarity between the user profile and the XML document depends upon the depth at which the dissimilarity occurs. A dissimilarity near the root, results in very less similarity values whereas dissimilarity near the leaf nodes, can still result in high similarity values.

Case 2: Assume $m = n$ i.e., the dissimilarity in the two documents occurs at the same level. From equation 5.5, it can be inferred that the similarity S now depends upon the degree of congruity, dc. That is, $S(Dx_1, userp_j) > S(Dx_2, userp_j)$ iff $dc(userp_j, level_m(Dx_1)) > dc(userp_j, level_n(Dx_2))$. Thus, higher the value of dc, better are the similarity values and vice versa.

The architecture for selective dissemination of XML documents based on the user model learnt using SAMGA and SVM is given in Figure 5.12. The user model is used for two purposes. First, it classifies the user profiles among the various user interest categories. The profile is then stored under the corresponding category. Second, for streaming XML documents it determines the user interest category. The similarity metric of Equation 5.2 is used to find the similarity between the user

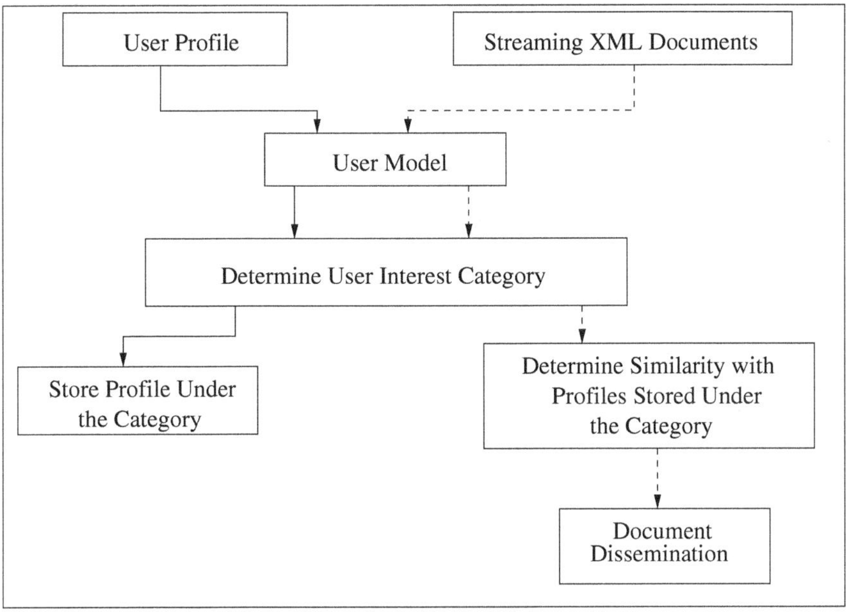

Fig. 5.12 Architecture of the Selective Dissemination System

profiles and the XML document. A high similarity value represents that the corresponding user is interested in the document. The document is disseminated to the top *k* users whose profiles have the greatest similarity with the input XML document.

5.11 Performance Analysis

The SAMGA used in our approach takes a small number of user queries and the documents adjudged as relevant by the user as input. The GA explores all possible tag combinations from the tag pool and tries to find the best tag combination which satisfies the maximum number of queries. This tag combination forms the profile of a user. Even for small XML document collections the tag pool is usually large. The time taken for the dissemination of the documents depends upon two factors: The number of stored user profiles and the number of user interest categories. The user interest categories are various divisions like *sports*, *books*, *politics*, *religion*, etc., to which the user profiles are assigned. It is important to have sufficient numbers of such categories. If the number of user interest categories is less, large number of profiles come under a single category and the time to find a matching profile increases. Thus, maintaining a optimal number of user interest categories results in good performance. The time taken for selective dissemination of XML documents is shown in Figure 5.13.

The number of user interest categories utilized also determines the accuracy of the selective dissemination task. The accuracy of the selective dissemination system

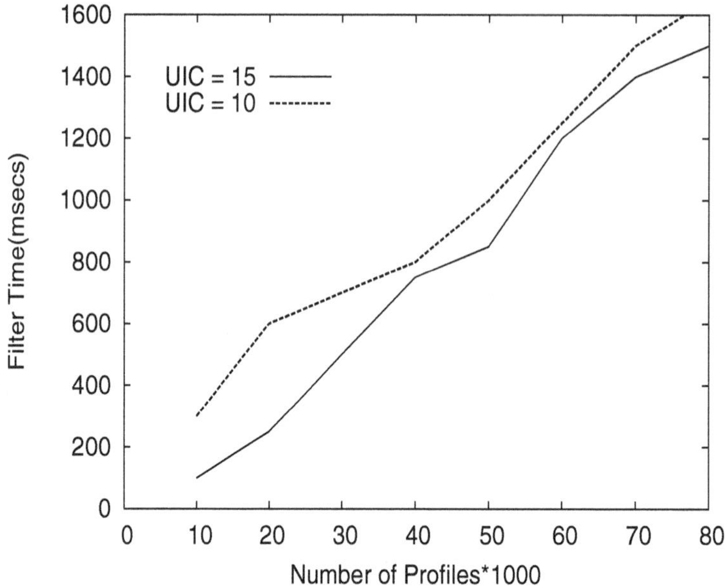

Fig. 5.13 Time for Selective Dissemination

is the proportion of disseminated documents that are relevant to the user. The variation of the accuracy with the number of User Interest Categories (UIC) is shown in Figure 5.14.

In order to validate the accuracy and efficiency of the proposed technique, we compare it with Multi-level Indexing Technique proposed in [19]. From Figure 5.15 it can be observed that the accuracy of selective dissemination increases with the number of profiles accessed. If an exhaustive search over the user profiles is performed, both the accuracy and time for dissemination increase. Since selective dissemination systems should be able to handle a large number of user profiles, the number of profiles accessed for an incoming document must be limited. From Figure 5.15, the accuracy of both the techniques is same when the percentage of profiles accessed is high. When the percentage of the profiles accessed is in the range of 30-40%, the proposed technique outperforms the Multi-level Indexing strategy in [19]. Thus the application of SAMGA and SVM helps in accurate and fast selective dissemination of XML documents.

Too many categories result in segmentation of the user interests and results in low accuracy. If the number of categories is less, it leads to superior performance with respect to accuracy but the time for selective dissemination increases. The intricate relationship among the number of profiles, the number of user interest categories, accuracy and time for selective dissemination, is given in Table 5.10 and it includes the following metrics; (i) Number of User Profiles (NUP), (ii) Number of User Interest Categories (NUIC), (iii) Accuracy (Acc), and (iv) Time for Selective Dissemination (TSD). From Table 5.10, it can be observed that the accuracy depends more

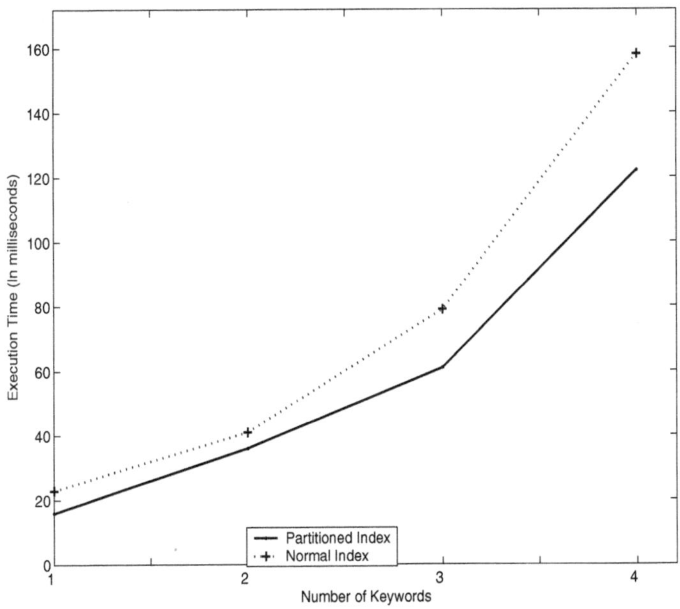

Fig. 5.14 Accuracy of Selective Dissemination

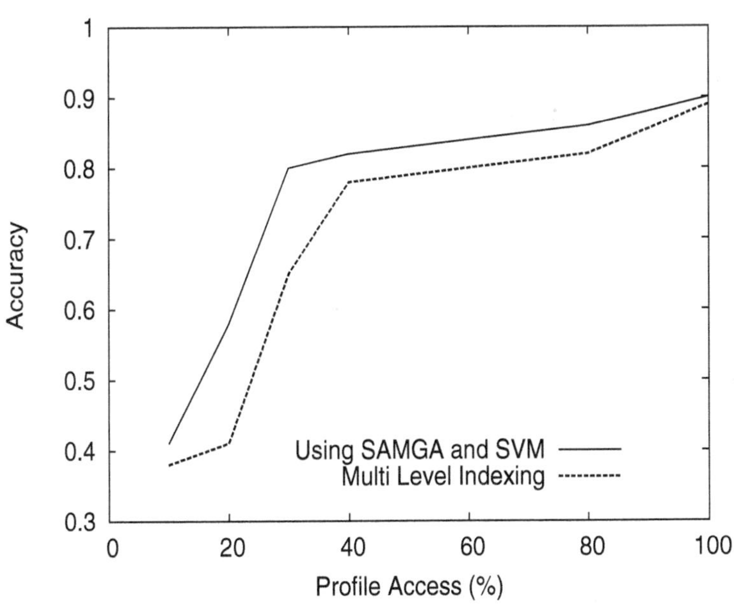

Fig. 5.15 Accuracy versus Number of Profile Access

Table 5.10 Variations of Accuracy and Time

NUPx1000	NUIC	Acc	TSD(msec)
5	10	0.92	162
5	15	0.87	143
10	20	0.89	201
10	25	0.81	189
20	30	0.74	331
25	15	0.87	547

on the number of user interest categories than on the number of profiles. Thus the system is potent in handling a large number of users. An optimal number of user interest categories should serve as a tradeoff between the accuracy and the efficiency and can result in good performance of the selective dissemination task.

5.12 Categorization Using SVMs

In this section, we have explored the possibility of topic categorization of XML documents using SVMs. The semantic information in the self describing tags is used for the purpose of categorization. A feature extraction strategy which can improve the efficiency of the SVM is also employed. Experimentation evaluations which compare the proposed technique with other classification techniques are also provided.

5.12.1 XML Topic Categorization

The topic categorization is the process of assigning a document to a specific category, from a predefined set of labels. Topic categorization of XML documents is of great interest as it can help in efficient XML data management. As the number of XML documents on the Internet is growing at an enormous rate there is a need for systematic organization of the documents. Such an organization must be capable of meticulously storing the XML documents according to the category to which they belong. The topic categorization can also help in effective information retrieval. A category specific knowledge discovery process discloses more relevant information compared to a general search.

The topic categorization problem has been extensively studied for text and HTML documents. In the case of text and HTML, a document is classified to belong to a specific category if it shares a common set of keyword terms with other documents under the same category. However, such a classification is often not accurate as the context of the keyword terms is not known. But in the XML environment, using self describing and extensible tags the context of the keywords can be precisely highlighted. Thus, the tags in the XML document can help in effective knowledge discovery.

The topic categorization of XML documents poses several new challenges. The hierarchical and the structural information in the XML document must be utilized in order to improve the quality of the knowledge discovered. A knowledge discovery framework which assigns equal priority to both the tags and the contents of an XML document is not able to exhibit any significant performance improvement when compared to knowledge discovery over text documents. The tags represent the semantics of the contents of an XML document and thus are more significant than the contents during the process of classification. Thus, a topic categorization framework with prominence to tags is highly efficient.

There has been a recent increase in interest regarding topic categorization of electronic documents. The topic categorization can be considered as a binary classification problem [21] or multi class classification problem [22]. A binary classification model, checks whether a document is relevant or not with respect to a category. However, for real world applications the binary model has limited applicability. In a multi class classification model the class to which a document belongs is judged from a set of predefined classes. In [23], distributed clustering is applied for document classification. Words are clustered into groups based on the distribution of class labels associated with each word. The classification of XML documents is discussed in [24]. The focus of this paper is to build effective feature space. Here, apart from the term frequency vectors, XML tag paths are used as features. In [25], an artificial neural network is used for categorization of HTML documents. The Principal Component Analysis (PCA) has been used to select the most relevant features for the classification. The most regular words that exist in each class are manually selected and weighted using an entropy weighting scheme. The fixed number of regular words from each class will be used as a feature vectors together with the reduced principal components from the PCA.

The weighing schemes can be used for feature selection in text mining applications. The weights are real numbers which can be used to judge the significance of terms. In [26], the *tf.idf* weighing scheme is discussed. The *tf* is the term frequency and the *idf* is the inverse document frequency. The *tf.idf* scheme assigns higher weights to more significant tags and vice versa. In [27], a Bayesian inference network model is used for term weighing. SVM is a machine learning approach for the task of classification which is capable of efficient multidimensional function approximation. It is based on the principle of structural risk minimization [28]. SVMs are widely used for text classification [29]. In [30], the standard feature representation of text is reviewed. The properties of text which make them suitable for classification using SVMs are discussed. Empirical results are also provided to prove that SVM text classification is more efficient than conventional text classification methods.

5.12.2 Feature Set Construction

The SVM is a machine learning approach to classification which is capable of low generalization error. For the application of SVM to the given XML data a feature vector must be constructed. The choice of the feature set determines the overall accuracy of the categorization task. Hence a robust feature extraction strategy is required. In the proposed technique, all the distinct tags from the training set XML documents are collected. This represents the initial tag pool. The tag pool can have rarely used tags and tags spread over all the categories apart from the more frequently used tags. Such tags can deteriorate the performance of SVM. The purpose of feature selection is to select the optimal feature subset that can achieve highest accuracy.

Consider a training set with n XML documents. Let $T_I = \{t_1, t_2, ..., t_l\}$ represent the initial tag pool, constructed using each of the distinct tags in the training set XML documents and l represents the total number of distinct tags. The t_i represent the ith tag in the tag pool. The purpose of feature selection is to obtain an optimal tag set $T_o = \{t_1, t_2, ..., t_m\}$, such that $m \leq l$. Feature selection has two advantages: First, the dimension of the feature vector is reduced. This can help in faster computations. Second, the features, or in the present context the tags, which are spread over all the categories are eliminated and importance is given to those tags that are representatives for a particular category. This can help in greater accuracy of the categorization task. In order to perform feature selection each tag in the tag pool is weighed using a standard weighing scheme. We analyze two weighing schemes: *tf.idf* [26] and *t*-statistics [31]. The weight of a tag t_i in T_I is given by,

$$w_i = tf_i.log(\frac{N}{df_i}) \tag{5.3}$$

where w_i is the weight of the ith tag in T_I , tf_i is the frequency of the ith tag in the document, N is the total number of documents in the collection and df_i is the number of documents in which the ith tag occurs. This scheme assigns weights to tags proportional to the number of times it occurs in document. However, the tags which are common among all the classes are assigned lower weights. But, in the case of XML documents only the information about the existence or the non-existence of a tag is sufficient to classify the documents and there is no necessity to calculate the tag frequency. Hence another weighing scheme based on *t*-statistics [31] is considered. Here, the initial expression vector for the training samples is given by $F^x = \{f_1^x, f_2^x, ... f_l^x\}$, where $1 \leq x \leq m$, m is the number of training samples and l is the number of features. Here F^x, represents the training sample x, and $\{f_1^x, f_2^x, ... f_l^x\}$represent all the features of this sample. Each sample is labeled with $D \in -1, +1$. All samples belonging to a particular class are considered as positives for the class and the rest of the samples are considered as negatives. The following values are computed for each tag:

- n^+ and n^-, the number of positive and negative training samples.
- μ_i^+ and μ_i^-, the mean values of the ith tag over the positive and the negative samples respectively.
- δ_i^+ and δ_i^-, the standard deviations of the ith feature respectively over the positive and the negative samples.

The weight of a tag is given by,

$$w_i = \frac{|\mu_i^+ - \mu_i^-|}{\sqrt{\frac{(\delta_i^+)^2}{n^+} + \frac{(\delta_i^-)^2}{n^-}}} \tag{5.4}$$

Equation 5.5, measures the normalized feature value difference between two groups. Higher weights are assigned to the tags that are discriminators for a class. Also, tags which are spread among all the categories are assigned low weights. In both the *tf.idf* and the *t*-statistics weighing schemes only the top $k\%$ of the tags with highest tag weights are chosen as dimensions for the optimal tag set T_o.

After the feature selection process, a feature vector for an input XML document must be constructed. The XML document is parsed and all the tags present in the document are extracted. Only binary values are assigned as dimensions of the feature vector. The feature vector of XML document consists of a 1 if the tag is present in the document and 0 otherwise. The feature vector thus constructed is the input to the multi class SVM.

5.13 SVM for Topic Categorization

SVM is a machine learning approach for the task of classification which is based on structural risk minimization. Here, the decision surface chosen must minimize the test error on unseen samples. The robust performance of SVM for both linearly separable and non-linearly separable data has lead to a rapid increase in the number of applications making use of SVM. First, we explain the binary and linearly separable classification problem and later extend the SVM to support multi class XML categorization. For a binary SVM, the training sample is $\{(x_1,y_1),(x_2,y_2),...(x_n,y_n)\}$, where x_i represents the feature vector for the ith sample and $y_i = \{-1,1\}$ i.e., a class label +1 or -1 is associated with each of the training samples. If we assume the profiles to be linearly separable then the equation of the hyperplane that does the separation is given by, $w^T x + b = 0$, where x is an input vector, w is an adjustable weight vector and b is the bias. Thus, $w^T x_i + b \geq 0,\ if\ y_i = +1\ and\ w^T x_i + b \leq 0,\ if\ y_i = -1$ or equivalently,

$$y_i(w^T x_i + b) \geq 1 \tag{5.5}$$

The training samples (x_i,y_i) for which Equation 5.5, is satisfied with a equality sign are called Support Vectors. Support Vectors represents the classification samples that are most difficult to classify. Hence, maximizing the margins between the Support Vectors results in a model which has good generalization properties. In order to take care of the non-separable data points a non-negative scalar variable ξ is introduced in Equation 5.5. The variable ξ is known as the slack variable. The resulting equation for a soft margin SVM is given by, $y_i(w^T x_i + b) \geq 1 - \xi$. A soft margin SVM solves the following optimization problem

$$min_{w,b,\xi_i} \frac{1}{2} w^T w + C\Sigma_{i+1}^n \xi_i \tag{5.6}$$

subject to $y_i(w^T x_i + b) \geq 1 - \xi, \xi > 0$

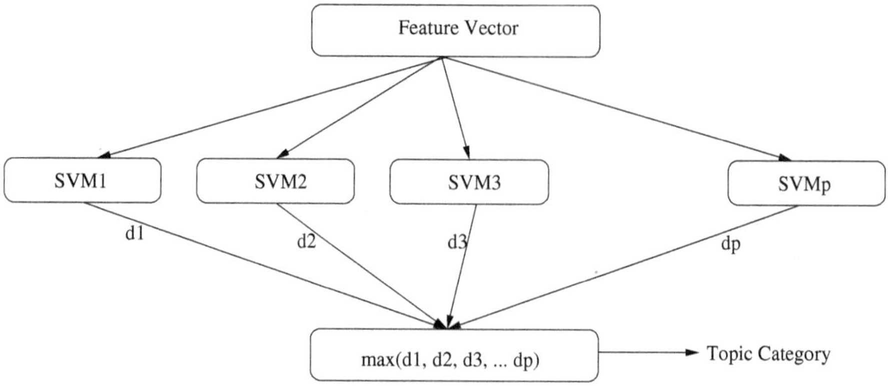

Fig. 5.16 All-vs-One SVM topic categorization

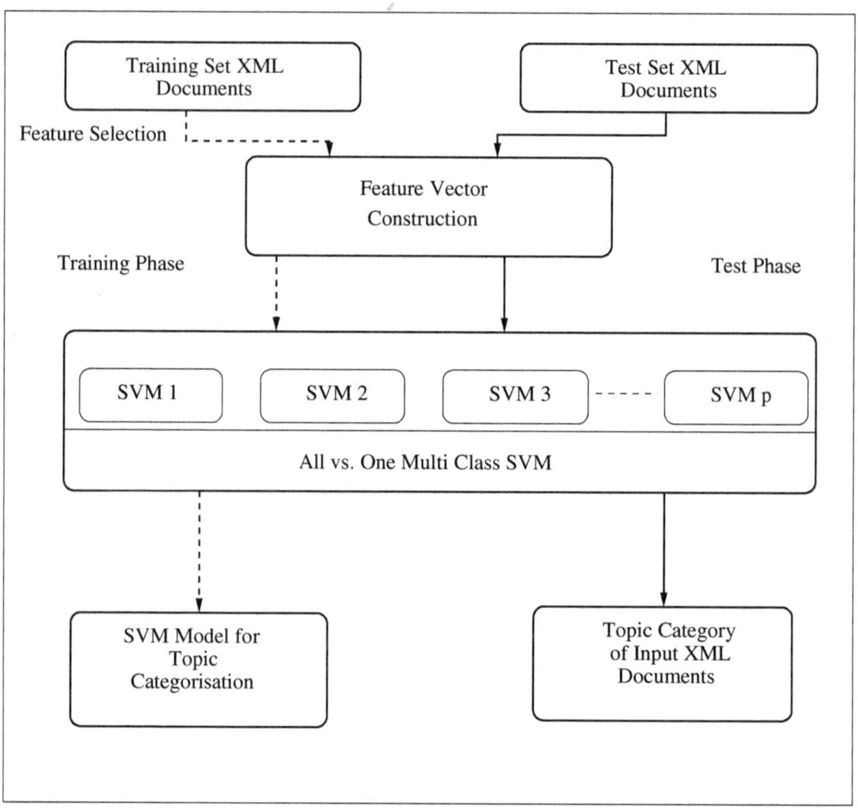

Fig. 5.17 Architecture of the Topic Categorization Framework

where C is a variable selected by the user. A quadratic programming algorithm is usually used to solve the constrained optimization problem of Equation 5.6. In order to classify samples that are not linearly separable the feature space can be mapped into a high dimensional linearly separable space using kernel tricks. However, in the present context where tags represent the features, the linear kernel is efficient compared to polynomial kernels.

The binary SVM can be extended to support multi class classification using the All-vs-One approach. Here P binary SVMs are used where, P is the number of predefined topics. Each of the P SVMs are trained using the training set documents. When a new XML document has to be categorized the feature vectors for the document are constructed and are individually provided as inputs to the P SVMs. Each SVM decides whether the given document is in its own topic or not. The All-vs-One approach for categorization of XML documents is shown in Figure 5.16. Here, each SVM calculates the distance to the optimal hyperplane. A document belongs to a particular category if the corresponding SVM representing the category yields maximum positive distance value. The architecture for topic categorization using SVM is shown in Figure 5.17. The dotted lines are used to represent the steps during the training phase and the solid lines represent the steps during the test phase. During the training phase, the feature vectors are constructed and given as inputs to the multi class SVM. The number of topic categories are predefined. Each SVM in the multi class setup is trained with the feature vectors for a particular category. The result of the training is a All-vs-One SVM model for topic categorization. In the test phase, the feature vectors are constructed for the input XML documents. The multi class SVM model is used to assign a topic category for the input XML document.

5.14 Experimental Studies

In this section, we analyze the efficiency and accuracy of the topic categorization of XML documents performed using the proposed approach. The experiments are performed with XML documents belonging to various categories collected from the Web as well as synthetic XML documents generated using the XML Generator [32]. The XML Generator is used to generate various XML documents conforming to a number of Document Type Definitions (DTDs). The performance experiments are done with 6,216 XML documents belonging to 33 topics.The feature selection performed prior to the application of SVM has a profound impact on the accuracy of the framework. Accuracy is the percentage of the correctly categorized XML documents.

The variations in the accuracy of the categorization with the values of k are shown in Figure 5.18. It can be observed that the accuracy of the t-statistics feature selection is greater than the *tf.idf* feature selection scheme for all values of k. This is because the t-statistics scheme is able to select an optimal number of features which are effective discriminators for the various classes. From Figure 5.18, it can also be seen that the accuracy is highest when k is in the range of 60-70%. When the value of k is low, the number of features selected is less. This affects the generalization performance of the SVM and it performs poorly over the test set. When the value of

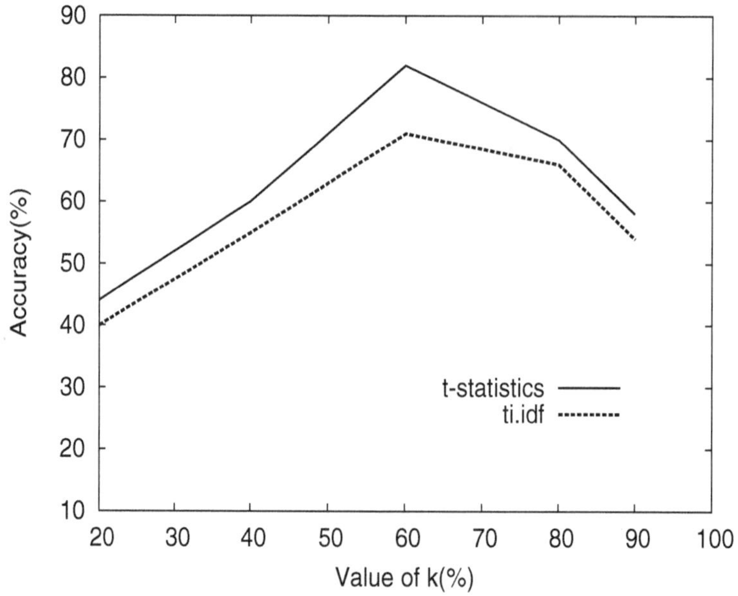

Fig. 5.18 Variation of accuracy with value of k

Table 5.11 F1-measure values for linear and Polynomial kernel

Category	F_1-Measure (linear)	F_1-Measure (polynomial)
Agriculture	0.5381	0.5416
Astronomy	0.6612	0.6709
Biology	0.5911	0.6215
Computer science	0.6617	0.6451
Entertainment	0.6910	0.7101
Health	0.7152	0.7093
News	0.5916	0.5911

k is high most of the features get selected. This results in poor classification accuracy of the SVM. A value of k in the range of 60-70% results in good classification accuracy as it chooses an optimal number of tags as features.

In addition to the accuracy of the system, precision, recall and F1-measure can be used to evaluate the classification system. Precision is the percentage of predicted documents for the given topic that are correctly classified. Recall is the percentage of total documents for the given topic that are correctly classified. Also, high precision values leads to low recall and vice versa. A classification framework which can exhibit optimum precision and recall values is required. Hence, the F1-measure, which is a weighted combination of precision and recall is used for evaluation. The F1-measure is given by,

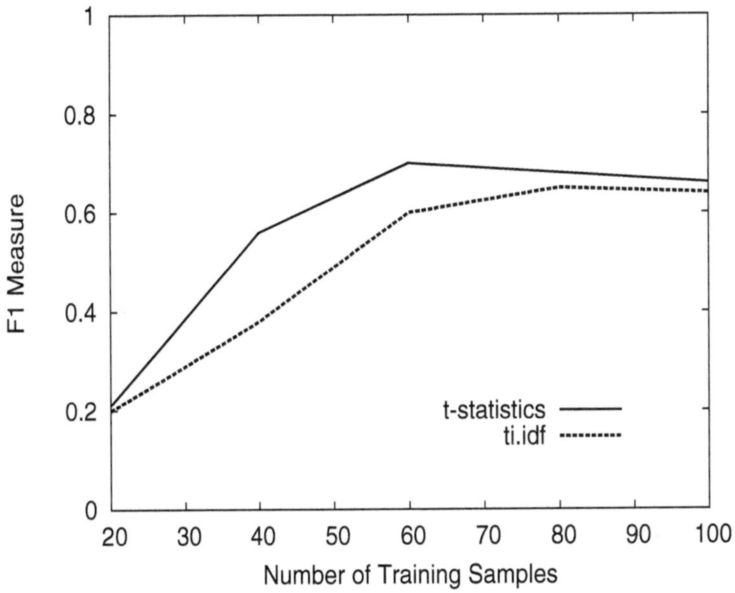

Fig. 5.19 F1 measure using the two schemes

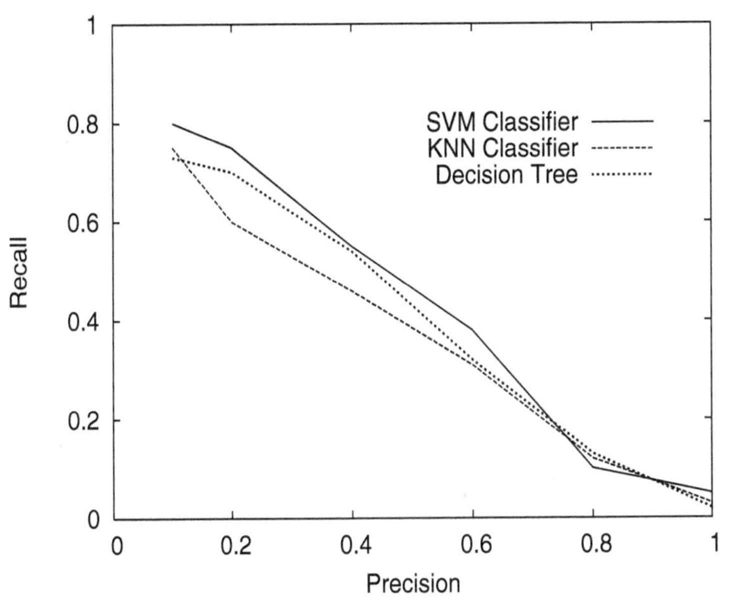

Fig. 5.20 Precision versus Recall

$$F_1 = \frac{2 * precision * recall}{precision + recall}$$

The F1-measure for the two feature selection schemes is shown in Figure 5.19. The T-statistics feature selection scheme results in better F1-measure values. The comparison of the linear kernel with Polynomial kernel for each topic category is shown in Table 5.11. It can be observed that the improvement in accuracy of the categorization task due to the application of kernel functions is not significant. Also, the time for application of the kernel function acts as an overhead during the categorization task. In the proposed technique though the accuracy of the linear kernel is slightly below that of the polynomial kernel, it is still utilized as it is more efficient.

In order to verify the accuracy of the proposed technique, we compare it with two traditional classification techniques: k-Nearest Neighbor and Decision Tree classifier. The average precision versus recall graph for all the three techniques is shown in Figure 5.20. It can be observed that the SVM classifier with the *t*-statistics feature selection strategy outperforms the KNN classifier and the Decision Tree techniques for the XML topic categorization problem.

5.15 Summary

A framework for information retrieval from XML documents that uses tag information to improve the retrieval performance is proposed in this chapter. Genetic Algorithms, which are efficient for search in large problem spaces, are used to learn the significance of the tags. A Self Adaptive Real Coded GA is used in particular because of its ability to perform a rapid exhaustive search over a large problem space. The notations for relationship strength and semantic relationship help in efficient retrieval of semantically interconnected results as well as ranking the search results based on the proximity of the keywords. Experiment on real data show that the SAGAXsearch is accurate and efficient. It has the flexibility of keyword query search, but the results obtained maintain accuracy values comparable to that of structured queries over XML documents.

The problem of Selective dissemination of XML documents that makes use of genetic algorithms to learn user interests is also discussed. A Self Adaptive Genetic Algorithm is used in particular as the problem space is large and a rapid exhaustive search is required to determine the user interests. Profiles which represent the user interests are built. A model which assigns user interest categories to the user profiles as well as the streaming XML documents is also proposed. A similarity metric between the user profile and the incoming XML document is used to determine the users to whom the document is to be disseminated. The efficiency and the accuracy of the selective dissemination of XML documents using the proposed technique is superior compared to the Multi Level indexing strategy.

The topic categorization of XML documents that makes use of Support Vector Machines for the purpose of classification is explained in this chapter. Feature selection schemes that can improve the accuracy of the SVM are presented. The All-Vs-One Multiclass SVM is used to assign a topic category to the input XML document from among the various pre-specified categories. Experimental evaluations are performed to test the efficiency of the feature selection schemes. The performance of the proposed framework is compared with other classification techniques like k-Nearest Neighbors and Decision Tree Classifiers. The comparisons show that the proposed technique outperforms the traditional classification techniques with respect to accuracy and efficiency.

References

1. Brin, S., Page, L.: The Anatomy of a Large-Scale Hypertextual Web Search Engine. In: Proceedings of International Conference on Seventh World-Wide Web conference (WWW7) (1998)
2. Luk, R., et al.: A Survey of Search Engines for XML Documents. In: SIGIR Workshop on XML and IR (2000)
3. Shanmugasundaram, J., et al.: A General Technique for Querying XML Documents using a Relational Database System, SIGMOD Record (2001)
4. Abiteboul, S.: On views and XML. SIGMOD Record 28(4), 30–38 (1999)
5. World Wide Web Consortium. XQUERY: A Query Language for XML W3c Working Draft, `http://www.w3.org/XML/Query`
6. Florescu, D., Kossmann, D., Manolescu, I.: Integrating Keyword Search into XML Query Processing. The International Journal of Computer and Telecommunications Networking 33(1), 119–135 (2000)
7. Gordon, M.: Probabilistic and Genetic Algorithms for Document Retrieval. Communications of the ACM 31(1), 1208–1218 (1988)
8. Yang, J., Korfhage, R.R.: Effects of Query Term Weights Modification in Annual Document Retrieval: A Study Based on a Genetic Algorithm. In: Proceedings of the Second Symposium on Document Analysis and Information Retrieval, pp. 185–271 (1993)
9. Yang, J., Korfhage, R.R., Rasmussen, E.: Query improvement in Information Retrieval using Genetic Algorithms: A Report on the Experiments of the TREC project. In: Proceedings of the First Text Retrieval Conference (TREC-1), pp. 31–58 (1993)
10. Pathak, P., Gordon, M., Fan, W.: Effective Information Retrieval using Genetic Algorithms based Matching Functions Adaptation. In: Proceedings of 33rd Hawaii International Conference on System Sciences (2000)
11. Kim, S., Zhang, B.T.: Genetic Mining of HTML Structures for effective Web Document Retrieval. Applied Intelligence 18, 243–256 (2003)
12. Hristidis, V., Papakonstantinou, Y., Balmin, A.: Key-word Proximity Search on XML Graphs. In: International Conference on Data Engineering (2003)
13. Guo, L., et al.: XRANK: Ranked Keyword Search over XML Documents. In: SIGMOD (2003)
14. Cohen, S., Mamou, J., Kanza, Y., Sagiv, Y.: XSEarch: A Semantic Search Engine for XML. In: VLDB 2003, pp. 45–56 (2003)
15. DBLP XML Records (2001), `http://acm.org/sigmoid/dblp/db/index.html`
16. Shanmugasundaram, K.T., Zhang, C., He, G., DeWitt, D.J., Naughton, J.F.: Relational Databases for Querying XML Documents: Limitations and opportunities. In: VLDB 1999, pp. 302–314 (1999)
17. Yan, T., Garcia-Molina, H.: The SIFT Information Dissemination System. ACM Transactions on Database Systems TODS 24(4), 529–565 (1999)

18. Altinel, M., Franklin, M.: Efficient Filtering of XML Documents for Selective Dissemination of Information. In: International Conference on Very Large Databases (VLDB 2000), pp. 53–64 (2000)
19. Stanoi, I., Mihaila, G., Padmanabhan, S.: A Framework for Selective Dissemination of XML Documents based on Inferred User Profiles. In: Proceedings of the Nineteenth International Conference on Data Engineering (ICDE 2003) (2003)
20. Joachims, T.: Text Categorization with Support Vector Machines: Learning with Many Relevant Features. In: Nédellec, C., Rouveirol, C. (eds.) ECML 1998. LNCS, vol. 1398, pp. 137–142. Springer, Heidelberg (1998)
21. Sebastiani, F.: Machine Learning in Automated Text Categorization. ACM Computing Surveys 34(1), 1–47 (2002)
22. Weiss, S.M., Apte, C., Damerau, F.J., Johnson, D.E., Oles, F.J., Goetz, T., Hampp, T.: Maximizing Text-Mining Performance. IEEE Intelligent Systems 14(4), 2–8 (1999)
23. Baker, K.D., McCallum, A.K.: Distributional Clustering of Words for Text Classification. In: Proceedings of ACM SIGIR Conference on Research and Development in Information Retrieval (SIGIR 1998), pp. 96–103 (1998)
24. Theobald, M., Schenkel, R., Weikum, G.: Exploiting Structure, Annotation, and Ontological Knowledge for Automatic Classification of XML Data. In: Proceedings of International Workshop on the Web and Databases (WebDB) (2003)
25. Selamat, A., Omatu, S.: Web Page Feature Selection and Classification using Neural Networks. Information Sciences-Informatics and Computer Science: An International Journal 158(1), 69–88 (1999)
26. Salton, G.: Automatic Text Processing. Addison-Wesley, Reading (1989)
27. Turtle, H., Croft, W.B.: Inference Networks for Document Retrieval. In: Proceedings of the Thirteenth International Conference on Research and Development in Information Retrieval, pp. 1–24 (1990)
28. Christopher, J.C.B.: A Tutorial on Support Vector Machines for Pattern Recognition. Data Mining and Knowledge Discovery 2, 121–167 (1998)
29. Joachims, T.: Transductive inference for Text Classification using Support Vector Machines. In: Machine Learning - Proceedings of Sixteenth Inernational Conference (ICML 1999), pp. 200–209 (1999)
30. Joachims, T.: Text Categorization with Support Vector Machines: Learning with Many Relevant Features. In: Nédellec, C., Rouveirol, C. (eds.) ECML 1998. LNCS, vol. 1398, pp. 137–142. Springer, Heidelberg (1998)
31. Liu, H., Li, J., Wong, L.: A comparative Study on Feature Selection and Classification Methods using Gene Expression Profiles and Proteomic Patterns, Genome Informatics (2002)
32. Diaz, A.L., Lovell.: XML Generator (1999), http://www.alphaworks.ibm.com/tech/xmlgenerator

Chapter 6
Soft Computing Based CBIR System

Abstract. Multimedia mining primarily involves, information analysis and retrieval based on implicit knowledge. The ever increasing digital image databases on the internet has created a need for using multimedia mining on these databases for effective and efficient retrieval of images. Contents of an image can be expressed in different features such as Shape, Texture and Intensity-distribution(STI). Content Based Image Retrieval(CBIR) is the efficient retrieval of relevant images from large databases based on features extracted from the image. Most of the existing systems either concentrate on a single representation of all features or linear combination of these features. The chapter proposes a CBIR System named STIRF (Shape, Texture, Intensity-distribution with Relevance Feedback) that uses a neural network for nonlinear combination of the heterogeneous STI features. Further the system is self-adaptable to different applications and users based upon relevance feedback. Prior to retrieval of relevant images, each feature is first clustered independent of the other in its own space and this helps in matching of similar images. Testing the system on a database of images with varied contents and intensive backgrounds showed good results with most relevant images being retrieved for a image query. The system showed better and more robust performance compared to existing CBIR systems.

6.1 Introduction

Multimedia mining is a subfield of data mining that deals with an extraction of implicit knowledge, data relationships, or other patterns not explicitly stored in the multimedia database. Relevant information analysis and retrieval is a key aspect of this field. The recent growth of the image databases on the internet has created a need for a more effective and efficient method for retrieval of images from a large database. The most common method of image retrieval is to use textual keywords to retrieve appropriate images. The major problem faced in context based textual image searches is the tedious manual process of image annotation. Content based image retrieval overcomes this problem by efficiently retrieving relevant images from large databases based on automatically derived imagery features like shape, color and texture features [1].

K.R. Venugopal, K.G. Srinivasa, L.M. Patnaik: Soft Comput. for Data Min. Appl., SCI 190, pp. 119–137.
springerlink.com

Image searches using text keyword often result in retrieving images that may be irrelevant to the user as the context in which the user uses the keyword may be different. For instance, if the user is looking for images of tea cups and hence gives the keyword as cups, he might get search results related to world cup as keyword matches. Further, the increasing size of image databases makes annotating images a tedious task. CBIR helps to solve these problems by using visual contents to retrieve relevant images.

Motivation: In classical CBIR the number of classes of inputs is very huge and not known before hand. Hence combining all the heterogeneous features and using a single SOM or Linear Vector Quantization(LVQ) over the entire set is not feasible or efficient. For individual features, the exact number of classes is not known and the number varies from feature to feature. Hence, LVQ would not help in dealing with heterogeneous features separately and combining them later to obtain good results [3, 4, 5].

The method proposed in this chapter is similar to method proposed in [2] only upto the the part, where the features are clustered independently. Even in clustering the topological ordering of SOM helps in the retrieval of images similar to query images easily. The system uses a non linear combination of the features with user relevance feedback to finally get merged results. This helps in not only retrieving the most relevant images, but also in taking users perspective into consideration.

The CBIR system proposed in this chapter, combines heterogeneous STI features in a nonlinear fashion to retrieve relevant images. An efficient comparison of the features is obtained by first clustering them independently in their own space. A static clustering system, along with a dynamic classification system using relevance feedback from users help in effective retrieval of images from large databases. The relevance feedback also helps the system to adapt itself to the user perspective.

6.2 Related Work

There has been considerable work done in the field of general purpose CBIR systems. Two of the pioneering CBIR systems are IBM QBIC [6, 7] and the MIT Photobook [8]. In these systems shape, color and texture features are generally represented in the form of histograms. The major drawback of this method is sensitivity to intensity variations, color distortions and cropping. Many approaches have been proposed to improve the retrieval by minimizing the effects of intensity variations and color distortions like the PicToSeek system that uses color model invariant of the object geometry, illumination and pose. There have also been systems which have concentrated on image retrieval based on a single set of features like shape features [9, 10] or texture features [11, 12] by finding appropriate way to extract the features. The performance of these system hint that rather than seeking a single representation for all features, it is better to use appropriate representation for each feature independently and finally use some way of combining them to achieve better results.

User relevance feedback can be a key to improve performance of CBIR. The Virage [13] and UIUC Mars [14] Systems allow user feedback and the systems learn to retrieve images better by learning from user feedback. To perform more relevant retrieval, systems like UCSB NeTra [15] and Berkley Blobworld [16] systems perform automatic image segmentation to obtain objects present in the image and use these for the image search. But both systems often partition a single object to several segments and none of the segments completely represent the object. In this chapter, we have attempted to overcome the above mentioned shortcomings.

6.3 Model

The image retrieval system proposed has two distinct phases. The first phase consists of creating the database of signatures used for image retrieval. Here the Shape, Texture and Intensity-distribution(STI) features are extracted from the image and each feature is individually clustered using Kohonen Self Organizing Map(SOM). The vector obtained by listing winner neuron's index for each of the clustered feature is then stored as signature for the corresponding image in the database.

The model of the final image retrieval system proposed which processes the image queries (query phase) is shown in Figure 6.1. The system accepts images as query key. The image is then preprocessed, followed by the feature extraction stage where the shape, intensity distribution and texture features are extracted from the image. Next, the trained Kohonen SOM used during the first phase is reused on the features to get signature of the query image. The classifier system, which is a backpropagation neural network, is used to compare this signature with the ones in

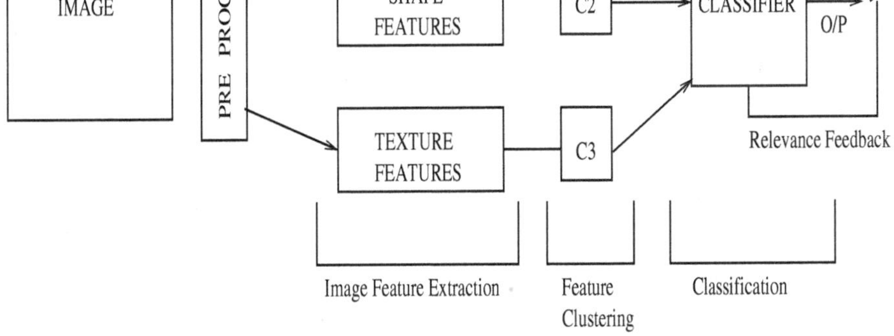

Fig. 6.1 Retrieval System Model

the database and retrieve appropriate images from the database. The users relevance feedback is used as the supervision for the backpropagation network to learn to retrieve the most relevant images. Thus, the system adapts to the users perspective.

6.3.1 Pre-processing

The preprocessing stage consists of first normalizing all the images to size 256x256. This makes the features computed to be independent of the original image size. Next, the histogram equalization of the image is done to distribute intensity across image uniformly. The preprocessing is done to make the system more robust to intensity variations and variable image size.

6.3.2 Feature Extraction

The system uses nine features for the purpose of image retrieval, four shape features, four texture features and one intensity feature.

(i) Shape Features: The four Shape features used are histogram of edge directions, Co-occurrence matrix of edge directions, Discrete Cosine Transform(DCT), and Hough transform.

Histogram of Edge Directions: The first step in finding Histogram of Edge Directions is converting RGB image to HIS(Hue, Intensity and Saturation) format using the formulae

$$I = \frac{1}{3(R+G+B)} \tag{6.1}$$

$$S = 1 - \frac{3}{R+G+B} min(R,G,B) \tag{6.2}$$

$$H = arccos(\frac{(R-G)+(R-B)}{2((R-B)^2+(R-B)(G-B))^{1/2}}) \tag{6.3}$$

Further, Sobels mask is used to obtain the direction edge maps for the eight directions 0,45,90,135,180,225,270, and 360 degrees. For shape analysis, Sobel edge detection provides detailed shape information about the image as compared to MPEG7. Combination of only intensity and saturation values of the image are used to obtain the edge maps as hue does not contribute to useful edges. The histogram of edge directions is computed using the eight direction edge maps. The edge histogram is translation invariant and captures the general shape features in the image. The edge histogram is computed by counting the edge pixels in each direction. The histogram is then normalized by the number of pixels in the image rather than number of edge pixels as done in [17]. To make the histogram more robust to rotation, smoothing of the edge histogram [17] (To make the histogram robust to rotation. Objects in the image when rotated should not affect the query results) is performed as follows:

$$H_e^S(i) = \frac{\sum_{j=i-k}^{i+k} H_e(j)}{2k+1} \tag{6.4}$$

where $H_e^S(i)$ is the histogram for direction i.

Co-occurrence Matrix of Edge Directions: To retain spatial features of the image, a more generalized edge direction histogram, a two dimensional edge histogram or co-occurrence matrix of edge directions is used as one of the features. The matrix is computed by taking every edge pixel pairs and enumerating based on their directions. The normalized 64 dimensional co-occurrence matrix H_{co} is obtained as follows :

$$H_{co}(i,j) = \frac{1}{NM} \sum_{x=0}^{N-1} \sum_{y=0}^{M-1} I_{e,i}(x,y) \sum_{(\hat{x},\hat{y})U(x,y)} I_{e,i}(\hat{x},\hat{y}, i = 1,...,8, j = 1,...,8. \tag{6.5}$$

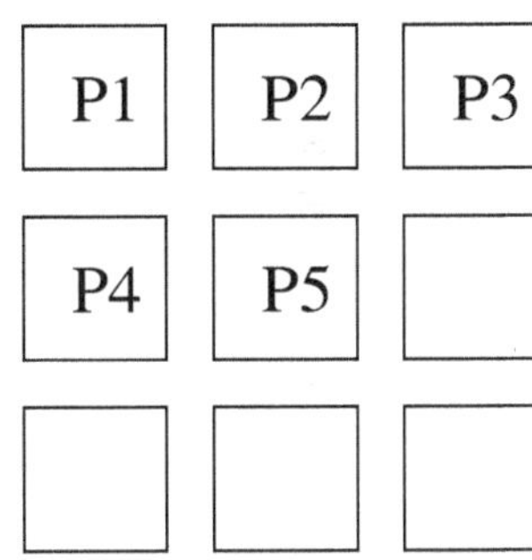

Fig. 6.2 Causal neighborhood set U(P5) = {P1,P2,P3,P4}

Where U(x,y) is the causal neighborhood set of the pixel (x,y)[Figure 6.2].

DCT of Edge map: Since edge map of an image contains all relevant shape features, DCT of edge map is used as one of the features. As the most significant information about the image is stored in the top left corner of the DCT matrix, only these values are used as one of the shape features. Image size becomes 256x256 after the normalization process is carried out during preprocessing. For every 256x256 image, 16x16 (because, 8*8 DCT coefficients for a large set of images were zero) DCT coefficients of the edge map are considered as feature vector and they are arranged row and column wise.

Hough Transform: The Hough transform is a technique which can be used to isolate shape features within an image. The main idea behind Hough transform is to apply the algorithm for detecting straight line to edge map of an image. In this

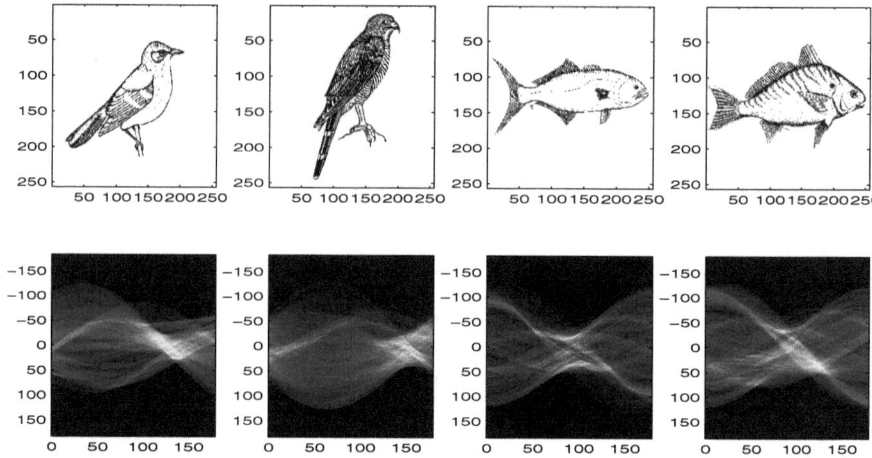

Fig. 6.3 Hough transform on objects

transform, each line is characterized by two parameters d and θ, d is the distance of line from origin and θ is the angle the normal of the line makes with the reference direction. Thus, the Hough transform of an image is again an image where the color of a pixel at a distance d and orientation θ from the origin is given by number of edge pixel that lie on the line. This transform is especially efficient when the image has a single distinct object. As Hough transform can be used to retrieve original images, it is evident that dissimilar images have dissimilar Hough transform. Moreover, the transform gives rotation, scale and transition invariant.

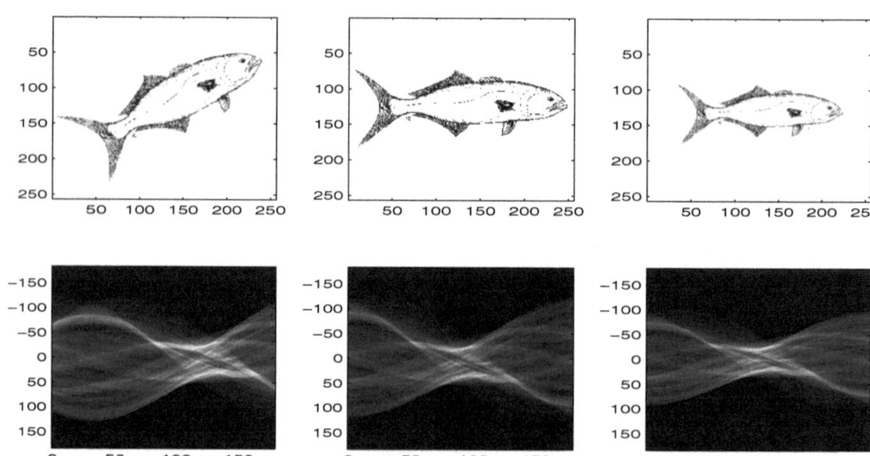

Fig. 6.4 Hough transform scaling and rotation invariance

The Figure 6.3 shows Hough's transform for two fish and two bird images. It can be seen that the fish images have similar Hough transform image and the bird figures have similar Hough transform images. The transform is rotation, scaling and translation invariant. This is seen in Figure 6.4 where the image produces similar Hough transform even after it is scaled or rotated.

(ii) Texture Feature: Texture feature is extracted using co-occurrence matrix [18] of the grey-scale image. The co-occurrence matrix is a two dimensional histogram of size GxG where G is number of grey levels. If I is a discrete image array and $d = (d_x, d_y)$ a displacement vector, The co-occurrence matrix H^{CO} is defined as

$$H^{CO}_{ij} = \{x|I(x) = i, I(x+d) = j\} \tag{6.6}$$

where the $(i, j)^{th}$ element of the matrix is the number of appearances of gray levels i and j in the distance and direction from each other determined by the displacement vector d, is the number of elements in the set and $x = (x, y)$ runs through the image array I. For extracting texture features, the co-occurrence matrix is computed for different displacement vectors and from these co-occurrence matrices energy, entropy, inertia and homogeneity terms are calculated and used as features. The computation of these terms from co-occurrence matrix reduces the feature vector size while retaining texture details. The Energy is computed as

$$Energy(H^{CO}) = \frac{1}{C^2}\sum_{i=1}^{G}\sum_{j=1}^{G}(H^{CO}_{ij})^2 \tag{6.7}$$

Entropy is computed as

$$Entropy(H^{CO}) = \frac{1}{C}\sum_{i=1}^{G}\sum_{j=1}^{G}H^{CO}_{ij}log(\frac{H^{CO}_{ij}}{C}) \tag{6.8}$$

Inertia is computed as

$$Inertia(H^{CO}) = \frac{1}{C}\sum_{i=1}^{G}\sum_{j=1}^{G}(i-j)^2 H^{CO}_{ij} \tag{6.9}$$

Homogeneity is computed as

$$Homogeneity(H^{CO}) = \frac{1}{C}\sum_{i=1}^{G}\sum_{j=1}^{G}\frac{H^{CO}_{ij}}{1+(i-j)^2} \tag{6.10}$$

where C is number of entries in co-occurrence matrix. The sets of different values of energy, entropy, inertia and homogeneity for different displacement vectors is used as the four feature vectors for texture feature.

(iii) Intensity Distribution Feature: Histogram of intensities is used as intensity distribution feature. Various CBIR systems have shown good results using histogram matching. Histograms contain some information about features like color, texture, geometry and shape of images. Here histogram for the image is computed based on Intensity map of image obtained using Equation 6.1. The smoothing of histogram (Equation 6.4) is applied as well for rotation invariance. This makes the feature more robust to rotation.

6.3.3 Feature Clustering

Once the heterogeneous feature vectors are obtained, each feature is clustered independent of the others using Kohonen neural network. Since the Kohonen SOMs are topology preserving, similar features are mapped close to one other. A Gaussian neighborhood function is used for training. As epochs proceed, the neighborhood size decreases and neurons far away from the winner neuron(neuron whose weights are closer to the input, taking Euclidian distance into consideration) are affected to a lesser extent by the winner neuron. After training the Kohonen neural network, for each image, the index of the winner neuron of each feature is taken as signature vector. The number of neurons used in the Kohonen neural network that cluster each of the extracted feature varies as each feature may require different cluster map size. Hence the signature is normalized by dividing each winner neuron index by corresponding number of neurons in the corresponding Kohonen neural network. The normalized signature is used for the image retrieval by comparing it with signatures of images in the database. Based on how well the images of different classes get separated by clustering each feature, the SOM size for each feature varies. Obviously the larger the cluster size better is the separation. But beyond a certain point, increase in the SOM size does not lead to improvement in clustering. Hence, the SOM size for each feature is fixed as the size beyond which not much improvement in results are seen. The SOM is trained on the part of the images available initially and as new classes of image are added to the system the SOM is retained. In the example, half of the images from each class is used for training.

6.3.4 Classification

Based on the clustered heterogeneous features, the image retrieval system retrieves relevant images. A back propagation neural network is used to compare query image signature with those in database and hence retrieve appropriate images. The input to the neural network is the signature of query image and signature of image in database to be retrieved. The number of signatures from the database fed to classifier for retrieval are pruned to increase the speed. The Eucliedian distance between signature of query image and images in the database is considered for pruning. This is possible because of the topological ordering of clusters in the SOM. If the output of the neural network is greater than 0.5, then the image is retrieved(Consider, an example of a database in which no images of a car are present. Hence, query of a car image should result in retrieving n images that are irrelevant just because they are the closest to query image in the database. On the other hand, if there are 100 images of bird in the database, a query of bird image should retrieve all of them, as all are relevant). The order of retrieved images is in the order of their outputs. The system accepts relevance feedback from the user to retrieve more relevant images the next time. The user relevance is used to set target value to supervise the output of the backpropagation neural network. The user relevance feedback can be -1, 0 or 1 corresponding to irrelevant, relevant and alike. This feedback f is converted to current target value as $target_{cur}$

$$target_{cur} = (f+1)/2 \tag{6.11}$$

During training the average target values is used as target for training the neural network using backpropagation of errors. The average target value $target_{avg}$ is obtained as

$$target^{t}_{avg} = (\alpha)target_{cur} + (1-\alpha)target^{t-1}_{avg} \tag{6.12}$$

where α is decay factor. The relevance feedback accounts for difference in the perspective of different users. During the testing and the use of system, set the target of the back propagation neural network to $target_{cur}$. During training per query per image retrieved use user feedback to update $target_{avg}$ and use it as target for the back propagation neural network. The neural network used here, contains one hidden layer and the input layer size is given by twice the number of cluster maps(each feature has a cluster map and hence a pair of 8 inputs, hence input vector size is 16). The user relevance feedback entails a nonlinear combination of features as the user feedback is assumed not to be just a weighted linear combination of features, but subject to perspective which is non-linear (Figure 6.5).

The clustering neural network uses static learning to cluster features of images in the database. The classifier on the other hand learns dynamically and becomes better after each retrieval. Since a total of nine features are considered the input is only a set of 18 values (a set of nine values of key image and another set of nine values of image in database under consideration.) Due to low dimension of classifier, the retrieval and learning is fast. When no example image is available in the beginning

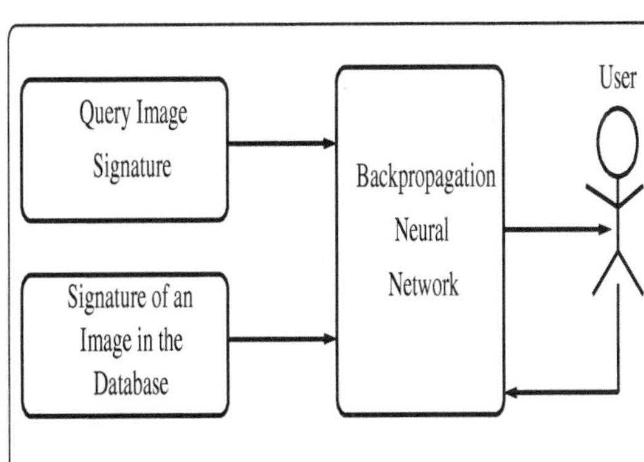

Fig. 6.5 Use of query and relevance feedback in BPNN

of the query, the query is matched with other images in the database and if any of them closely resemble the query image only those are retrieved. If none of the images resemble the query image then the search retrieves no image.

6.4 The STIRF System

The description of the STIRF(Shape, Texture, Intensity-distribution features with Relevance Feedback) system is presented in Table 6.1.

Problem Definition: Given a large image database consisting of *n* images, the objective of the system is to retrieve the most relevant images from the database, based on Shape, Texture and Intensity-distribution features and Relevance Feedback.

The STIRF System: The query image is accepted as an input in RGB format and normalized to 256x256 pixels by scaling the image. The image is represented as three 256x256 RGB matrices. Histogram equalization is then performed on the image to distribute intensities uniformly across the image. Four shape features, an 8x8 co-occurrence of edge direction matrix, a vector of length eight as edge direction histogram, a 16x16 matrix as the DCT of edge map and a 180x362 matrix as Hough

Table 6.1 Algorithm for STIRF System

1. Input the query image
2. Normalize the image to size 256x256
3. Perform histogram equalization on the image
4. Extract the shape features Hough transform, co-occurrence matrix of edge direction, edge direction histogram and DCT of the edge map.
5. Extract the texture feature by first calculating co-occurrence matrices for different displacement vectors and calculating energy, entropy, inertia and homogeneity for the different co-occurrence matrices.
6. Extract the intensity distribution feature by calculating histogram of intensities.
7. Use Kohonen SOM to cluster each feature individually.
8. Create signature for the image by taking a vector made up of winner neuron index normalized by dividing them by the number of neurons in the neural network used to cluster the feature.
9. For each of the *n* images in the database, *do*
   a. Use the signature of the query image and images in the database as input to classifier and if output is greater than selected threshold, retrieve the image.
10. Rearrange the images in the order of output from classifier.
11. If relevance feedback is given by user for any of the retrieved images, use relevance feedback to train the classifier.

transform are extracted from the processed image. In the next step, the texture features are computed as Energy, Entropy, Homogeneity and Inertia vectors of size 24, one each for displacement vectors. Finally, the histogram of intensities, a vector of size 256 is obtained as intensity-distribution feature.

Once the feature extraction is completed, it is fed as input to a Kohonen neural network that clusters each feature independently. The number of inputs to the Kohonen neural network is equal to the feature size. Once clustering is performed, we get eight winner indices, one for each feature as output of clustering. Next, the signature for the image is created, using the clustered features. For each image in the database, the signature of the image along with signature of the query image is fed as input to the classifier.

The output of the classifier is a value between zero and one. All images in the database, that give output of more than a selected threshold for the query image are retrieved. The threshold value decides the number of images retrieved. The retrieved images are presented to user after they are rearranged according to the output from the classifier. The user relevance feedback can be used to train the classifier.

6.5 Performance Analysis

Retrieval tests are conducted with different combinations of texture, intensity and shape features. Figure 6.6 shows the first ten retrieval results based on Hough transform alone. The first image to the top left corner of the figure is the query image. Here a total of three misclassifications (images balloon(6),cloud(7) and cloud(9)) can be seen. The misclassifications are due to the fact that, certain edges in the misclassified images are responsible for producing Hough transform for these images that are similar to Hough transform of query image. Figure 6.7 shows retrieval result for another query of a plane image based on the remaining three shape features

Fig. 6.6 Hough transform based retrieval

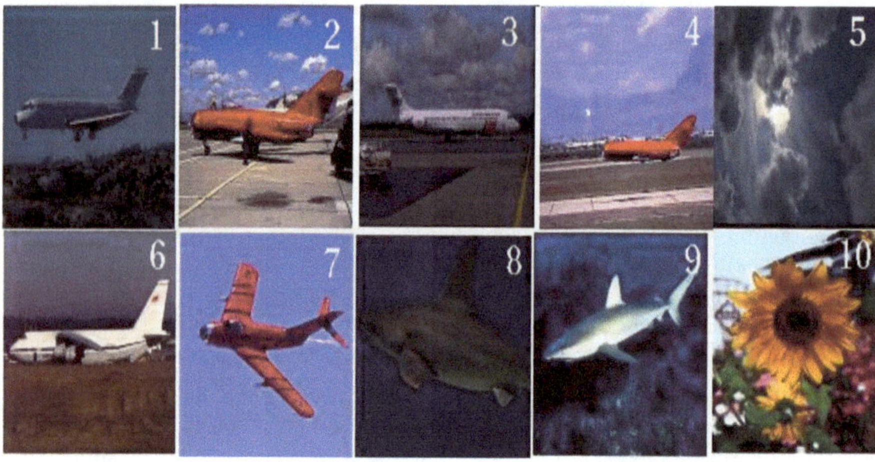

Fig. 6.7 Based on remaining three shape features

(Edge direction histogram, edge direction co-occurrence matrix and DCT of edge map). Here four (images cloud(5), shark(8), shark(9) and flower(10)) out of the top ten are misclassified due to similarity in certain edges present in the images and due to similar edge histogram being produced for the images, even though the objects in the image are different.

Figure 6.8 shows top ten query result of a cloud image based on texture feature alone. There is only one misclassification here (image 6(airplane)). The misclassification is due to the cloud background of the misclassified image which gives it similar texture feature as query image. Finally, Figure 6.9 shows top ten query results of a bird image based on intensity histogram. Here, we see seven

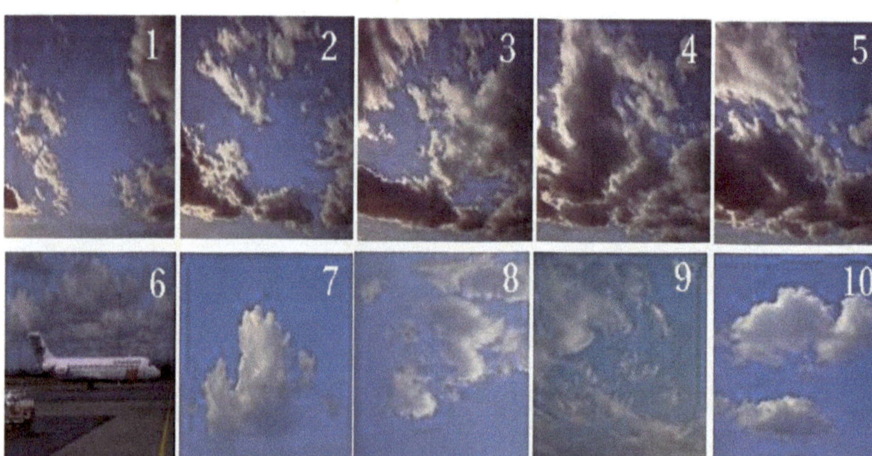

Fig. 6.8 Retrieval based on Texture feature

Fig. 6.9 Retrieval based on Intensity distribution

misclassifications (images polar bear(3), sharks(4), airplane(5), building(6), airplane(8), tiger(9) and clouds(10)) due to similarity in intensity histograms of dissimilar images. In all these queries, the query image chosen was that image which gave best results with the chosen feature. Similarity matching of each feature is based on some factor independent of the other.

When all the heterogeneous features are combined together and similarity matching is done then distance between images is compared in an n dimensional space where n is the number of features. Hence the result obtained is much better and complete. Figure 6.10 shows the top 16 query result when all the features are considered together. Figure 6.11 shows improvement in the performance when relevance feedback is used in retrieval. It can be seen that relevance feedback has a significant impact on the rank of the retrieved images. Further, in the top 16, one misclassification (image 16) is resolved by the use of relevance feedback.

Fig. 6.10 Retrieval before Relevance feedback

Fig. 6.11 Retrieval after Relevance feedback

Tests were conducted on the effect of distortions in the image on the retrieval of the image to show the robustness of the system. Figure 6.12 shows the effect of contrast variations on the rank of the retrieved image. The x axis shows the variation in contrast of the image (each step by a factor of 5 units) and the y axis shows the change in the rank of the retrieved image. An interesting fact that can be observed is that lowering of contrast does not affect retrieval as much as increasing contrast of the image beyond a certain point ($x = 2$). This phenomenon, is due to the fact that, increasing the contrast by a large extent induces edges that previously were not detected in the low contrast image. For instance, when we consider an image of a cloud, when contrast is further increased, the sky becomes distinctly bright blue and clouds distinctly white. Hence the edge detection gives clear outline of the cloud while edge map of the low contrast image does not have such a distinct outline. From the figure, it can be observed that our system is more robust to lowering of contrast of image than the CBIR system that uses local histogram matching. Both methods however deliver similar performance when contrast is increased.

Figure 6.13 shows effect of sharpness variation of an image on retrieval. The steps used to increase sharpness or blur the image is Gaussian filter of size 5. Positive value for variation indicates that the image is sharpened and negative value indicates that it is blurred. It can be observed that there is a symmetry in the graph about y axis. That is, sharpening and blurring of the image have similar effects. The rank initially increases and then returns back to rank of original image. Initially, when the image is sharpened or blurred, the distortion in the image, plays a role in increasing the rank of the image. As image is further sharpened or blurred, the system considers a larger area of the image as feature, and hence maps general shape of object with query image and hence performance improves. One more interesting observation in the graph is the sharp rise in the rank when the image is sharpened beyond a particular point ($x = 15$). This is due to the roughly pixelating effect that occurs due to increasing sharpness of image excessively. Overall, the effect of sharpness of image has little effect on retrieval compared to other distortions as the change in rank of image is in the range of about zero to ten as can be seen in Figure 6.13.

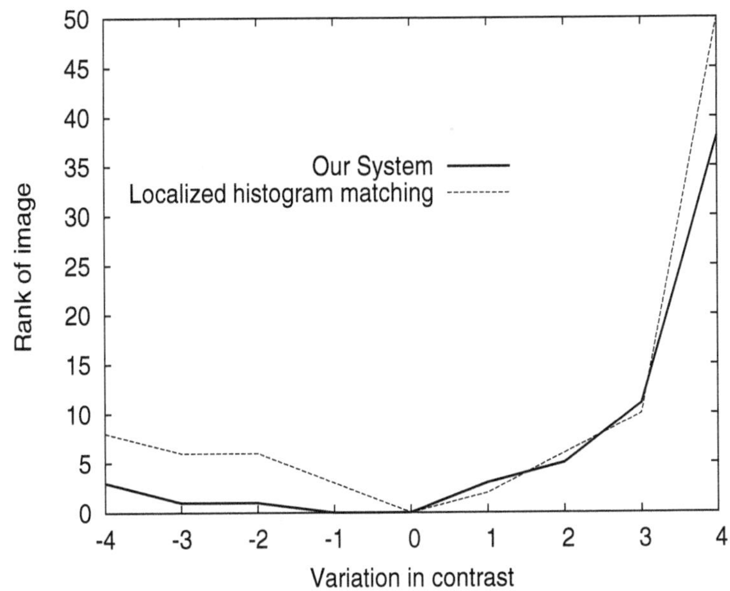

Fig. 6.12 Effect of contrast variations on retrieval

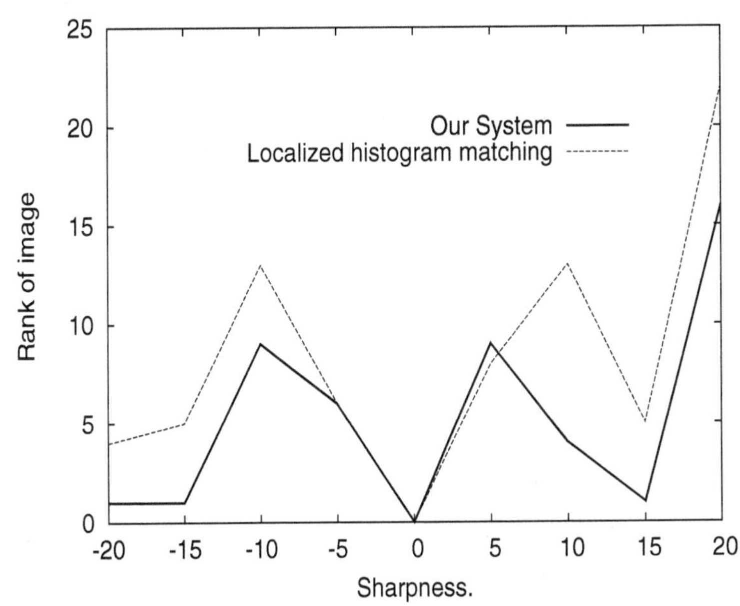

Fig. 6.13 Effect of sharpness variations on retrieval

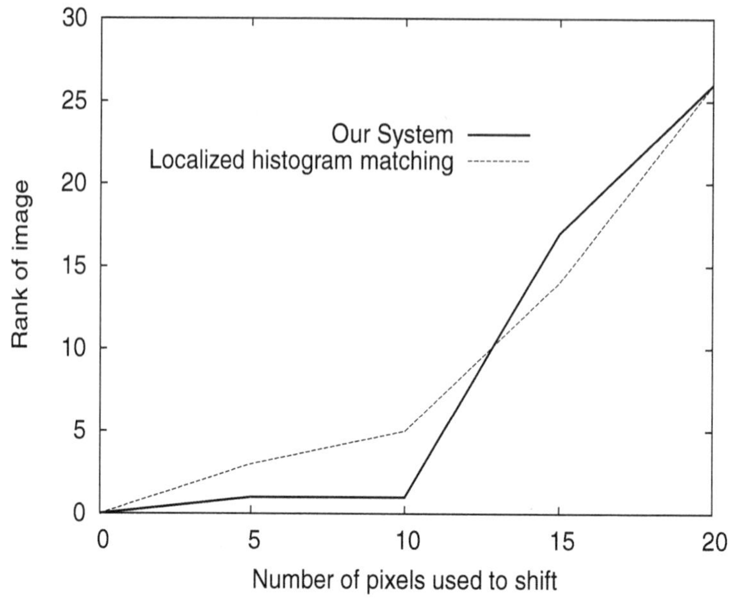

Fig. 6.14 Effects of image shift on retrieval

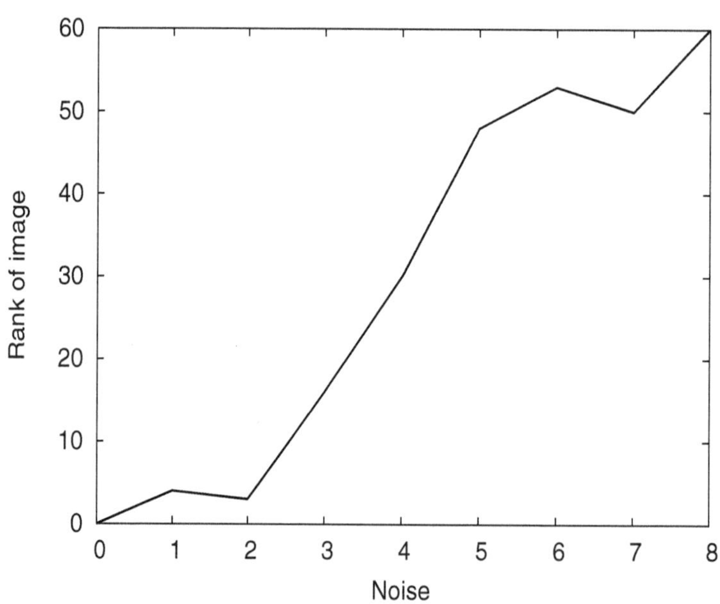

Fig. 6.15 Effects of noise on retrieval

Here, better results are obtained by our system compared to CBIR system using local histogram matching.

Often a shift effect is seen, especially in the images captured by a moving camera. Figure 6.14 shows the effect of shift in image on the rank of retrieved image. In the experiment, image was shifted in steps of 5 pixels. Upto a critical point($x = 10$) shift effect does not affect the rank of the image retrieved. Beyond the critical point, as shift effect distorts the shape of the image and hence there is a variation in the rank. It can be observed that our system is more robust to shift in image than the local histogram matching method for image retrieval.

Figure 6.15 shows the effect of noise in image on the rank of retrieved image. The x axis represents the amount of intentional noise added to the image. The noise added is in steps of 0.2 Gaussian noise pixels being added per pixel of image in all red green and blue channels. It can be observed from Figure 6.15 that low noise has insignificant effect on the rank of the image. As the noise increases, there is a linear increase in the rank of the image. Beyond a particular point ($x = 5$), addition of noise to the image does not affect the rank of the image significantly. Thus, the rank of the image stabilizes as adding noise to an already distorted image does not affect the image. In all the above experiments, the step of variation is large, so that the effect on image is clearly distinguishable to the human eye.

Description of the Database: The dataset consisting of the classes mentioned in Figure 6.16 plus some assorted images. This is considered as our test environment. The training was on less than the half the testing data. The table in Figure 6.16 shows the general performance of the system on image database where images are classified by humans roughly based on objects present in them. It can be observed that the performance of the system depends widely on the kind and variety of objects in the image.

The recall is defined as the proportion of relevant documents retrieved and precision is defined as the proportion of retrieved documents that is relevant. The

Fig. 6.16 Performance of the STIRF system

Sl. No.	Category	Average Precision	Precision	Recall	Accuracy
1	Birds	0.468	0.734	0.93	0.468
2	Sea Birds	0.88	0.94	1.00	0.88
3	Buildings	0.36	0.68	0.86	0.36
4	Clouds	0.94	0.97	1.00	0.94
5	Flowers	0.716	0.858	0.97	0.716
6	Hot air ballons	0.63	0.815	1.00	0.63
7	Planes	0.467	0.7335	0.75	0.467
8	Sharks	0.342	0.671	0.71	0.342

accuracy is defined as the difference between the proportion of retrieved documents that are relevant and the proportion of retrieved documents that are irrelevant. The recall, precision and the accuracy for the database is shown in Figure 6.16.

The average precision is calculated as Average Precision = Average (No. of relevant pictures retrieved - No. of irrelevant pictures) / No. of relevant pictures in the database, the average is calculated over query of all the images in the class.

6.6 Summary

We have developed a CBIR system based on the combination of heterogeneous STI features. The system is self-adaptive and improves image retrieval based on the user feedback. Unlike traditional systems, which combine features in a linear fashion, this system learns to appropriately combine features in a nonlinear fashion based on relevance feedback. Since features are clustered independently, the system retrieves images based on various similarity factors like shape, texture and intensity variation. Since the dimension of the signature obtained from the cluster is small, the retrieval process is efficient. The key to the good performance of the system, is that the slow and expensive process of clustering is made static and the classification which is a relatively less intensive process is made dynamic which helps in increasing effectiveness of the system without compromise in efficiency. Although the proposed model is for CBIR, it can be easily extended to other type of Content Based Multimedia retrieval like audio or video by choosing appropriate features.

References

1. Chen, Y., Wang, J.Z.: A Region-Based Fuzzy Feature Matching Approach to Content-Based Image retrieval. IEEE Transactions on Pattern Analysis and Machine Intelligence 24(9) (2002)
2. Sheikholeslami, G., Chang, W., Zhang, A.: SemQuery: Semantic Clustering and Querying on Heterogeneous Features for Visual Data. In: Proceedings of ACM Multimedia, pp. 3–12. ACM Press, Bristol UK (1998)
3. Li, J., Wang, J.Z., Weiderhold, G.: IRM:Integrated Region Matching Approach for Image retrieval, ACM Multimedia, pp. 147–156 (2000)
4. Yu, L., Gimel'farb, G.: Image Retrieval Using Color Co-occurrence Histograms. Image and Vision Computing, NZ, pp. 41–46 (2003)
5. Srisuk, S., Fooprateepsirr, R., Petrou, M., Waraklang, S., Sunat, K.: A General Framework for Image Retrieval Using Reinforcement Learning. Image and Vision Computing, NZ, 36–41 (2003)
6. Faloutsos, C., Barber, R., Flinkner, M., Hafner, J., Niblack, W., Petkovick, D., Equitz, W.: Efficient and Effective Querying bi Image Content. Journal of Intelligent Information Systems: Integrating Artificial Intelligence and Database Technologies 3(3-4), 231–262 (1994)
7. Niblack, W., Barber, R., Equitz, W., Flickner, M., Glasman, E., Petkovick, D., Yanker, P., Faloutsos, C., Taubin, G.: The QBIC Project: Querying Images by Content Using Color, Texture and Shape. In: Proc. of SPIE, in Storage and Retrieval for Image and Video Database, vol. 1908, pp. 173–187 (February 1993)

8. Picard, R.W., Kabir, T.: Finding Similar Patterns in Large Image Databases. IEEE, Minneapolis V, 161–164 (1993)
9. Kadyrov, A., Petrou, M.: The Trace Transform and Its Applications. IEEE Transactions on Pattern Analysis and Machine Intelligence 23(8), 811–828 (2001)
10. Wolf, C., Jolion, J.M., Kropatsch, W., Bischof, H.: Content based Image Retrieval using Interest Points and Texture Features. In: Proc. of International Conference on Pattern Recognition, pp. 4234–4237 (2000)
11. Distasi, R., Nappi, M., Tucci, M., Vitulano, S.: CONTEXT: A Technique for Image Retrieval Integrating CONtour and TEXTure Information. In: Proc. of International Conference on Image Analysis and Processing, pp. 224–229 (2001)
12. Manjunath, B.S., Ma, W.Y.: Texture Features for Browsing and Retrieval of Large Image Data. IEEE Transactions on Pattern Analysis and Machine Intelligence (Special issue on digital libraries) 18(8), 837–842 (1996)
13. Gupta, A., Jain, R.: Visual Information Retrieval. Communications of ACM 40(5), 70–79 (1997)
14. Malhotra, S., Rui, Y., Ortega-Binderberger, M., Huang, T.S.: Supporting Content-based Over Images in MARS. In: Proc. IEEE Int. Conf. on Multimedia Computing and Systems, pp. 632–638 (1997)
15. Ma, W.Y., Manjunath, B.: NeTra: A Toolbox for Navigating Large Image Databases. In: Proc. IEEE Int. Conf. Image Processing, pp. 568–571 (1997)
16. Carson, C., Thomas, M., Belongie, S., Hellerstien, J.M., Malik, J.: Blobworld: A System for Region-Based Image Indexing and Retrieval. In: Proc. Visual Information Systems, pp. 509–516 (1999)
17. Jain, A.K., Vailaya, A.: Image Retrieval Using Color and Shape. Pattern Recognition 29(8), 1233–1244 (1996)
18. Haralick, R., Shanmugam, K., Dinstein, I.: Textual Features for Image Classification. IEEE Transactions on Systems, Man, and Cybernetics (2000)

Chapter 7
Fuzzy Based Neuro - Genetic Algorithm for Stock Market Prediction

Abstract. Stock market prediction is a complex and tedious task that involves the processing of large amounts of data, that are stored in ever growing databases. The vacillating nature of the stock market requires the use of data mining techniques like clustering for stock market analysis and prediction. Genetic algorithms and neural networks have the ability to handle complex data. In this chapter, we propose a fuzzy based neuro-genetic algorithm - Fuzzy based Evolutionary Approach to Self Organizing Map(FEASOM) to cluster stock market data. Genetic algorithms are used to train the Kohonen network for better and effective prediction. The algorithm is tested on real stock market data of companies like Intel, General Motors, Infosys, Wipro, Microsoft, IBM, etc. The algorithm consistently outperformed regression model, backpropagation algorithm and Kohonen network in predicting the stock market values.

7.1 Introduction

Financial forecasting involves the gathering and processing of enormous amount of data. The future stock market values depend upon various market variables. The market variables like past returns, dividend yields, default spreads, term spreads, level of short term interest rates, and value line predictions, are usually considered for prediction. Even parameters like gross exposure, net exposure, concentration, and volatility in real-time affect the future stock values. Thus the data required for predicting stock market values are high dimensional in nature and reducing its dimensionality results in loss of information. For a human expert, predicting the market using the stock variables is a laborious and error prone task. Stock market prediction involves extraction of information on the correlations between the stock market values from the complex and enormous stock market data, and presents an interesting and challenging data mining task.

Data mining is a process of extracting nontrivial, valid, novel and useful information from large databases. Clustering is a promising approach to mine stock market data and it attempts to group similar objects based on features in actual data [1].

K.R. Venugopal, K.G. Srinivasa, L.M. Patnaik: Soft Comput. for Data Min. Appl., SCI 190, pp. 139–166.
springerlink.com

Machine learning techniques like neural network and genetic algorithms have been successfully used to cluster large amount of data.

In backpropagation networks, errors are propagated back during training and using these errors weights are adjusted. Errors in the output determine the errors in hidden layer, which are used as a basis for adjustment of connection weights between the input and hidden layers. Adjusting two sets of weights between the pairs of layers and recomputing the outputs is an iterative process that is carried out until the errors fall below a tolerance level. Learning rate parameters scale the adjustments to weights and a momentum parameter is used to overcome local minima [2].

The objective of a Kohonen network is to map input vectors (patterns) of arbitrary dimension N onto a discrete map with one or two dimensions. The Kohonen layer works on the idea of neighborhood. Each node has a set of neighbors; when a node wins a competition, the weights of the winning node and its neighbors are changed. Further the neighbor is from the winner, smaller is the change in its weight. The Kohonen layer is composed of neurons that compete with each other. The weights of the winner and its neighbors are brought closer to input pattern during training and hence the output obtained are clusters, that are topologically ordered [2].

Genetic algorithms are examples of evolutionary computing methods which are highly nonlinear, multifaceted search process. It searches for the best solution in a large set of candidate solutions. Genetic algorithms can be considered as computational models consisting of starting set of individuals(P), crossover technique, mutation algorithm, fitness function and an algorithm that iteratively applies the crossover and mutation techniques to P using fitness function to determine the best individuals in P. The fitness function is determined by the goal of the genetic algorithm [3]. Fuzzy Inference Systems apply logics on fuzzy sets, and use a defuzzifier to obtain crisp outputs. In cases where the required output is not always crisp and precise, fuzzy inference systems provide good results [4].

7.2 Related Work

A survey of all the different techniques used for forecasting is briefed in [5]. In paper [5], all the models like Exponential Smoothing, ARIMA, Kalman Filters, Neural Nets, Regime Switching Models, State Space and Structural Models, Univariate Models, etc., for efficient and accurate time series forecasting is discussed in brief. The various measurements like Robustness, Seasonability, Accuracy Measures, etc. for determining the accuracy of the forecasting model is also explained. A state of the art paper on usage of Artificial Neural Networks(ANNs) for better forecasting is given in [6]. The paper [6], presents a summary of all the modeling issues of ANNs forecasting. The paper presents a relative performance of ANNs with traditional statistical models. A method called automated ANNs to develop an automatic procedure for selecting the architecture of an ANNs for forecasting purpose is explored in [7]. An analytical model for an efficient market hypothesis is given in [8]. The role of model specification uncertainity, the effect of dynamic learning and the effect of feedback on return predictability is discussed in [8]. A comparison of classification techniques and level estimation models for an efficient forecasting of stock indices is discussed in [9]. A set of threshold trading rules driven by the probabilities estimated by the classification models are also derived. A hybrid system of neural networks guided by genetic algorithms for better customer targeting is used in [10].

Pioneering systems for the prediction of time series uses classical mathematical models like autoregression. These models are static and hence not very effective in prediction of stock market values [11]. The use of artificial neural networks in prediction of time series [12] offers an attractive solution to better and more adaptable prediction model. A primitive stock market prediction model using a basic Artificial Neural Network is discussed in [13]. A methodology analysis on the applications of neural networks in stock market data analysis is explained in [14]. A prediction model, based on neural networks, for time series with origin in chaotic systems is used in [15]. Application of a hybrid/expert system for database mining in market survey data is addressed in [16]. This approach uses backpropagation method for market survey. However, the backpropagation neural network cannot efficiently extract useful correlations from the raw stock market values. Clustering of data before prediction often provides valuable information on correlations required for prediction. An analysis and prediction of stock market data using SOM and its recent developments are discussed in [17]. The use of genetic algorithms along with neural networks [18,19] has shown better performance with noisy input sets. A combination of recurrent neural network and genetic algorithms is used for predicting stock market values in [20]. The advantage of this model is the temporal preservance which is very important in any time series model. However, here [20] again the neural network is fed with the raw input without any clustering. This brings down performance of system.

7.3 Model

Classical economic models like multiple regression and rolling average models are used to predict stock market values. The multiple linear regression model uses the mathematical equation of the form

$$y = \beta_0 + \beta_1 x_1 + \beta_2 x_2 + \ldots + \beta_n x_n + e \tag{7.1}$$

where x_1, x_2, ..., x_n are the n previous stock market values, y is the predicted value, β_0, β_1, ..., β_n are the coefficients of regression and e is the error factor to predict the stock market. This model tries to find multiple linear correlations between the inputs and the output. The model uses the previous stock values to solve the simultaneous equation to obtain correlations to predict next stock value.

The regression model has many shortcomings. First, stock market is highly non-linear for a linear model of single equation to predict its values accurately. Further, the results are not generalizable on account of a high degree of multicollinearity, as the parameter estimates in the model may not be stable owing to the high variance of the estimated coefficients.

The backpropagation neural network overcomes the shortcomings of the classical economic models by using the concept of feedback. In the backpropagation neural network each layer represents a feature with the number of neurons representing the number of dimensions of the feature. It is difficult to determine the features required for predicting the stock market.

The problem of determining the features for predicting share market is overcome by using Kohonen algorithm along with backpropagation neural network. Here the Kohonen neural network learns to cluster by choosing a winner and using competitive learning. The learning process of the Kohonen algorithm can be summarized as (i) initialize the weights of all the neurons participating in the competition. (ii) train the Kohonen network by changing the weights of the winner neuron and its neighboring neurons appropriately. The cluster map obtained from the Kohonen network preserves the topological order.

Kohonen algorithm does not learn effectively but keeps oscillating about the learning peak when similar input patterns are far away from each other as noticed in the stock market data. The Kohonen network in conjunction with genetic algorithms helps in achieving outputs nearer to the learning peak.

In this chapter, we propose a five stage model to predict the stock market values, as shown in Figure 7.1. In the first stage, the ordered stock market values are fed as input. The preprocessing of raw stock market data is performed in the second stage. The third stage uses a Kohonen network to output a cluster map. The Kohonen layer uses evolving genetic algorithms to obtain weights of its winners. In the cluster map, the winners are assigned values in the order of their ranks forming a multichromatic cluster map unlike the usual cluster maps which have only zero or one as outputs. This is a unique feature which helps in accurate prediction of stock market data. This layer also uses a fuzzy inference system to obtain fuzzily the learning rate using number of epochs and rank of winner under consideration. The use of fuzzy logic helps achieve a faster convergence rate. The fourth stage uses a simple backpropagation algorithm to extract share values from the cluster map. In the final stage, the output of the backpropagation algorithm is processed to obtain the absolute share values. The predicted stock market value can be used in the next input set for future predictions.

Mathematical Model: The Kohonen layer can be considered as a mapping from input space I of n dimensions to output space of clusters O of m dimensions using the transformation weight matrix W. Consider an input vector, $\hat{I} = \ <I_1, I_2, ..., I_n>$ of n inputs and a weight matrix $W = \{w_{ij} | i < n \ and \ j < m\}$. The output vector is computed as $\hat{O} = \hat{I} \text{ x } W = \ <O_1, O_2, ..., O_m>$.

Let $P = \{P_1, P_2, ..., P_m\}$ be a set of population of the m neurons. Then P_i^j represents the j^{th}individual of population P_i and P_{ik}^j is the kth chromosome of individual P_i^j. A set of neurons are selected in order of their outputs as winners. For these winners, the corresponding weights w_{ik} are computed as, $w_{ik} = P_{ik}^j$, $1 \leq k \leq n$, where P_i^j is the best individual from the population P_i of neuron i.

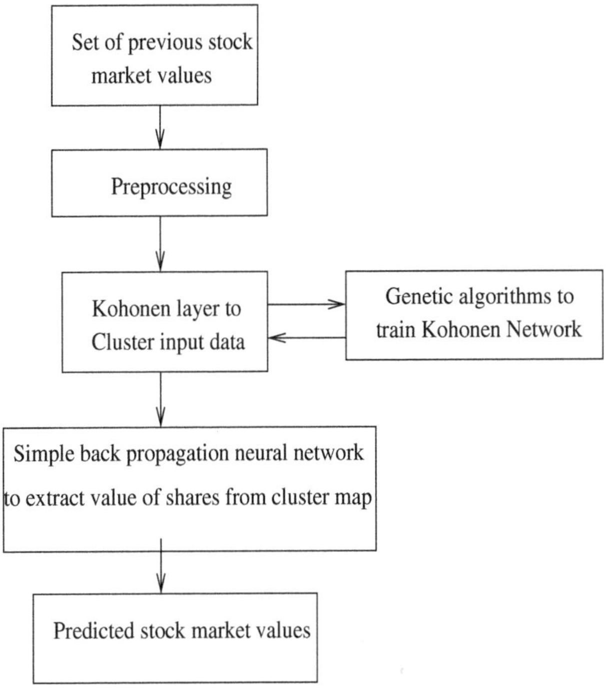

Fig. 7.1 Model of the FEASOM

Once the weights of winner neurons are fixed using genetic algorithm, for the remaining neurons, the change in weights of the neurons is given by:

$$\Delta w_{ik} = \begin{cases} 0 & \text{if } i \text{ is a winner} \\ \eta(\kappa,\gamma)\lambda(i,\phi,\gamma)(\mathrm{old}w_{\phi k} - w_{\phi k}) & \text{otherwise} \end{cases}$$

where ϕ is the winner under consideration, κ is its rank and γ is the number of epochs(an epoch is a learning cycle of neural network) and $\lambda(i,\phi,\gamma) = e^{-d^2/2\sigma^2(\gamma)}$ is the Gaussian distribution function, where $\sigma(\gamma) = 1/1 - e^{\gamma}$. $\eta(\kappa,\gamma)$ is the fuzzy inference system that fuzzily fixes the learning rate using number of epochs and the rank of the winner.

The fuzzy inference system uses the plot shown in Figure 7.2 to find the memberships μ_κ^H, μ_κ^M and μ_κ^L to fuzzy sets High(H), Medium(M) and Low(L) of rank of winners. Similarly, it uses the plot in Figure 7.3 to find the memberships μ_γ^H, μ_γ^M and μ_γ^L to fuzzy sets High(H), Medium(M) and Low(L) of number of epochs.

Further the memberships μ_η^H, μ_η^{HM}, μ_η^M, μ_η^{LM} and μ_η^L to fuzzy sets High(H), High Medium(HM), Medium(M), Low Medium(LM) and Low(L) of learning rate to be determined is obtained using the fuzzy rule table and the corresponding logic is shown in Figure 7.5.

Fig. 7.2 Fuzzy sets of rank of winners

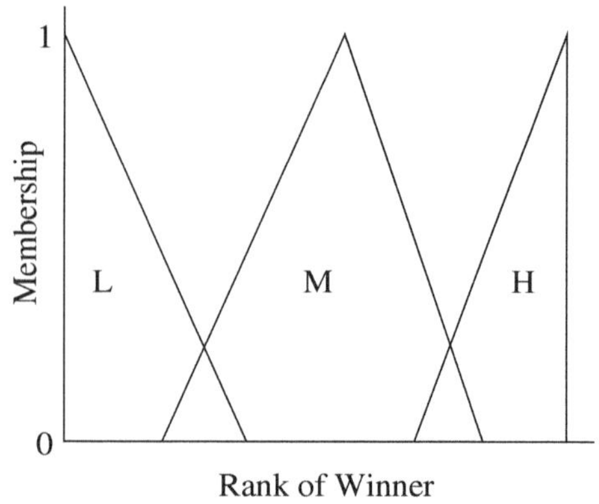

Fig. 7.3 Fuzzy sets of number of epochs

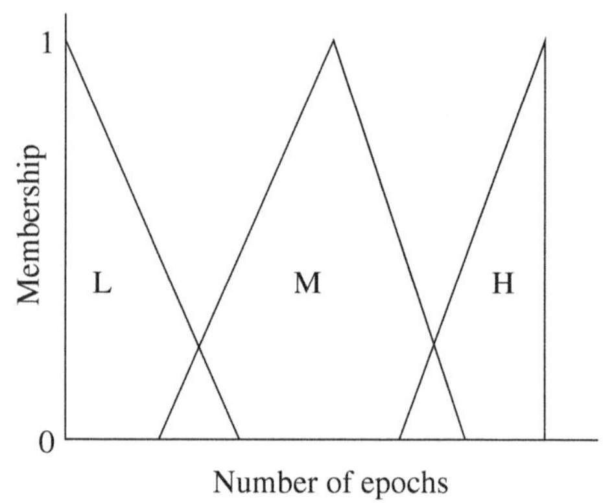

Fig. 7.4 Fuzzy sets of learning parameter

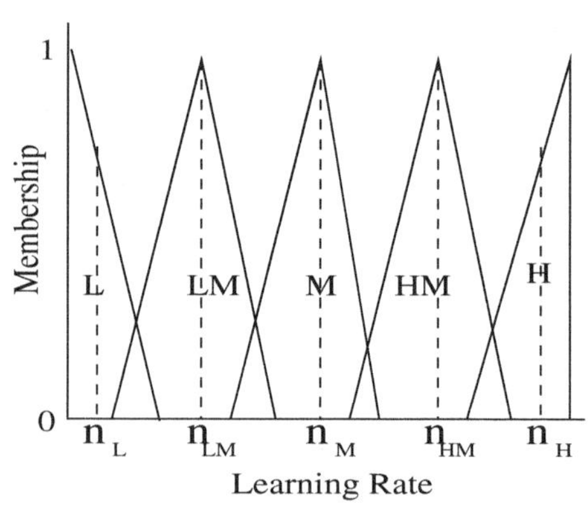

Fig. 7.5 Fuzzy Rules Table

Rank of Winner	Number of Epochs	Learning Rate
L	L	H
L	M	HM
L	H	M
M	L	LM
M	M	M
M	H	LM
H	L	M
H	M	LM
H	H	L

Finally using the plot in Figure 7.4 and the memberships to fuzzy sets of learning rates, the crisp learning rate is obtained as

$$\eta(\kappa,\gamma) = \eta_0 + \frac{\mu_\eta^H \eta_H + \mu_\eta^{HM} \eta_{HM} + \mu_\eta^M \eta_M + \mu_\eta^{LM} \eta_{LM} + \mu_\eta^L \eta_L}{\mu_\eta^H + \mu_\eta^{HM} + \mu_\eta^M + \mu_\eta^{LM} + \mu_\eta^L} \tag{7.2}$$

where η_H, η_{HM}, η_M, η_{LM} and η_L obtained from the Figure 7.4 are used to calculate the centroid which is the required learning rate. η_0 is a bare minimum learning used to offset learning rate parameter.

The entire population P_i is cultivated using a genetic algorithm with fitness function

$$f(P_{ik}^j) = \sum_{k=1}^{n} I_k^2 - \sum_{k=1}^{n} (P_{ik}^j * I_k) \tag{7.3}$$

As the better offspring of the current generation are carried over to the next generation, fitness of the best individual of the next generation is at least as good as the current generation. In general as the genetic algorithm evolves through the neural network epochs, all the individuals of a population lie around the ideal solution. The contributions of this work are as follows.

1. Genetic algorithms that evolve through the epochs of the Kohonen network is used. The Genetic algorithms bring the weights of the Kohonen neural network closer to the learning peak and hence prediction is better.
2. Use of multivalued winner neurons where the winner neurons are assigned output values varying according to the rank of the winner. Thus the cluster formed is multichromatic and hence gives more discernable output.
3. The neural network assumes that the neurons closer to the winner neurons are similar to winner neurons. Hence the calculation of change in weights of the neurons need to be done only for the winner neurons. Thus the time required for computation is less and the cluster map formed is more distinguishable.

4. The learning rate for an individual neuron is fuzzily decided using number of epochs and the rank of winner under consideration. This helps in achieving faster convergence rate.

7.4 Algorithm

Problem Definition: Given a large data set of ordered share values s_1, s_2, s_3, ..., s_n, the objective is to effectively predict m future share values, s_{n+1}, s_{n+2}, ..., s_{n+m} using the n previous share values.

7.4.1 Algorithm FEASOM

The FEASOM algorithm uses a modified Kohonen algorithm for the first layer which uses genetic algorithms to obtain weights of its winners. The output of this layer is a cluster map, which is fed to a backpropagation algorithm for prediction. The algorithm FEASOM can be summarized as follows:

1. Initialize Kohonen network with n input units, m output units and a weight matrix W.
2. Initialize for each neuron in Kohonen layer, a set of populations P.
3. Initialize a backpropagation algorithm with random weights.
4. Obtain the input vector Î for the Kohonen layer by subtracting an arbitrary value v from each element of S, where $v \leq min(S)$.
5. Use modified Kohonen algorithm with Î as input.
6. Set target t as (s_{n+1} - v)/f, where f is some arbitrary normalizing factor.
7. Feed the backpropagation algorithm with output cluster map of Kohonen layer as input and t as the target.
8. Multiply f to the output of the backpropagation algorithm and add v to get the predicted stock market value.

7.4.2 Modified Kohonen Algorithm

Consider a Kohonen layer defined by an input vector Î=$< I_1, I_2, ..., I_n >$ where n is the number of input units, an output vector Ô=$< O_1, O_2, O_3, ..., o_m >$ where m is the number of output units and a weight matrix $W = \{w_{ij} \mid i < n \; and \; j < m\}$.

1. Initialize the weight matrix W with small random values.
2. For each epoch from Zero to the maximum number of epochs,
   do

   a. Accept input Î from the training set.
   b. Obtain output Ô as Ô = Î x W.
   c. Select the l winner neurons with the l largest outputs.
   d. For each winner neuron with index r,
      do

- Use Genetic algorithms to fix the weights of winner neuron.
- For each neuron in the Kohonen layer with index i,
  do
  - Distance $d = abs(r - i)$;
  - Use fuzzy inference system to obtain learning rate fuzzily using rank of winner and number of epochs.
  - Change the weights of neurons by a factor of learning weight obtained using change in the weights of the winner neuron by a factor θ, where θ is the Gaussian of distance d which is the neighborhood function.
- Endfor

e. Endfor

3. Endfor

In this algorithm, we first initialize all the weights to some small random value (≤ 1). Next, for each input pattern, the outputs of every neuron i is calculated as

$$output = \sum(w_{ij}I_i). \tag{7.4}$$

The weights and the inputs are usually normalized, which means that the magnitudes of the weight and input vectors are set equal to one. The l neurons with the largest outputs are the winners. These neurons have a final output of one, for their corresponding input patterns.

Consider a vector, $A = ax + by + cz$. The normalized vector $\hat{A}$ is obtained by dividing each component of A by the square root of the sum of squares of all the components. In other words, each component is multiplied by $1/\sqrt{(a^2 + b^2 + c^2)}$. Both the weight and the input vectors are normalized with respect to the input during the operations of the Kohonen feature map. The reason for this is that the training law uses subtraction of the weight vector from the input vector. Using normalized values while subtraction reduces both input and weight vectors to a unit-less state, and hence makes the subtraction of like quantities possible.

In the next step, genetic algorithms are used to obtain the weights of the Kohonen network. The fitness function of the genetic algorithm is obtained from the training law of Kohonen algorithm given by, $W_{new} = W_{old} + \alpha * (Input - W_{old})$. The training law takes the weight at each epoch closer to the input. Consider the old weight vector W_{old} and the new weight vector W_{new} in a three dimensional plane. In Figure 7.6, it is evident that at each epoch, the weight is pushed closer to the input pattern by a factor of α. In higher dimensions, similar process occurs in the hyperplane.

The change in the weight of the winner neuron given by the genetic algorithm is used to adjust the weights of neighboring neurons (a factor of Gaussian of their distance from the winner neuron is used). This is done under the assumption that neurons closer to winner neurons resemble the winner. Since initially all neurons are initialized with small random weights, though the assumption may not be true

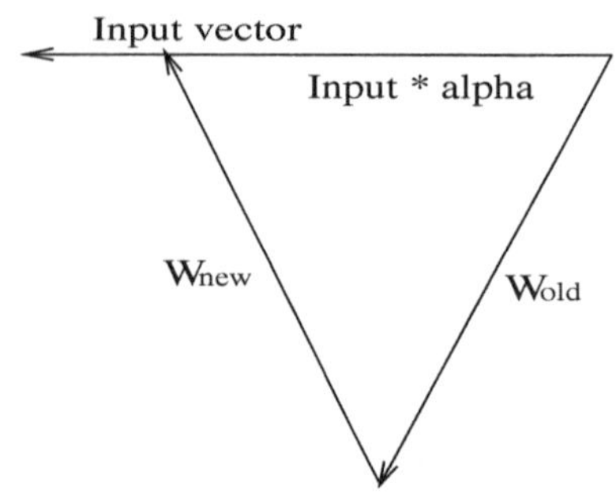

Fig. 7.6 Kohonen learning rule

initially the small weights of neurons ensures that the cluster map is properly formed. As the distance from the winner neuron increases, the Gaussian of the same decreases. Hence the change in the weight of the neuron far away from winner neuron reduces with its distance from the winner. The Gaussian function also decreases at each epoch and thus as the training proceeds, the size of the neighborhood decreases.

7.4.3 The Genetic Algorithm

Consider an ecosystem having m populations $P_1, P_2, P_3, ..., P_m$ of u individuals where each individual is a set of weights for the winner neuron. That is each individual is a vector of floats. Let $P^{`}$ be the new improved population for each *epoch.*

Since the genetic algorithm is used to adjust weights of the winner neurons of the Kohonen layer, the fitness function of the genetic algorithm follows the Kohonen learning rule. The genetic algorithm tries to move the weights of winner neurons closer to the input pattern. Hence the fitness function of the genetic algorithm for an individual is given by equation 7.3.

The algorithm for obtaining weights of the winner neurons for each epoch is as follows;

1. For each generation from Zero to maximum number of generations per epoch, *do*

   a. For each population P_i *do*
      i. $P^{`} = \{\ \}$
      ii. N = $|P_i|$
      iii. Sort P_i according to the fitness of the individuals
      iv. For $z = 0$ to some constant B of the better individuals *do*
         - $P^{`} = P^{`} \text{ U } P_i^z$
      v. Endfor
      vi. For the next C (some constant) individuals do
         - Select any individual P_i^j from P_i and mutate P_i^j
         - $P^{`} = P^{`} \text{ U } P_i^j$
      vii. Endfor

viii. For every few generations crossover and mutate between populations
ix. Until $|P^{'}| = N$
x. $P_i = P^{'}$

b. Endfor

2. Endfor

The genetic algorithm starts with a set of populations used previously during each epoch of the neural network to obtain the weights of the winner neurons. Its populations are preserved through the neural network epochs for each neuron. The populations of winners during each epoch are cultivated from the populations used last. The crossover used is a single point crossover. During mutation a small random value is added or subtracted to some randomly chosen weight in the individual.

7.4.4 *Fuzzy Inference System*

Input : κ the rank of the winner and γ the number of epochs.
Algorithm of fuzzy inference system:

1. Find the degree of membership of κ to each of the fuzzy sets High(H), Medium(M) and Low(L) of rank of winner.
2. Find the degree of membership of γ to each of the fuzzy sets High(H), Medium(M) and Low(L) of number of epochs.
3. Use fuzzy logic to obtain the degree of membership of learning parameter to be obtained to fuzzy sets High(H), High Medium(HM), Medium(M), Low Medium(LM) and Low(L).
4. Use centroid method to obtain crisp learning rate parameter from memberships to fuzzy sets.

7.4.5 *Backpropagation Algorithm*

Consider a backpropagation network with one input layer, multiple hidden layers and an output layer. Initialize all the weights to some small random values. For every input in the training set, update each network weight using the errors that are calculated for every layer, until the termination condition is met. The algorithm propagates back the square of the error and adjusts the weights accordingly.

7.4.6 *Complexity*

Consider a pure Kohonen network. Let n be the number of input units, m be the number of neurons and l be the number of winners. Computation of output requires nm operations per epoch and obtaining l winners requires lm steps per epoch. Finally, the adjustment of weights of the winner neuron and neurons surrounding each winners requires ln operations as after first few epochs, only winners and few other surrounding neurons weights are changed. Therefore, the computational complexity is typically of the order $O((n+l)m)$ per epoch. Generally, l is small and hence the complexity of the Kohonen network is of the order $O(nm)$.

Let p be the number of populations, u be the number of individuals and g be the number of generations per epoch in the genetic algorithm used to train Kohonen network. For the genetic algorithm to calculate the fitness function it takes n steps. Fitness function of n individuals are to be computed. Sorting of the u individuals is of the order $u * log(u)$. These steps are to be performed for p populations and g generations. Hence $nupg$ number of steps are to be performed per neural network epoch per winner by the genetic algorithm. Hence the complexity of our algorithm is of order $O(nm+lm+wnupg)$.

For a complex input set with many possible classes m is much greater than $nupg$. Hence, the order of computational complexity of the proposed algorithm is same as Kohonen, that is $O(nm)$, which is linear and the training complexity of the backpropagation network is exponential and higher than FEASOM.

7.5 Example

Consider the simple network shown in Figure 7.7, with two input units, four neurons in clustering neural layer and a single backpropagation neuron. Consider the network to be partially trained. Let the weights of clustering layer be

$$W_c = \begin{bmatrix} 0.09 & 0.21 & 0 & 0.2 \\ 0.19 & 0 & 0.21 & 0.1 \end{bmatrix}$$

Consider an input vector Î=$< 0.1, 0.2 >$ and let the expected output be 0.3. Now we obtain the outputs of the clustering neurons by normalizing the vector Î x W_c with respect to the input, by dividing each element in the input vector by the summation of squared values is and given by (0.1^2 + 0.2^2). Hence the outputs of the clustering network is $< 0.94, 0.42, 0.84, 0.8 >$

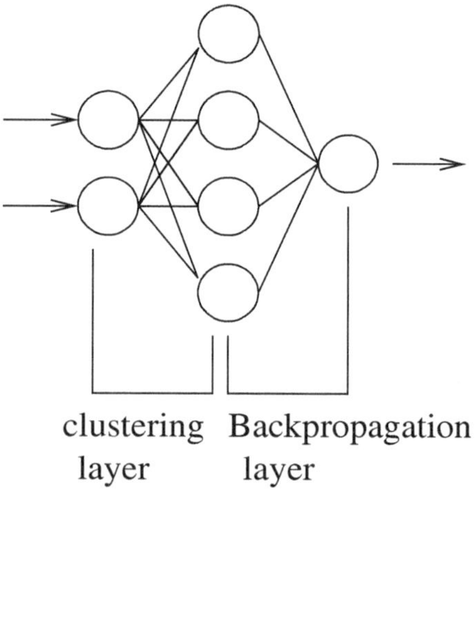

Fig. 7.7 Example Neural Network

Choosing the winner with the largest value from the output of the clustering network and assigning it with value one and the rest of the neurons with value zero, the output of the clustering layer is $\hat{O} = <1,0,0,0>$

Let the weights of the backpropagation neuron be

$$W_b = \begin{bmatrix} 0.29 \\ 0 \\ 0.4 \\ 0.01 \end{bmatrix},$$

then the output of the neural network is $\hat{O}$ x W_b = 0.29 + 0 + 0 + 0 = 0.29.

Now for training, consider the learning rate η to be 0.3. For the backpropagation network, the change in weights is given by $\Delta w_{ik} = \eta o_i(1-o_i)(t_i - o_i)Input_k$. Here if the target t is set to 0.3, the output o obtained is 0.29. Hence the change in weight is given by

$\Delta w_{ik} = 0.3 * 0.29 * (1-0.29) * (3-0.29)Input_k = 0.0062Input_k$

The new weights for the backpropagation neuron are

$$W_b = \begin{bmatrix} 0.29 + (0.0062 * 1) \\ 0 + (0.0062 * 0) \\ 0.4 + (0.0062 * 0) \\ 0.01 + (0.0062 * 0) \end{bmatrix} = \begin{bmatrix} 0.2962 \\ 0 \\ 0.4 \\ 0.01 \end{bmatrix}$$

In the clustering layer for simplicity let the neighborhood size be one, that is only the weights of the winner neuron are changed during every epoch. Consider a population with four individuals from which the weights of the winner neuron are to be selected. Let the four individuals be P_1^1, P_1^2, P_1^3 and P_1^4.

$P_1^1 = <0.09,0.19>$, $P_1^2 = <0.08,0.189>$, $P_1^3 = <0.09,0.01>$ and $P_1^4 = <0.02,0.19>$

Now for the next generation, the value of the best individual is retained. Hence, the new population has an individual $P_1^1 = <0.09,0.19>$. Mutation is performed by choosing some random chromosome and altering its value slightly for the second individual to obtain the individual $P_1^2 = <0.09,0.189>$. Let the third and fourth individuals crossover at the second position giving individuals $P_1^3 = <0.09,0.19>$ and $P_1^4 = <0.02,0.01>$. Slightly mutating them, we get $P_1^3 = <0.092,0.193>$ and $P_1^4 = <0.024,0.008>$

Now applying the fitness function given by $\sum input^2 - \sum(gene * input)$, the fitness of P_1^1 is 0.003, P_1^2 is 0.0032, P_1^3 is 0.0022 and P_1^4 is 0.0046. Choosing the best individual as P_1^3 which has the lowest fitness value and hence is closest to the input pattern, the new Kohonen weight matrix is as follows,

$$W_c = \begin{bmatrix} 0.092 & 0.21 & 0 & 0.2 \\ 0.193 & 0 & 0.21 & 0.1 \end{bmatrix}$$

In the next epoch, repeating the input vector $< 0.1, 0.2 >$ we obtain the prediction value of 0.2962 which is better than the previous result and the winner is pushed closer to the given pattern, thus assuring it the winning position in the next epoch.

Similarly, consider an input vector $< 0.2, 0.1 >$. For this, the winner is the fourth neuron. Applying backpropagation, the output is 0.01. It is almost a straight line with decreasing slope of 0.1. Thus, with this simple model of neural network with trained weights we could predict two sets of linear values with a fixed slope.

7.6 Implementation

The algorithm FEASOM is implemented to take in eight previous stock values and predict three values. The Kohonen layer had eight input units and six hundred neurons with two hundred winners. The winners' are assigned outputs varying from zero to one, such that the first winner has one as its output and subsequent winners have lower outputs. To train the weights of the Kohonen layer, the genetic algorithms used, evolved through one hundred generations per epoch of the neural network. The genetic pool consists of a collection of five populations and inter-population crossovers are carried out for every thirteen generations. Each population consists of seventeen individuals and the best four individuals of a population are retained for the next generation. The populations are preserved between neural network epochs for better results. The fuzzy inference system divided inputs rank of winners and number of epochs into three fuzzy sets each as Low [0 : 75], Medium [50 : 150],and

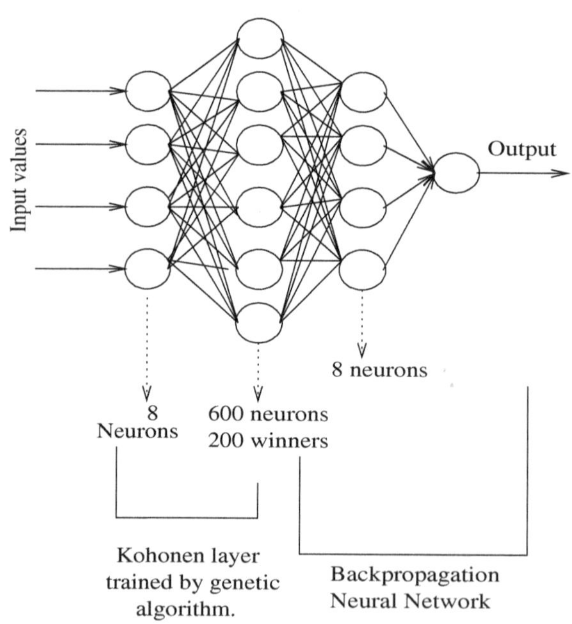

Fig. 7.8 Implementation model

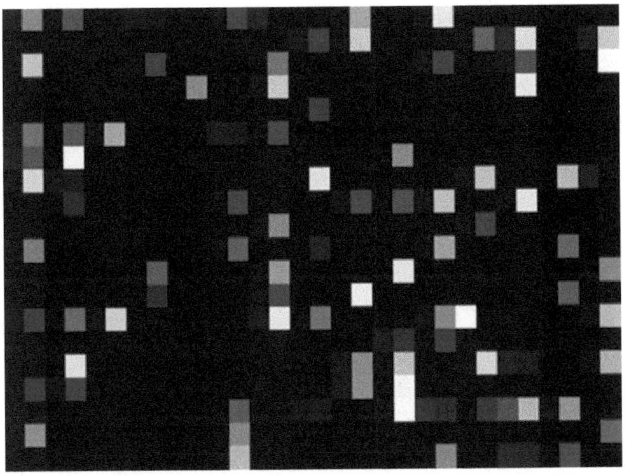

Fig. 7.9 Cluster 1

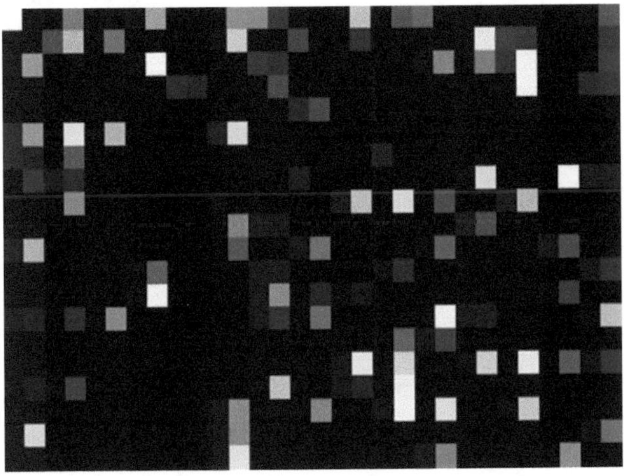

Fig. 7.10 Cluster 2

High [125 : 200]. Output learning rate is divided into six fuzzy sets and centroid method is used to obtain crisp learning rate value.

To extract stock market prediction values from the cluster map, a simple two layer backpropagation neural network is used, which has eight neurons in the first layer and one neuron in the next layer to predict the share value. Momentum is used to avoid settling of the neural network at local minima. The cluster map obtained from the Kohonen layer shows complex but distinct traits for different features in the input stock market values. The backpropagation neural network is used to discern these features and predict the share value.

The inputs to the model are the eight previous stock market values subtracted by a value less than the least share value. The output of neural network is a value which when added to the least value initially subtracted gives the predicted value. The predicted value is used as the input to the next two predictions and thus three future predictions are made from eight past values.

The entire implementation model is summarized in Figure 7.8. Figure 7.9 and Figure 7.10 show example illustrations of two cluster maps. The different shades of gray show the corresponding values of output varying from zero to one. This multichromatic nature of the cluster map helps in better prediction of stock market value by the backpropagation algorithm.

7.7 Performance Analysis

To analyze and compare the efficiency of the proposed algorithm with those of the other algorithms, we trained our network(FEASOM) with a synthetic dataset taken from a linear function. The training set consists of one-fourth the test set which is evenly spread. The Kohonen, FEASOM and backpropagation networks are trained with the same training set. The difference between predicted and actual values is plotted in Figure 7.11 for all the three models. From the graph, it can be seen that FEASOM is closer to the base line and has less error than Kohonen and backpropagation neural networks. The sample taken from the Figure 7.11 showing errors of the three models is tabulated in Figure 7.12. It can be observed that the rate of increase

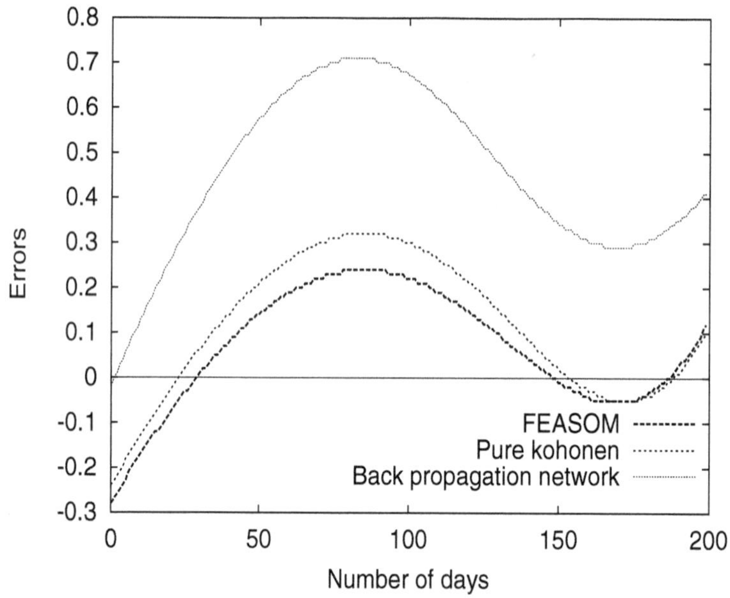

Fig. 7.11 Error graph on a set of linear data

Fig. 7.12 Errors generated on prediction of linear dataset

EASOM	Kohonen network	Back propagation network
0	0.06	0.37
0.01	0.06	0.38
0.02	0.07	0.39
0.02	0.08	0.09
0.03	0.04	0.1
0.05	0.11	0.41
0.05	0.11	0.42
0.06	0.12	0.43
0.07	0.13	0.44
0.07	0.14	0.45

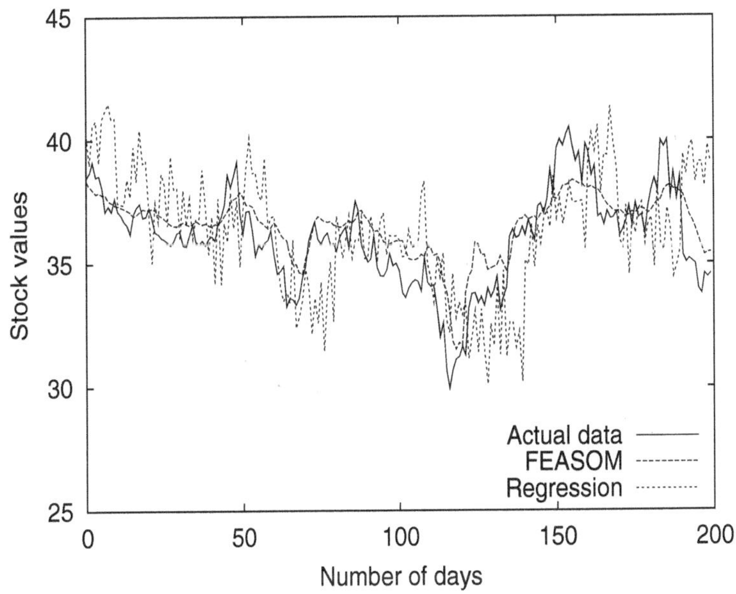

Fig. 7.13 Predicted values for GM stock market

of error in backpropagation method is more than that in the Kohonen or FEASOM methods.

One striking feature seen in the graph is the oscillation of error about the base line that can be, observed in both Kohonen and FEASOM methods. From the Figure 7.11, it is evident that the use of Genetic algorithms in FEASOM restricts

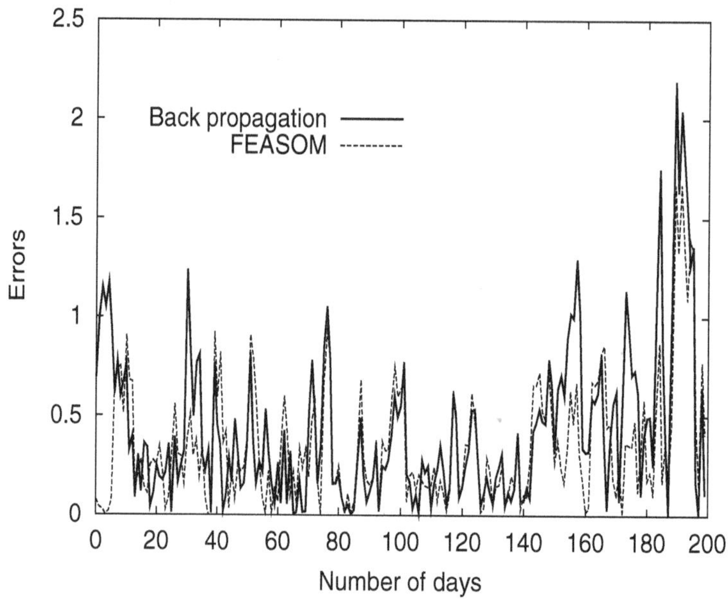

Fig. 7.14 Error generated for GM stock values from Figure 7.13

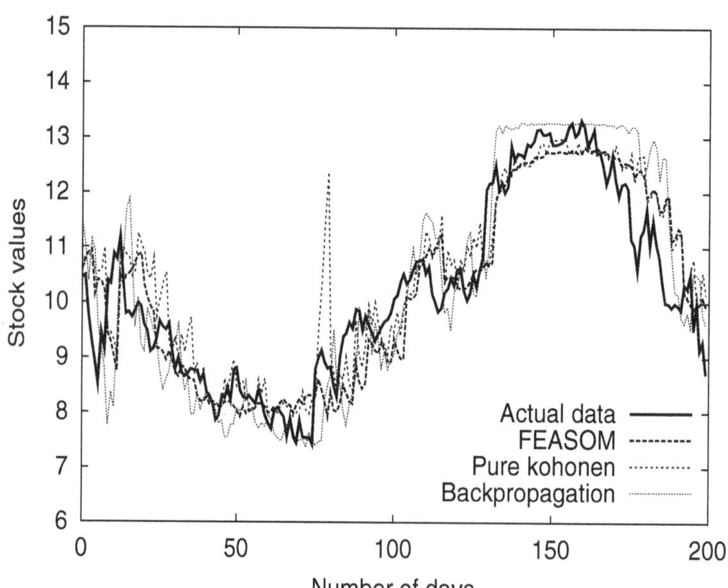

Fig. 7.15 Predicted values for Satyam stock values

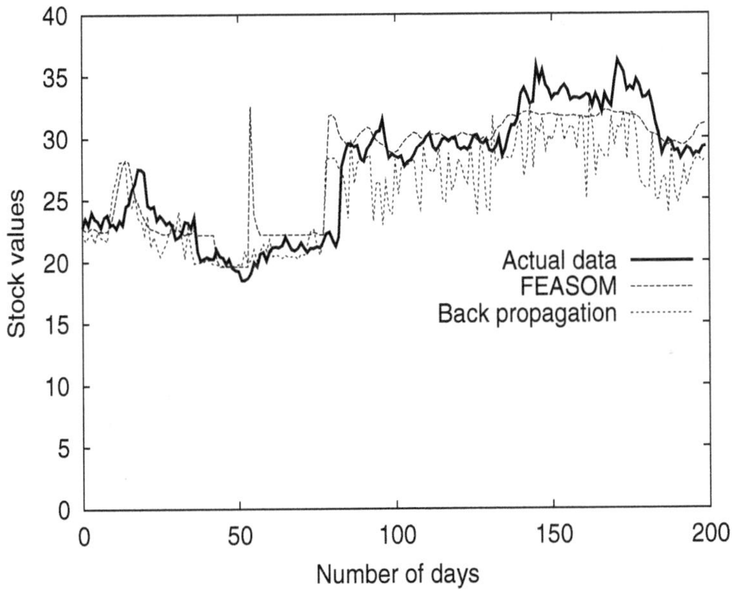

Fig. 7.16 Predicted values for Wipro stock values

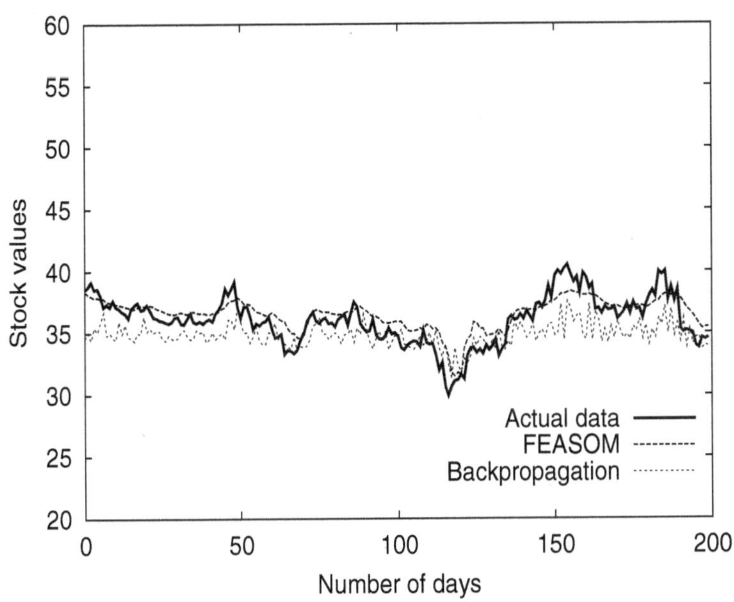

Fig. 7.17 Predicted values for General Motors stock values

the range of error. Better predictions are achieved because, during the later epochs of the neural network, all individuals lie within a small area round the learning peak.

To test the effectiveness of FEASOM, the algorithm is tested on several real datasets. The stock market data of all the companies are collected from BSE(Bombay Stock Exchange) on request. The training and test sets taken are disjoint. A comparative study is conducted to juxtapose regression model, backpropagation algorithm and Kohonen algorithm.

Figure 7.13 shows the plot of predictions of GM stock market by both classic regression model and FEASOM. It is clear from Figure 7.13 that FEASOM outperforms the regression model. The error graph in Figure 7.14 displays the errors in predictions of FEASOM and regression model while predicting GM stock market data. It is observed that the predictions of the regression model are erroneous compared to those of FEASOM.

The three algorithms, FEASOM, Kohonen and backpropagation, are trained to predict the actual stock values of a leading Indian software company Satyam Inc. The training set consists of 40 datapoints of Satyam stock values from the year 1999 to mid 2001. Finally when the neural networks trained using the three algorithms are tested, FEASOM shows the best performance amongst the three methods. The graph plotted in Figure 7.15 shows the predictions of the three algorithms on Satyam stock market values. As the graph indicates, the stock values are highly fluctuating, yet it is observed that FEASOM is able to predict the share values well.

The graph in Figure 7.16 shows the predictions of stock values for Wipro software company, another Indian IT giant. Here the share prices have been more stable and it can be seen that FEASOM gives good prediction results while the backpropagation algorithm gives highly fluctuating result. Figure 7.17 shows the prediction of GM(General Motors) stock market values. The graph shows much better predictions by FEASOM than by the backpropagation algorithm. Figure 7.18 shows the error plots of prediction of FEASOM and backpropagation algorithms. The error considered here is difference between actual and predicted values. It is evident that FEASOM consistently performs better than backpropagation. Figure 7.19 shows the prediction of Microsoft stock market. It can be noticed that the stock market values are more or less stable except towards the end where the values are fluctuating. Figure 7.20 shows the comparison of errors in predicting Microsoft stocks by FEASOM and Backpropagation. Here at most places FEASOM has much less errors. Figure 7.21 shows the prediction of Motorola stock market. The prediction by FEASOM here is close to the actual value. Figure 7.22 shows the errors in prediction of the Motorola stock market by FEASOM and backpropagation methods. The errors in the predictions of FEASOM are less than that of errors in predictions by the backpropagation algorithm. Figure 7.23 shows the prediction of US Steels stock values by FEASOM. The prediction is very close and only one place where the market value had a very sudden sharp rise, the prediction is not possible. The predictions of stock market values of Infosys, Intel, Lee and IBM are shown in Figure 7.24, Figure 7.25, Figure 7.26 and Figure 7.27 respectively and it is observed that FEASOM performs well consistently in all the four cases. As already mentioned eight previous values are used to predict next three stock market

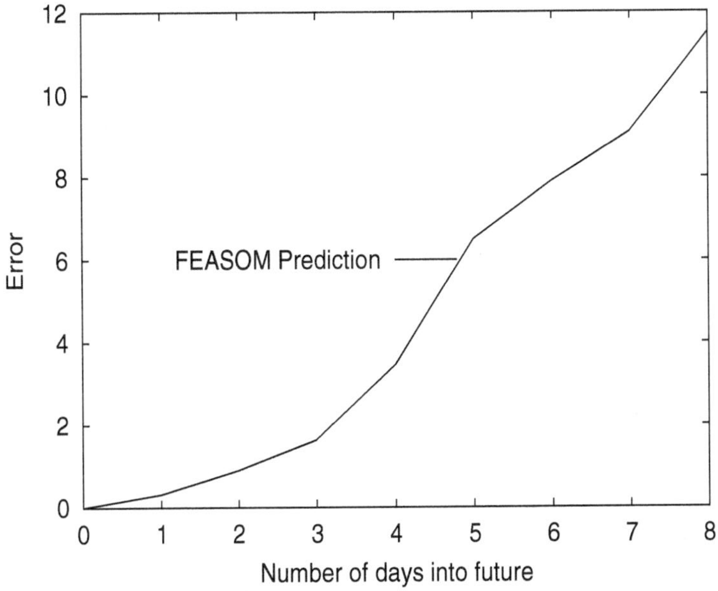

Fig. 7.18 Error generated for GM stock values from Figure 7.17

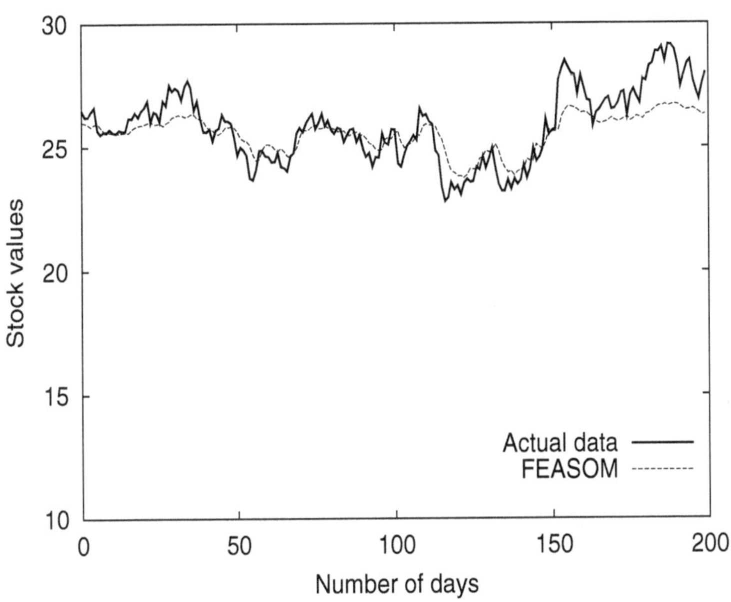

Fig. 7.19 Predicted values for Microsoft stock values

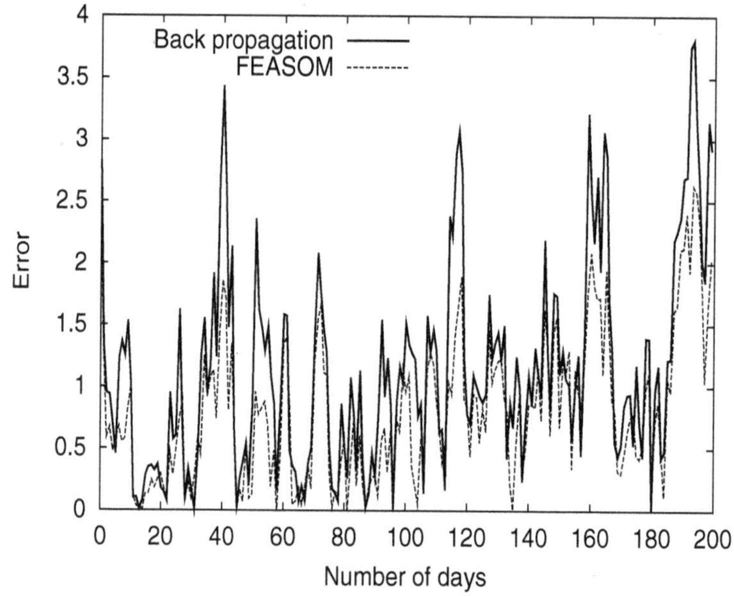

Fig. 7.20 Error generated for Microsoft stock values from Figure 7.19

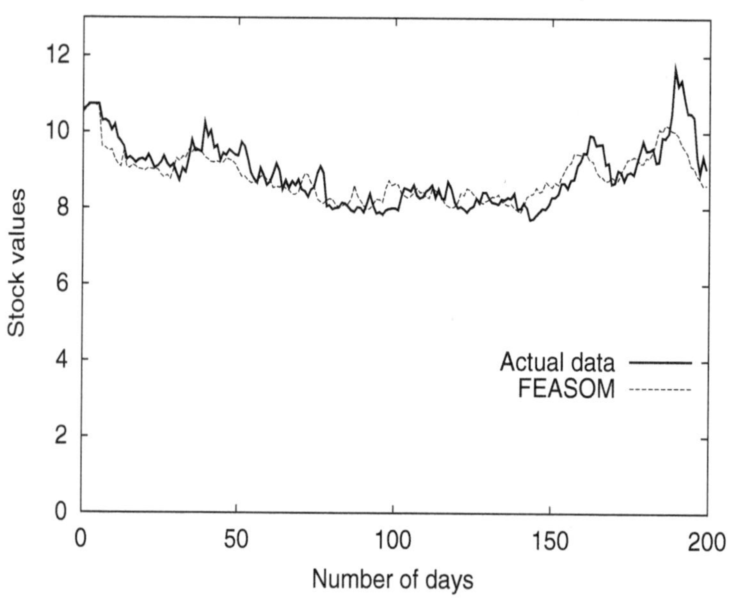

Fig. 7.21 Predicted values for Motorola stock values

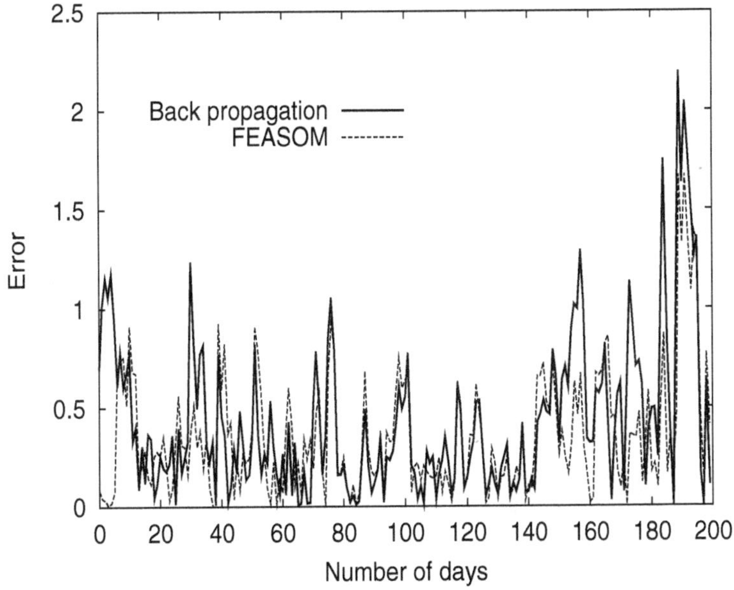

Fig. 7.22 Error generated for Motorola stock values from Figure 7.21

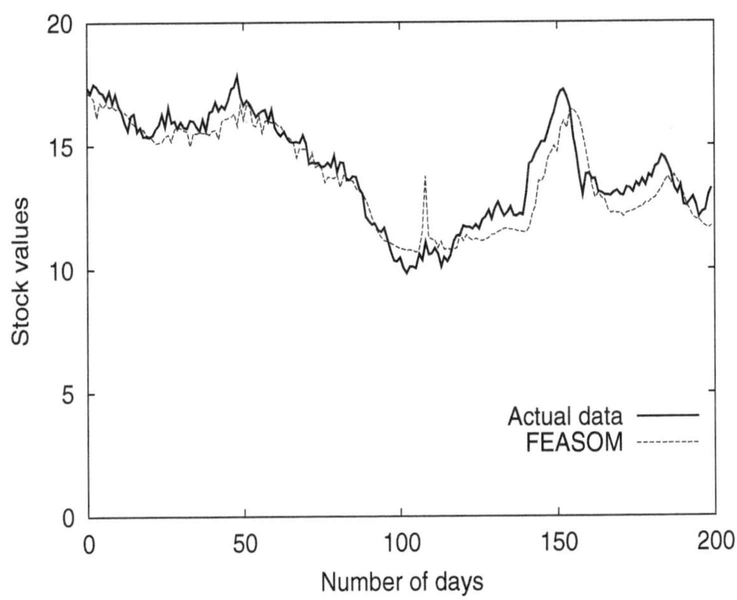

Fig. 7.23 Predicted values for US Steels stock values

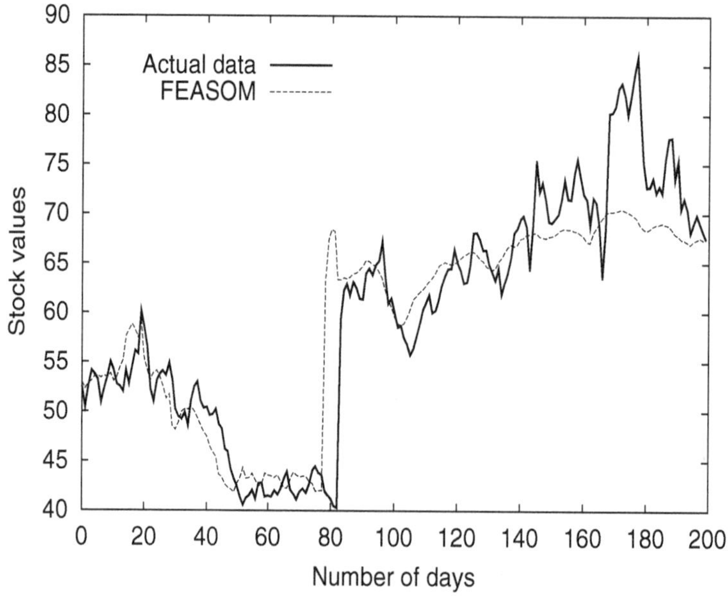

Fig. 7.24 Predicted values for Infosys stock values

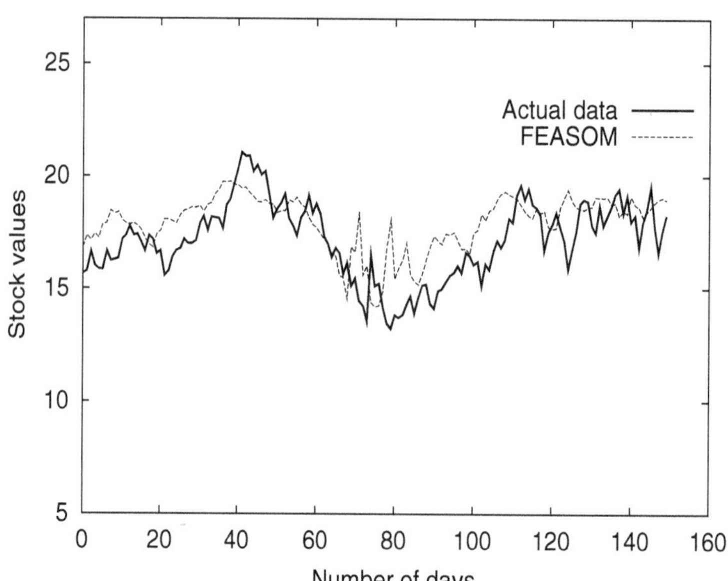

Fig. 7.25 Predicted values for Intel stock values

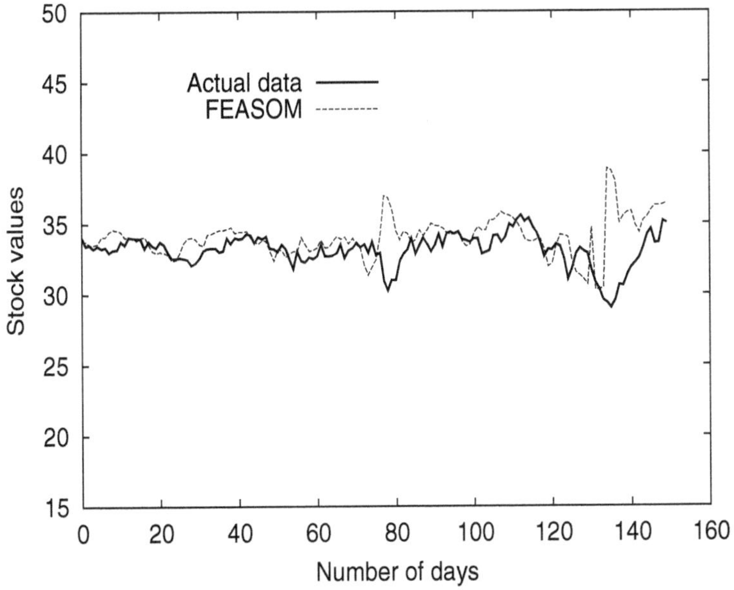

Fig. 7.26 Predicted values for Lee stock values

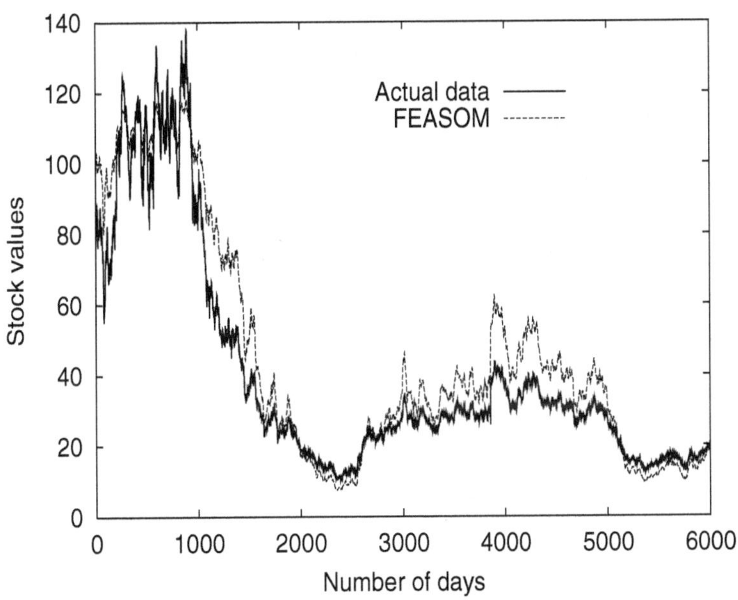

Fig. 7.27 Predicted values for IBM stock values

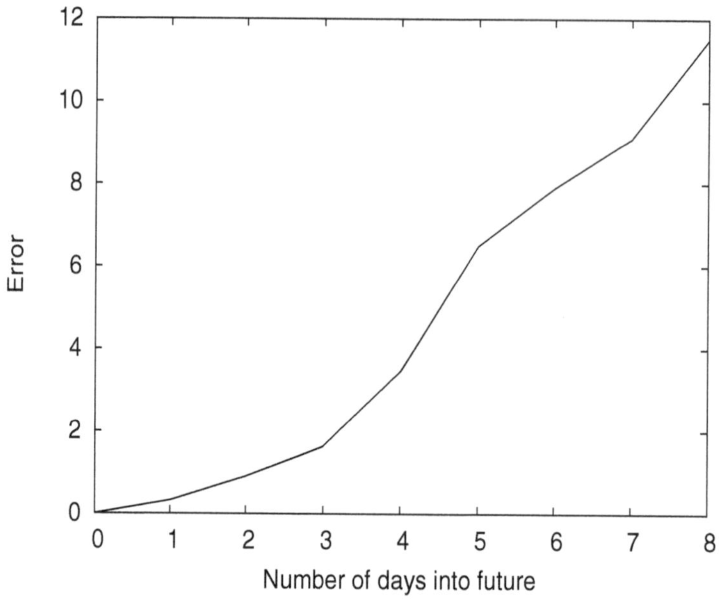

Fig. 7.28 Performance degradation of prediction

Method	MAD	SSE	MSE	RMSE	MAPE
FEASOM	1.115	446.9	2.235	1.495	0.005
Backpropagation ANN	1.335	578.7	2.894	1.701	0.017
Regression	1.884	1113.6	5.668	2.381	0.056

Fig. 7.29 Performance Measures

closing prices. Figure 7.28 shows plot of how performance degrades as the same eight values are used to predict closing values of stocks further into the future. The X axis has days into the future and the Y axis the average error of prediction. It can be seen that predicting three days into the future is most appropriate, as graph increases steeply after that. Certain standard performance measures are used to evaluate the performance of the method and compare it with other methods. These performance measures include

- the mean absolute deviation (MAD) = $\frac{\sum |e_t|}{N}$
- the sum of squared errors (SSE) = $\sum (e_t)^2$
- the mean squared error (MSE) = $\frac{\sum (e_t)^2}{N}$

- the root mean squared error (RMSE) = $\sqrt{MSE}$
- the mean absolute percentage error (MAPE) = $\frac{1}{N}\sum|\frac{e_t}{y_t}|(100)$

The table in Figure 7.29, shows the values of these performance measure for autoregression, backpropagation and FEASOM predictors and it is seen that the performance of FEASOM is superior.

7.8 Summary

In this chapter, we examine the issue of predicting stock market values using an effective fuzzy based neuro-genetic algorithm. The algorithm is efficient and effective in predicting share values. The time complexity of FEASOM is linear which is more efficient than other neural network algorithms which are usually of nonlinear complexity. The use of genetic algorithms to train the weights of the Kohonen network and the multivalued winner approach makes the prediction of stock market more accurate. The algorithm FEASOM outperforms both backpropagation and Kohonen networks in predicting accurate share values. Extensive simulations on real data varying from software industries to steel industries shows the consistent better performance of FEASOM. However sharp rise and fall in the stock market values due to unforeseen circumstances and external factors cannot be predicted by the algorithm. The algorithm can also be extended to heterogeneous and distributed databases.

References

1. Shenoy, P.D., Srinivasa, K.G., Mithun, M.P., Venugopal, K.R., Patnaik, L.M.: Dynamic Subspace Clustering on Very Large High-Dimensional Databases. In: Liu, J., Cheung, Y.-m., Yin, H. (eds.) IDEAL 2003. LNCS, vol. 2690, pp. 850–854. Springer, Heidelberg (2003)
2. Haykin, S.: Neural Networks, A Comprehensive Foundation. Pearson Education Inc., London (1999)
3. Shenoy, P.D., Srinivasa, K.G., Venugopal, K.R., Patnaik, L.M.: Evolutionary Approach for Mining Association Rules on Dynamic Databases. In: Whang, K.-Y., Jeon, J., Shim, K., Srivastava, J. (eds.) PAKDD 2003. LNCS, vol. 2637, pp. 325–336. Springer, Heidelberg (2003)
4. Cox, E.: Fuzzy Modeling and Genetic Algorithms for Data Mining and Explorations. Morgan Kaufmann, San Francisco (2004)
5. de Gooijer, J.G., Hyndman, R.J.: 25 Years of IIF Time Series Forecasting: A Selective Review, TI 2005-068/4, Tinbergen Institute Discussion Paper (2005)
6. Zhang, G., Patuwo, B.E., Hu, M.Y.: Forecasting with Artificial Neural Networks: The State of the Art. International Journal of Forecasting 14, 35–62 (1998)
7. Balkin, S.D., Ord, J.K.: Automatic Neural Network Modeling for Univariate Time Series. International Journal of Forecasting 16, 509–515 (2000)
8. Timmermann, A., Granger, C.W.J.: Efficient Market Hypothesis and Forecasting. International Journal of Forecasting 20, 15–27 (2004)

9. Leung, M.T., Daouk, H., Chen, A.-s.: Forecasting Stock Indicies: A Comparision of Classification and Level Estimation Models. International Journal of Forecasting 16, 173–190 (2000)
10. Kim, Y., et al.: Customer Targeting: A Neural Network Approach Guided by Genetic Algorithms, Technical Report, Management Sciences Department, University of Iowa, USA (2000)
11. Bansal, A., Kauffman, R.J., Weitz, R.R.: Comparing the Modeling Performance of Regression and Neural Networks as Data Quality Varies: A Business Value Approach. Journal of Management Information Systems 10, 11–32 (1993)
12. Dorffner, G.: Neural Networks for Time Series Processing. Neural Network World 6(4), 447–468 (1996)
13. Egeli, B., Ozturan, M., Badur, B.: Stock Market Prediction Using Artificial Neural Networks. In: Hawaii International Conference on Business (June 2003)
14. Zekic, M.: MS Neural Network Applications in Stock Market Prediction - A Methodology Analysis. In: Proc. of 9th Intl' Conf. Information and Intelligent Systems, pp. 255–263 (September 1998)
15. Lu, R., Mois, M., Pires, F.M.: Prediction Model, Based on Neural Networks, for Time Series with Origin in Chaotic Systems. In: Workshop on Artificial Intelligence for Financial Time Series Analysis (2001)
16. Ciesielski, V., Palstra, G.: Using a Hybrid Neural/Expert System for Database Mining in Market Survey Data. In: Proc. of Intl' Conf. on KDD 1996, pp. 36–43. AAAI Press, Menlo Park (1996)
17. Ivakhnenko, A.G., Miiller, J.: Recent Developments of Self Organizing Modelling in Prediction and Analysis of Stock Market, `http://www.inf.kiew.ua/GMDH-home`
18. Balakrishnan, K., Honavar, V.: Evolutionary Design of Neural Architectures: Prelimnary Taxonomy and Guide to Literature, Technical Report CS TR95-01, Department of Computer Science, Iowa State University (1995)
19. Yao, X.: Evolutionary Artificial Neural Networks. Encyclopedia of Computer Science and Technology 33, 137–170 (1995)
20. Armano, G., Murru, A., Roli, F.: Stock Market Prediction by a Mixture of Genetic-Neural Experts. IJPRAI 16(5), 501–526 (2002)

Chapter 8
Data Mining Based Query Processing Using Rough Sets and GAs

Abstract. The optimization of queries is critical in database management systems and the complexity involved in finding optimal solutions has led to the development of heuristic approaches. Answering data mining query involves a random search over large databases. Due to the enormity of the data set involved, model simplification is necessary for quick answering of data mining queries. In this chapter, we propose a hybrid model using rough sets and genetic algorithms for fast and efficient query answering. Rough sets are used to classify and summarize the datasets, whereas genetic algorithms are used for answering association related queries and feedback for adaptive classification. Here, we consider three types of queries, i.e., select, aggregate and classification based data mining queries. Summary tables that are built using rough sets and analytical model of attributes are used to speed up select queries. Mining associations, building concept hierarchies and reinforcement of reducts are achieved through genetic algorithms. The experiments are conducted on three real-life data sets, which include KDD 99 Cup data, Forest Cover-type data and Iris data. The performance of the proposed algorithm is analyzed for both execution time and classification accuracy and the results obtained are good.

8.1 Introduction

The process of classification involves finding a set of models that describe and distinguish data classes or concepts for the purpose of being able to use the model to predict the class of objects whose class labels are unknown. The derived model is based on the analysis of a set of training data. The derived model may be presented in various forms such as classification rules, decision trees, mathematical formulae, etc.. The interestingness of patterns are measured by support and confidence. Objective measures for association rules of the form $X \rightarrow Y$ are support and confidence [1], that represent the percentage of transactions that satisfy the rule and degree of certainty of the detected association, respectively. Pattern classification problems have been widely used as traditional formulation of machine learning problems and researched with various approaches including statistical methods [2], neural networks [3] [4], genetic algorithms [5], etc..

K.R. Venugopal, K.G. Srinivasa, L.M. Patnaik: Soft Comput. for Data Min. Appl., SCI 190, pp. 167–195.
springerlink.com

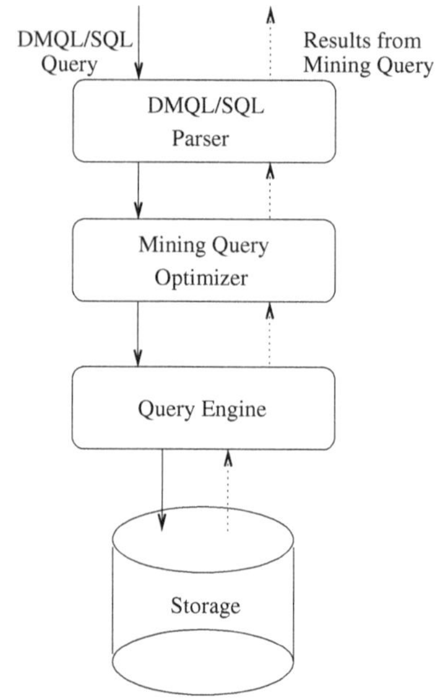

Fig. 8.1 Database model

Major issues in data mining are the mining methodology and user interaction issues. The mining methodology is concerned with coverage of wide spectrum of data analysis and user interaction deals with interactive mining of knowledge at multiple levels of abstraction with reference to the domain knowledge. Just as relational query languages allow users to pose ad-hoc queries, Data Mining Query Languages(DMQL) [1] have been developed to describe ad-hoc mining tasks by facilitating the specification of the relevant knowledge, kinds of knowledge to be mined and the condition and constraints to be enforced on discovered patterns. Some of the commands used in knowledge discovery to specify the kinds of knowledge to be mined include (i) Generalized relations (ii) Characteristic rules (iii) Discriminant rules (iv) Classification rules and (v) Association rules. The typical Database model using DMQL is shown in Figure 8.1. The DMQL parser validates the user given commands and builds the query trees. The optimizer determines the optimal query evaluation path. Finally, the query engine transforms the query algebra and executes over the database. The results obtained after the execution of query are presented to the user in a representable form. The syntax of DMQL is close to that of SQL and is generally of the form,

```
use database <database_name>
{use hierarchy <hierarchy_name> for <attribute>}
<rule_spec>
related to <attr_or_agg_list>
from <relation(s)>
[where <conditions> ]
[order by <order list>]
{with [<kinds of>] threshold= <threshold_value>
[for <attribute(s)>]}
```

The optimization of queries is critical in aspect of database management and the exponential time complexity involved in finding optimal solutions has led to the development of heuristic approaches. Here, we experiment with the concepts of rough sets and genetic algorithms in order to reduce the query execution time with approximate reasoning of data, without going into exact statistical measures [6], [7], [8].

Since its inception, rough set theory has proved itself of being valuable in the context of intelligent information systems. Developed by Pawlak [9] for classification analysis of data tables, it seeks to synthesize approximation of information through focus on discernibility of objects in the domain. It bears on the assumption that in order to define a set, a prior information about the elements of the universe is needed. This information about the elements is presented through an attribute-value system. All objects are classified into higher level concepts. The classification itself is in the form of lower and upper approximations, where the lower approximation consists of all the objects that *surely* belong to the concept and those in the upper approximation *possibly* belong to the concept. The primary focus is on extracting information *granules* and *reducts*. The information granule is a group of objects grouped together through similarity or functionality. Data is reduced by identification of equivalence classes through information granules, thereby reducing the number of attributes that are needed to represent the entire class. The retained attributes form the reducts. However, the concept of rough set is more general and can be extended to situations where underlying equivalence relation is not available [10]. Attempts to extend rough set theory to handle complex objects have also been made through case-based-reasoning systems [11].

On the other hand, there has been tremendous developments in the field of Genetic Algorithms(GA) over the last two decades. Their origin is attributed to Holland's work on cellular automata [12]. The applications of the theory include diverse areas such as training objective optimization, job scheduling, neural nets, classification, image extraction, etc.. A genetic algorithm is a search process that follows the principles of evolution through natural selection. It is efficient, adaptive and robust search process producing near optimal solutions and enjoys the advantage of implicit parallelism. GAs are executed iteratively on a set of coded solutions, called population with three basic operators: selection/reproduction, crossover and mutation.

8.2 Problem Definition

Assume that the entire database is represented as a single relational table R with attributes $A_1, A_2, A_3, \ldots A_n$. Let the number of tuples in R be N with each tuple identifiable by a *TupleID*, where t_i represents the ith tuple. The objective of this chapter is to efficiently answer the queries belonging to the following three categories.

1. An information retrieval query whose purpose is to search the tuples satisfying one or more conditions usually specified by the WHERE clause
2. An aggregate query which involves the extraction of statistical information that does not exist as it is, but have to be derived from the existing attributes.

3. The third type of queries are those that exist for the discovery of knowledge and patterns in an information system. These involve mining characteristics, identifying associations, dynamic classification defining hierarchies and visualization among others.

Assumptions

1. The proposed framework works in conjunction with the existing information systems.
2. Only the queries belonging to the categories as mentioned above are executed by the proposed framework, whereas, other types of queries are executed by the existing information systems.
3. The entire dataset is represented as a single relational table.

8.3 Architecture

A data mining query language that allows ad-hoc mining is used for this purpose. The DMQL adopts SQL like syntax so that it can be seamlessly integrated with relational query language SQL.

Figure 8.2 shows the block diagram of the proposed frame work {Upper Section(a)} in conjunction with the existing information system{Lower Section(b)}. Here, a classification based information rich(compressed), alternate data catalog is built using roughsets on the primary data source. Only the essential features that best describe the database are stored in the catalog in the form of summary tables. Genetic algorithm-based approach is used for association and concept hierarchy

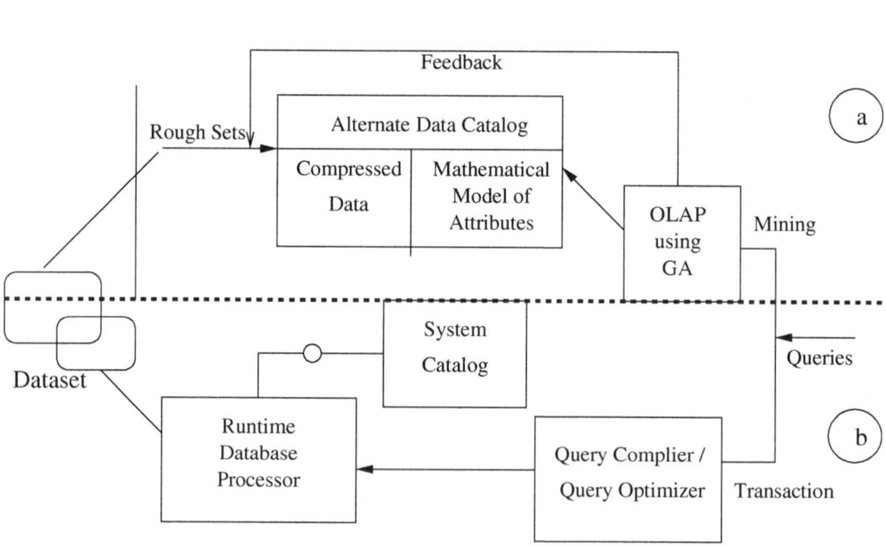

Fig. 8.2 Architecture of the Proposed Model

queries, in which the summarized models are used in order to reduce the execution time. Genetic algorithms are also used as feedback to improve the accuracy of classification. Only the queries belonging to the three categories as mentioned in the problem definition are answered by the proposed framework (Figure 8.2, section (a)), whereas, the remaining queries are answered by the existing system (Figure 8.2, section (b)).

The proposed framework as shown in Figure 8.2, seeks to speed up the processing time required to produce results for all the three types of queries, where the results obtained from the proposed system slightly differ from those obtained by querying the actual database.

For the first type of query, the performance is measured by the commonality between the results obtained. If T is the set of $tuple-id$s that are retrieved for an actual query and T' is the set obtained from the proposed framework; the quality can be measured as, $\alpha = |T \cap T'|/|T|$ and $\beta = |T - T'|/|T|$ where α represents the relevance factor and β signifies the margin of error. For other types of queries, the deviation of the obtained result from the actual one is quantifiable and can be flexibly changed by the user. Mining related queries such as finding association, characteristics, concepts hierarchies are evaluated based on the percentage of records in the data sets that are correctly classified. The approach given here focuses on efficient results with lower query-execution times.

8.3.1 Rough Sets

The rough set theory, despite being relatively simple, has the capability to deal with imperfections, such as noise and unknown values. Some of the concepts that are relevant to this article are briefed here. Details can be referred from [13] and [14].

Information System: An Information System is a set of objects with attribute related to it. By definition, *An Information System is an ordered pair* $\mathscr{A}=(U, A)$ *where U is a nonempty finite set of objects - the Universe, and A is a non-empty, finite set of elements called Attributes. Every attribute* $a \in A$ *is a total function* $a : U \rightarrow V_a$, *where* V_a *is the set of allowed values for the attribute*

A Decision System IS $\mathscr{A}=(U, A)$ *for which the attributes in A are further classified into disjoint sets of condition attributes C and decision attributes D.* $(A = C \cup D, C \cap D = \phi)$

Indiscernibility: *With every subset of attributes* $B \subseteq A$ *in the IS* $\mathscr{A} = (U, A)$, *an equivalence relation IND(B) called an Indiscernibility Relation is associated: which is defined as follows:*

$$IND(B) = \{(x,y) \in U^2 | a(x) = a(y) \forall a \in B \}$$

By definition, $U/IND(B)$ is the set of all equivalence classes in relation $IND(B)$. The equivalence classes induced by Indiscernibility relation are known as granules.

The partition induced by equivalence relation can be used to build new subsets of the universe.

The Lower Approximation $\underline{B}X$ and the Upper Approximation $\overline{B}X$ of a set of objects $X \subseteq U$ with reference to a set of attributes $B \subseteq A$ (defining an equivalence relation on U) may be defined in term of the classes in the equivalence relation, as follows:

$$\underline{B}X = \bigcup\{E \in U/IND(B)|E \subseteq X\}$$

$$\overline{B}X = \bigcup\{E \in U/IND(B)|E \cap X \neq \phi\}$$

Reducts: Indiscernibility relation reduces the data by identifying equivalence classes, using the available attributes. Only one element of the equivalent class is needed to represent the entire class. The minimal set of attributes min_A are taken from initial relation A, such that min_A induces same partition on the domain of DS as done by A. The above set of attributes are called reducts. Reducts have been appropriately characterized in [14] by *discernibility matrices* and *discernibility functions*. For a set of attributes $B \subseteq A$ in $A = (\mathscr{U}, A)$, the Discernibility Matrix $M_D(B) = m_D(i,j)_{nxn}$ $1 \leq i,j \leq n = |U/IND(B)|$, where

$$m_D(i,j) = \{a \in B|a(E_i) \neq a(E_j)\} for\ i,j = 1,2,\ldots,n.$$

The entry $m_D(i,j)$ in the discernibility matrix is the set of attributes from B that discern object classes $E_i, E_j \in U/IND(B)$. The Discernibility Function $f(B)$ of a set of attributes $B \subseteq A$ is

$$f(B) = \bigwedge_{i,j\in\{1...n\}} \bigvee \overline{m}_D(E_i,E_j)$$

where $n = |U/IND(B)|$, and $\overline{m}_D(E_i,E_j)$ is the disjunction taken over the set of boolean variables $m_D(i,j)$ corresponding to the discernibility matrix element $m_D(i,j)$.

The relative discernibility function $f'(B)$ computes the minimal sets of attributes required to discern any equivalence class from all the others. Similarly, the relative discernibility function $f'(E,B)$ computes the minimal sets of attributes required to discern a given class E from the others.

Dispensibility : An attribute a is said to be dispensable or superfluous in $B \subseteq A$ if $IND(B) = IND(B - \{a\})$, otherwise the attribute is indispensable in B.

An Example: An example of decision system is shown in Table 8.1, with income being the decision attribute.

From Table 8.1, it can be observed that,

$$U/IND(\{studies, education, works\}) = \{\{1,2\},\{3\},\{4,5\}\}$$

The objects that are grouped together cannot be discerned between one another when using the selected set of attributes. The equivalence class is formed with such a group. In Table 8.2, it can be observed that, class E_1 comes from objects 1 and 2, class E_2 from object 3, while class E_3 comes from objects 4 and 5.

The following relative discernibility functions can be calculated:

Table 8.1 An Example of Decision System

	studies	education	works	income
1	no	good	yes	high
2	no	good	yes	high
3	yes	good	yes	none
4	no	poor	no	low
5	no	poor	no	medium

Table 8.2 Equivalence classes

	studies	education	works
E_1	no	good	yes
E_2	yes	good	yes
E_3	no	poor	no

Table 8.3 Discernibility Matrix

	E_1	E_2	E_3
E_1	-	studies	education works
E_2	studies	-	studies education works
E_3	education works	education works studies	-

$f'(E_1, C) = studies \wedge (\ education \vee works\ )$
$f'(E_2, C) = studies \wedge (\ studies \vee education \vee works\ )$
$f'(E_3, C) = (\ education \vee works\ ) \wedge$
$\quad (\ studies \vee education \vee works\ )$

From Table 8.3, it can be noted that, $IND(C) = IND(C - \{works\}) = IND(C - \{education\})$. The only dispensable attribute is *studies*.

It is clear that all the attributes are not needed to determine the classification of the dataset. *Dispensable* attributes should be mapped to the attributes from which its value can be derived. The attribute mapping table (Table 8.4) is constructed so that the query containing any of these attributes can be translated if needed, before execution. For example, the query

```
select ... from <table> where works="yes";
```

can be translated to

```
select ... from <table> where education="good";
```

The above methodology is described for discrete attributes, whereas slight modifications [9] are required to make the rough set to work for continuous attributes.

Table 8.4 A mapping of attributes to reducts

$\mathscr{A}$	$\mathscr{R}$
A_1	R_1
A_2	R_1
A_3	R_1
A_6	R_2
A_7	R_2

The difficulty of continuous-type attributes arises due to the fact that they cannot be used to partition the data set into classes based on their values since the number of values they assume are nearly the same as the number of tuples present in the dataset. The usual approach taken during the cleaning process is to quantize the values into classes and then treat the attributes like a discrete valued function [15]. However in this case the complexity is increased by the fact that the classifier needs to produce efficient results for user-invoked SQL-like queries. The approach taken here is predominantly analytical in nature described in section *V*.

8.3.2 *Information Streaks*

TupleID: Assume that each of the tuples in the data table is identified by a unique identifier called *TupleID*. The primary attribute(primary key) can also be used as unique identifier. However, for the explanation we assume *TupleID* to be just an integer identifying the position of the tuple. The objective of finding information streaks(consecutive tuples) is to identify *TupleID*-ranges such that tuples within the range belong predominantly to an information class. Each range value contains the summarized information about the attributes it is built upon.

The pseudocode used to find the information streaks is given in Algorithm 1. Its purpose is to find tuple-ranges of size $\geq l$ in the entire dataset such that tuples in the range can be considered to belong to an information class(E). In order to avoid fragmentation, w is used as a weighing factor which proportionally increases with the length of the current streak. The sensitivity of classification can be considered by constants β and β'; p denotes the number of samples taken at each iteration and thus determines the resolution of the classification.

It can be observed that the above algorithm resembles clustering with proximity consideration. Clustering takes into account all the classes and hence tuples are bound to be well distributed in the entire data space. Accessing tuples belonging to one cluster would mean many blocks of data being read, causing time overhead. Information Streaks would overcome this problem with some sacrifice being made to the accuracy of classification. Figure 8.3 depicts a typical scenario. Left side of Figure 8.3 shows the data block accesses by the clustering algorithm. Consider the highlighted tuples belonging to a particular information class. The clustering algorithm would retrieve all the tuples resulting in four block accesses, whereas only one block is accessed using the information streaks.

Algorithm 1. Obtaining tuple-ranges

p: number of samples to be taken at each iteration
l: the minimum length for a tentative information range to be accepted
α: the tolerance level for a sample belonging to a different class to be included in the current streak
β, β′: constants to determine the nature of exponential averaging
w: information to store the weight associated with each class
E_k: information class to be included in the current streak
tl: store the tentative length of the streak
E_c: current information class
p𝒜: a pseudo tuple fragment

```
s ← 1
tl ← 0
while s ≤ lastTupleID do
   sample tuples with TupleID's from s to s + p
   produce p𝒜 whose attribute values are obtained from averaging over the sample set
   Classify p𝒜 based on rules deducted earlier(say E_k)
   if tl = 0 then
      w_k ← β + (1 − β)w_k
      E_c ← E_k
   else if E_c ≠ E_k then
      w ← (1 − β)w_k − β′
      if w_k > α then
         Add the current sample range to the streak
         tl ← tl + p
      else
         if tl > l then
            current streak to the range table with summarized information
         else
            s ← s + tl
            tl ← 0
         end if
      end if
   end if
end while
```

8.4 Modeling of Continuous-Type Data

Continuous-type attributes which are not included in the reduct table are individually represented as mathematical functions. This might seem like an unreasonable assumption to make since real-world data scarcely lend themselves to be fit into analytical deductions. However, it is a feasible concept to use when attributes change gradually over a period of time such as daily temperature, traffic flow, stock index, etc.. Attributes from a couple of data sets from Time-series Library are shown in

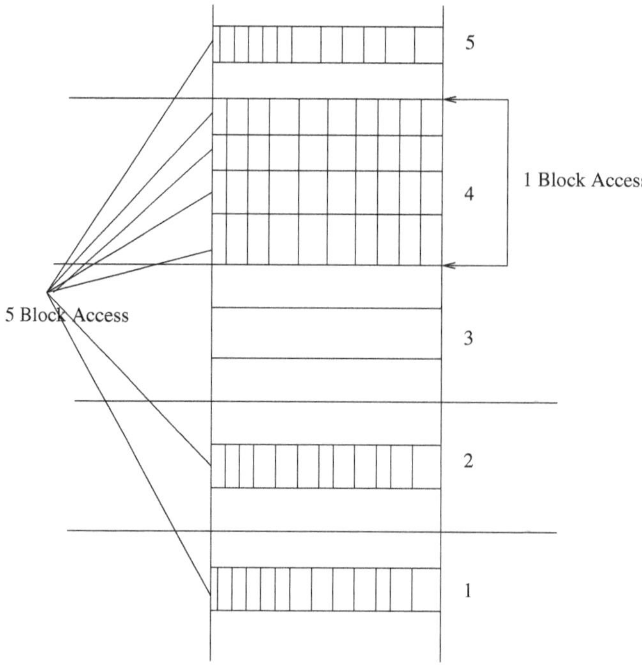

Fig. 8.3 Comparison of Clustering against Information Streaks

Figure 8.4 and 8.5. Figure 8.4 shows variation of daily stock closing prices. By observation, it can be roughly approximated to an exponential distribution. Figure 8.5 shows the quarterly variation of Sales and Purchase rates. Similarly, Figure 8.5 is found to be closely approximated to a function involving *sine* components with five different frequencies.

If A is an attribute that has been represented by a mathematical function $f(t_{id})$, then any query involving condition checking on values of A can be answered by solving $f(t_{id})$, to find the corresponding tuple-id ranges.

All the functions shown in Figures 8.6 through 8.9 are obtained by approximating the exchange rates to function of sine and cosine components of three frequencies. The Figure 6 depicts single valued selection for *tupleID* versus exchange values. The points of intersection give the *tupleID* satisfying the condition for exchange-rate = 1.2. In Figure 8.7, the query involving the range function is given. The corresponding tuple ranges are $t_1 - t_1'$, $t_2 - t_2'$ and $t_3 - t_3'$(shaded regions in Figure 8.7) for satisfying the range between 1.2 to 1.4.

A combination of interval-based sampling and analytical treatment can be obtained if patterns are found in *classes* [15]. The smoothness of data can be achieved by taking the mean for each of τ values, where τ is an user defined value for smoothening the data values. Some of the parameters involved in projection of tuples are given below:

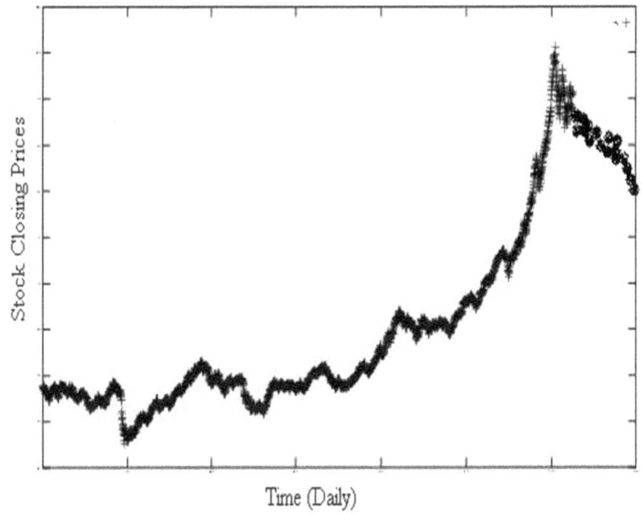

Fig. 8.4 Stock Market Index

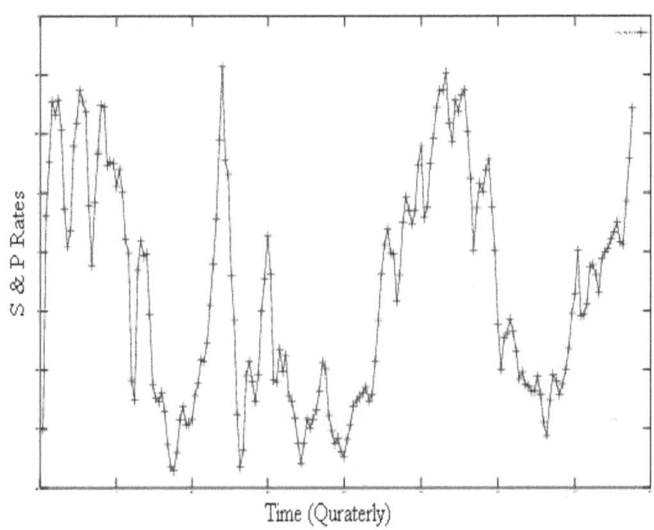

Fig. 8.5 Quarterly S and P rates

(i) Marginal error width (m_e): Since the projections are not accurate enough to pick single-values, the number of adjacent tuples to be included is determined by m. Selection in Figure 8.7 cannot be accurate. Therefore, adjacent tuples around the point of selection are also considered. The width of the selection defines the marginal error. This scenario is shown in Figure 8.8.

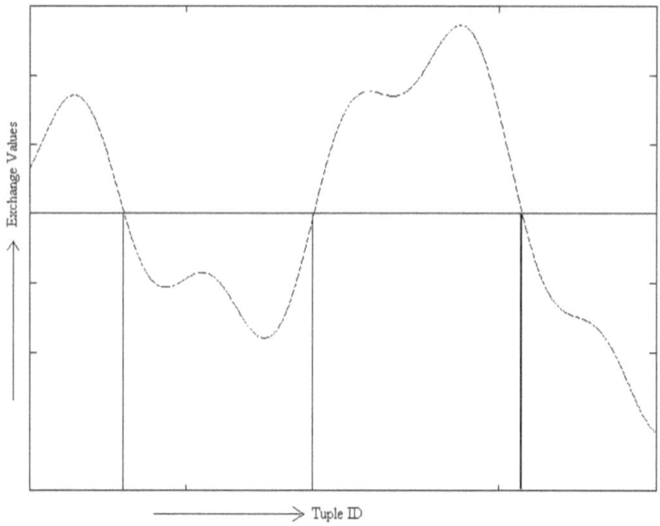

Fig. 8.6 A single-valued selection

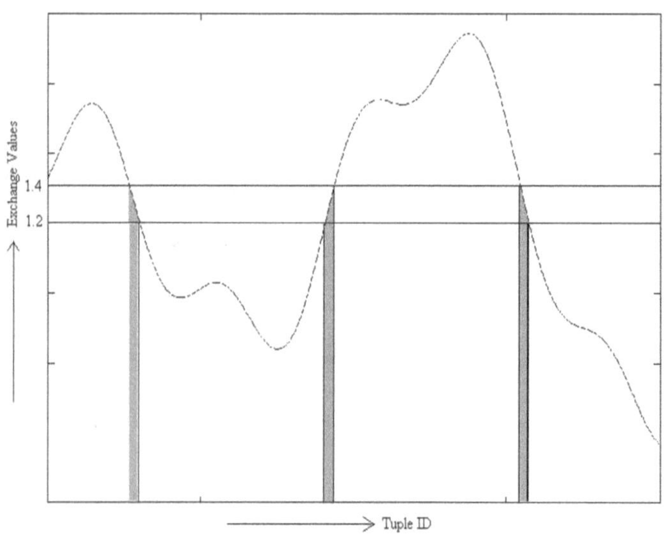

Fig. 8.7 Range query involving " $SELECT\ldots WHERE 1.2 <= EXCHANGE < 1.4$

(ii) Snap factor (s_n): Snap factor is a width lower than which two adjacent projection ranges can be merged, s_n appears to achieve the same function as that of τ in attempting to coalesce closely split tuples. However, a closer look would reveal that τ is a parameter used when the attribute model is *built* and hence a change

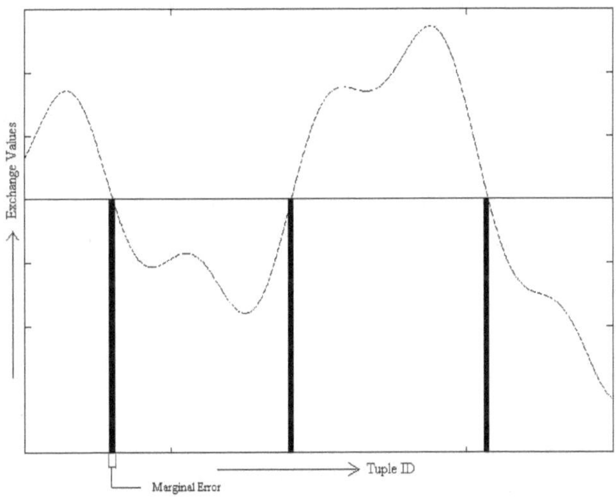

Fig. 8.8 Marginal error selection

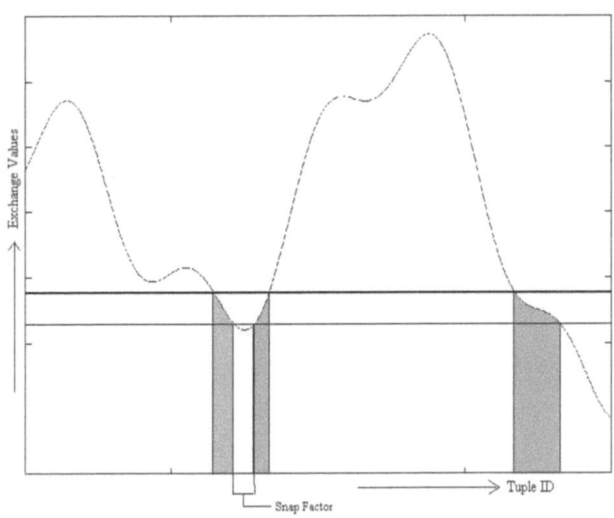

Fig. 8.9 Snap factor

in τ would require that data blocks are re-read, whereas s_n is a tunable factor that can be set at the time of execution of a query. Figure 8.9 shows the effect of snap factor, where the two tuple ranges with the distance less than s_n are merged into a single range.

8.5 Genetic Algorithms and Query Languages

The search bias during genetic search depends on the kind of problem solved, the structure of search space and the genetic operators. There are two possible ways to build meaningful blocks

1. Search through possible encodings for a good one while searching for a solution.
2. Expand the number of good encodings by increasing the types of building blocks considered meaningful.

Here, task-based specific queries are posed in form of DMQL queries. Some of the popular mining characteristics sought are comparison, classification, associations and concept hierarchies.

A typical specification of each query in DMQL may be generalized as:

```
<Mine_Knowledge_Specification> ::= <Mine_Char>  |
                                   <Mine_Discr> |
                                   <Mine_Assoc> |
                                   <Mine_Class>
```

In general initial population is created consisting of randomly generated rules. Each rule represented by a string of bits. For example, if the training set consist of two attributes A_1 and A_2 and a class C, then the rule "IF A_1 AND NOT A_2 THEN C" can be encoded as 101. Attributes with more values can be encoded using more bits, while continuous attributes can be encoded after interval-based classification. The fitness of a rule is assessed by its classification accuracy on the dataset. The general structure of GA is as follows:

```
t=0;
P(t) = Initialize Random Population
        (no_attributes, attribute_domains);
while ( t < max_generations )
        Evaluate fitness (P(t), dataset)
        t = t+1
        P(t) = select(P(t-1))
        Crossover(P(t));
        Mutate(P(t))
end while
```

the modified version is,

```
t=0;
P(t) = Generate biased Population
        based on query attributes
while (t<max_generations)
    Evaluate fitness(P(t), summarized data tables)
    t = t+1
    P(t) = select(P(t-1))
```

```
    Crossover(P(t));
    Mutate(P(t))
  end while
```

8.5.1 Associations

Each rule can be represented as a string. In generation of rules along with the support of a pattern, the number of *set* positions are also important. A pattern full of don't cares will gain a support of 100% but has no meaning in terms of knowledge discovery. Each discovered rule in the rule set is usually represented in the form

$$IF < condition_1 > \& < condition_2 > \ldots \& < condition_n >$$

$$THEN < action >$$

There are various representation methods for conditions and actions in terms of rule properties (fuzzy or non-fuzzy) and the attribute properties(continuous or discrete). A rule set supposed to be the solution for a classification problem. Processing a query which is of the form,

```
<Mine Assoc> ::= mine associations
                 [as <pattern name>]
                 [match <meta-pattern>]
```

involves the extraction of rules consisting of two parts: searching for hidden patterns and generation of rules based on those patterns. After the validation of candidates, the rules are selected based on expected values of support and confidence. Let the attributes selected to form a classification rule R_i are $< A_1, A_2, \ldots >$, out of which b attributes are of boolean type, and k of continuous and p_i denotes the number of intervals of A_i^{th} attribute and d the number of discrete attributes with each taking at the most d_i values. The length of the binary chromosome is given by :

$$l_b = b + log\left(\sum_{i=0}^{k} p_i\right) + log\left(\sum_{j=0}^{d} d_j\right) + m$$

where m is the length of the consequent.

The fitness of a chromosome reflects the success rate achieved and the corresponding rule set is used for classification. The GA operators use this information to evolve better chromosome over the generations. The fitness function actually measures the collective behavior of the rule set. The evaluation of the fitness can be speeded up using the summarized table built from the reduct rules and information streaks. In case of rules exactly matching the reduct rules, the support and confidence measures of earlier classification can be directly used with a new threshold value. The decomposition of a query in order to find rules already computed is a part of query relaxation. Query relaxation involves the rewriting of query to form a new query. Some of the main reasons for relaxation are

- the query is too general or too specific
- some of the query terms may not be used in the database
- one or more query terms have multiple meanings subject to context.

One approach used to deal with the above scenarios is generalization which involves rewriting of a query to a more generalized term based on information found in association mappings and attribute transformation. It is also possible to reduce associated clauses in the query to a single generalized attribute based on reduct table.

8.5.2 Concept Hierarchies

A frequently used and an important type of query is the concept hierarchy query, which usually takes the form

```
<Concept_Hierarchy_Definition_Statement>
        ::= define hierarchy< hierarchy_name>
            [for <attribute_or_dimension> ]
            on <relation_or_cube_or_hierarchy>
            as <relation_description>
            [where <condition>]
```

Concept hierarchies allow the mining of knowledge at multiple levels of abstraction. They define a sequence of mappings for a set of a low-level concepts to higher-level more general concepts. A concept hierarchy is represented as a set of nodes organized in a tree, where each node in itself, represents a concept. The common types of hierarchies are

- schema hierarchies which express semantic relationship between attributes.
- set-grouping hierarchies that organizes values for a given attribute or dimension into groups of constants, typically used for defining small sets of object relationships.
- operational-derived hierarchies is based on operations specified by users, experts or the data mining systems.
- rule-based hierarchies that occur when either the whole concept hierarchy or a portion of it defined by a set of rules.

Since problem of concept hierarchies involve classification of attribute values with multiple levels of abstraction genetic algorithm can be used with a slight modification of including multiple types of chromosomes each defining the strength of classification for each level of abstraction.

It can be observed that the number of features that have to be used to describe a particular class in the hierarchy may be different from one another. Consider the example of classification of connection types in KDD99-Cup Dataset [16], the number of attributes(features) that are needed to predict a normal connection is only four, while predicting the exact class of attack requires more than six attribute values. The concept hierarchy for KDD99-Cup dataset is shown below.

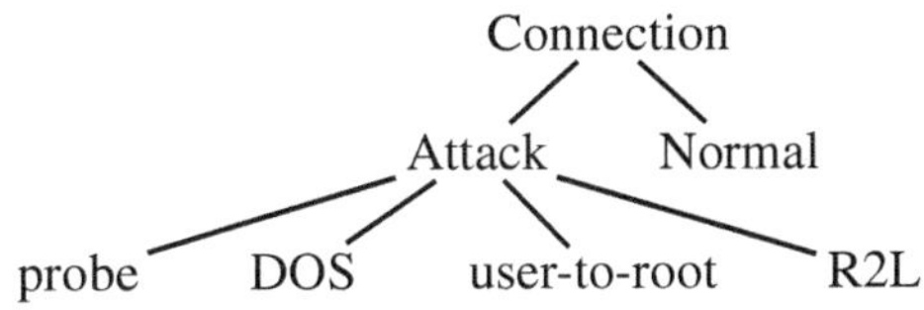

The difference in the number of bits to represent features demands the use of chromosomes with different length and a non-conventional type of crossover. A simple approach to take is to consider each path of the concept hierarchy as a series of classification where the support for a lower level concept p is based on the percentage of data objects belonging to its immediate higher level concept. The automatic building of concept hierarchies is close to concept maps and ontologies [17]. Figure 8.10 shows the first few actions of GA, with concepts being {PLANT, ROOT, STEM} and relation {has part}. Based on concept map(CM) and ontology the GA determines initial populations of CMs. Later it selects individuals better selected to be parents based on fitness values. Genetic operators are applied to these individuals. The fitness of the gene is calculated based on binary relation it expresses. The taxonomy that allows assigning of values to binary relations is described in [18]. The root of the hierarchy that has been denoted by a schema serves as a template

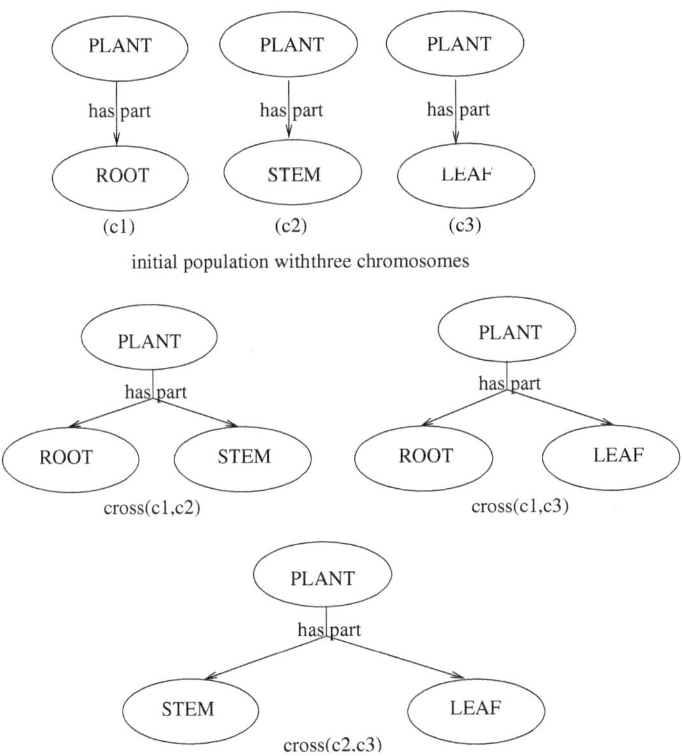

Fig. 8.10 Hierarchical Crossover

for all its descendants, which more specific instances of the root schema. At each stage GA selects in the individual population better fitted to be parents of the next population. It then submits these individuals to genetic operators of crossover and mutation. The new population is formed by the ancestors and descendants.

Since the initial set of population is biased to include genes of specific classes a efficient way to deal with over-specification and underspecification has to be used. In order to achieve this the concept of semantic hierarchy [19] of chromosomes can be used. The identification of points of similarity within two chromosomes and the new hierarchical operator, that can be used to generate meaningful offspring from parent solutions from within GA, suit the requirements of construction of concept hierarchies. The semantic hierarchy is simply a 'part-of' hierarchy, which defines the semantics of a genotype and not the syntax. An example of semantic hierarchy of 16 bit, 2 gene chromosome is shown in Figure 8.11.

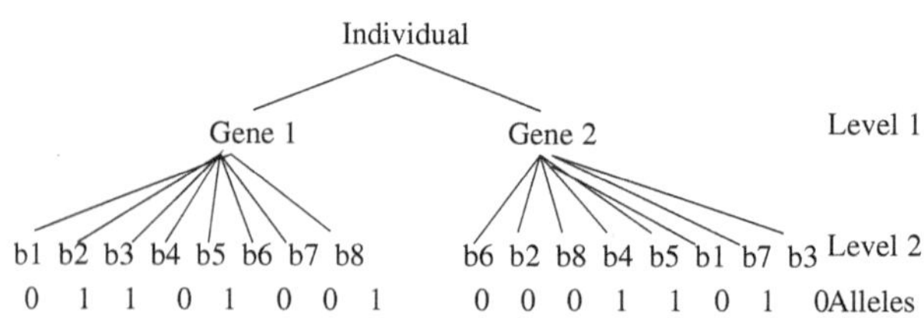

Fig. 8.11 Semantic Hierarchy of 16 bit, 2 gene chromosome

Binary encoding scheme works reasonably well for the above described problems, however, a different representation of chromosomes is needed when the solution space is huge.The chromosomes are represented as tentative solutions to sub-problems. Some of the queries that involve mining of characteristics or analysis of patterns, identification of key-attributes values for finding an optimal result can be solved using this approach.

The basic idea is to represent the solution as a combination of sub-problem each of which is solved using a subset of chromosomes. As the generations evolve the fittest chromosomes from each range are taken and the cross-over function may be adapted to fit the new populations. This avoids the clustering of fittest chromosomes in one region and explores the solution space in an efficient way. The structure of a chromosome is given in Figure 8.12.

Fig. 8.12 Structure of the Chromosome

Range	Strength	Cross-over function	Value

One of the direct applications of the above method of GA is to find the number of minimal reducts. Due to the exponential complexity of finding minimal reducts a model simplification has to be made unless it results in unacceptable loss of accuracy [20]. Order-based genetic algorithm for searching minimal reducts can be used with the fitness evaluated on already available summary tables rather than on the dataset. The new reducts, if discovered, can be integrated to the system thus acting as a feedback.

8.5.3 Dealing with Rapidly Changing Data

The methods discussed so far are based on the assumption that the dataset is not rapidly changing. When the dataset is incrementally changed the new data that is added can affect the classification in the following ways:

- Increase in the number of categories in classification
- Refinement of existing solutions due to data being added
- New rules being formed by added attributes

Incremental approach used for learning tasks as described in [21] can be used to overcome these problems. Each classifier agent has a current solution based on the attributes, classes and training data that are currently known. When new attributes, classes or data are acquired, the GA is used to learn the new changes and evolve into a reinforced solution.

There are many ways in which new chromosomes can be obtained from old ones. Either the best rule set chromosome can be used as a seed for the entire population or the whole group of chromosomes can be used if the current solution is available. New rules can be formed either by evolving new rule sets from the added attributes or by appending randomly selected new elements to old rule sets. The pseudocode to evolve new chromosomes is given by

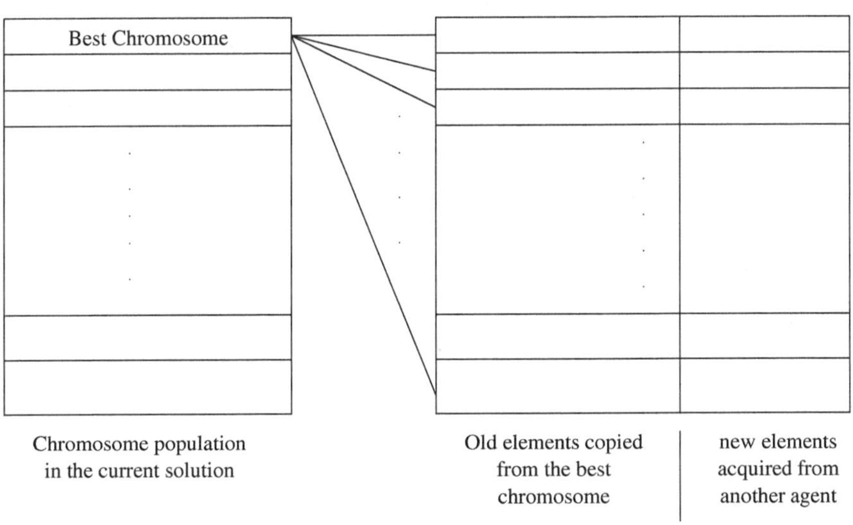

Fig. 8.13 Appending Additional Rules to Chromosomes

```
if(new attributes are to be integrated)
   select group chromosomes as seeds from current
   solution if available;
else
   select chromosomes with best fitness as seed;
if(new chromosomes for new attributes are available)
   integrate new rules to old rules;
else
   expand old chromosomes by randomly created elements;
   perform GA on the resulting population;
```

Figure 8.13 shows the pictorial representation of appending new rules to old ones. Various initialization schemes with different techniques for appending can be found in [21].

8.6 Experimental Results

Experiments are performed on three real-life data sets taken from UCI Machine Learning Archive [22]. The characteristics of data sets are summarized below:

1. *KDD 99 Cup data:* The competition task was to build a network intrusion detector, a predictive model capable of distinguishing between "bad" connections, called intrusions or attacks, and "good" normal connections. The attacks fall into four categories as DOS(denial of service), R2L, U2R and probing. The datasets contain a total of 24 training attack types. The attributes describe the basic features of TCP connections, content features within the connection suggested by domain knowledge and traffic features.
2. *Forest Cover-type:* It is a Geographical Information System data representing forest cover type like pine, fir, etc., found in US. The variables are cartographic and remote sensing measurements. It contains 10 dimensions, seven classes and 586,012 samples. All the attribute values are numerical.
3. *Iris:* A data set with 150 random samples of flowers from the iris species setosa, versicolor, and virginica. There are four features all of which are numeric.

Classification: The classification accuracy is defined as the percentage of data points whose class labels are correctly predicted. The classification accuracy of the datasets considered above is shown in Table 8.5. The size of summarized tables

Table 8.5 Classification Accuracy of all datasets

Dataset	Accuracy(%)
KDD 99	98.3
Covertype	64.2
Iris	97.6

Table 8.6 The sizes of summarized tables

Dataset	n_a	*size*
KDD 99	12	7.1%
Covertype	6	6.8%
Iris	4	2.6%

expressed as percentage of size of the original dataset is shown in Table 8.6. n_a denotes the number of attributes that are the part of the summarized tables.

Relation between l and classification accuracy: Figure 8.14 shows the variation of classification accuracy versus minimum streak length l for KDD99 Cup data, Forest Cover-type data and Iris databank. As the minimum length of the information streak is increased the resolution of the classification decreases, hence the drop in accuracy.

Aggregate functions: A random set of queries are generated involving aggregate functions count, aggregate and max/min to test the performance of the proposed methods. For count and aggregate operations the error is the difference between the values obtained, while for max/min operations it is the percentage of times the correct max/min value was returned. Table 8.7 shows the comparison of execution times for three aggregate functions. t denotes the average time that was taken to execute the queries directly on the database and t_s is the average time of execution when summarized tables are used. The last column shows the % error which is the deviation of the numerical value obtained from the actual value.

Accuracy of Concept Hierarchies: The accuracy of the hierarchies is calculated with respect to each level. If p^{ij} denotes the set of tuples at j^{th} class in the ith level and c^{ij} the corresponding number that have been classified correctly, then the accuracy of ith level is

$$\frac{\sum_{j=1}^{n} C_{ij}/P_{ij}}{n}$$

where n is the number of classes at that level. Overall hierarchy accuracy is expressed as average of all the levels. The experiments are performed on only two of the data sets, the results of which is shown in Table 8.8.

Analysis of improvement with GA feedback: The datasets are increased by 10% and experimented with GA feedback. A slight improvement is seen in two data sets and there is no change observed in Iris dataset due to the absence of formation of new rules. The results are tabulated in Table 8.9.

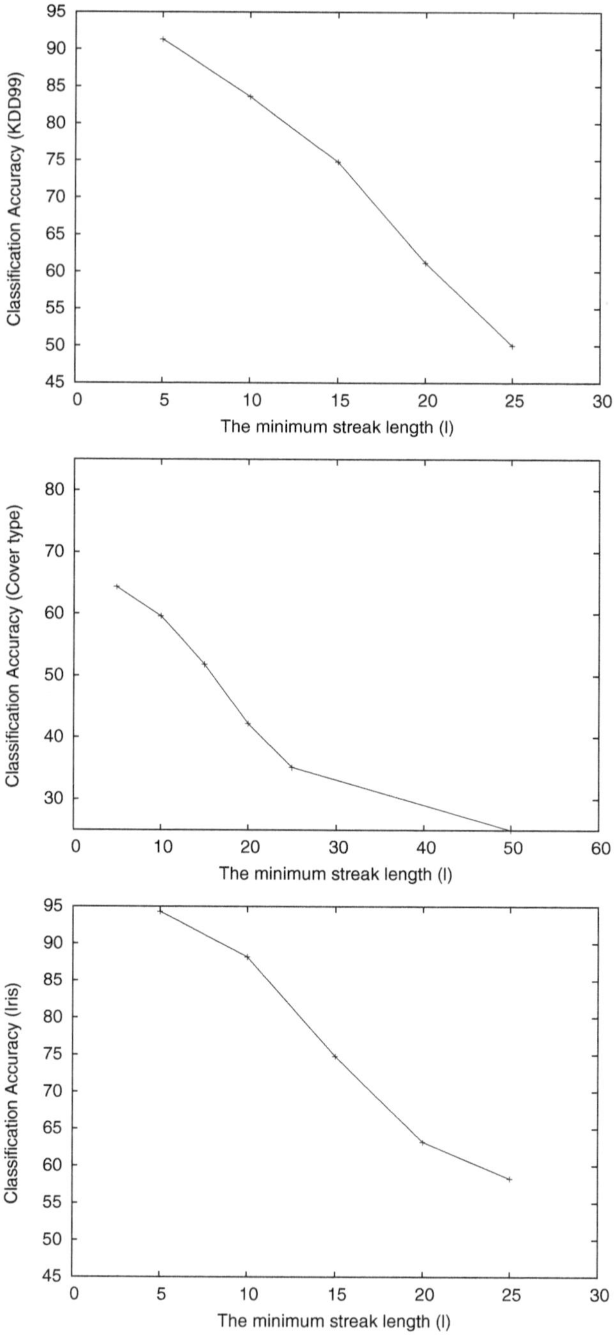

Fig. 8.14 Variation of Accuracy versus minimum streak length

Table 8.7 Comparison of execution times for Aggregate Function

Dataset	query-type	t (secs)	t_s (secs)	% Error
KDD 99	count	40.3	2.1	10.2
	avg	51.3	2.2	11.6
	max/min	43.3	2.1	23.0
Cover	count	63.1	5.1	14.3
	avg	71.3	6.2	13.6
	max/min	61.3	5.2	29.6
Iris	count	0.5	0.02	10.1
	avg	0.45	0.02	9.6
	max/min	0.32	0.02	14.1

Table 8.8 Average concept hierarchy accuracy

Dataset	Accuracy
KDD 99	95.9%
Covertype	61.2%

Table 8.9 Classification Accuracy with GA feedback

Dataset	Classification Accuracy
KDD 99	98.9
Covertype	66.2
Iris	97.6

8.7 Adaptive Data Mining Using Hybrid Model of Rough Sets and Two-Phase GAs

Extraction of knowledge and information retrieval presents a formidable challenge when huge databases are involved. Various techniques and algorithms are being continually experimented upon, with varying results. In this section, we provide a framework for efficient ad-hoc-query answering with quick response times, using hybridization of adaptive rough set approach and two phase genetic algorithm. To overcome the drawbacks of rough sets and to include user adaptiveness to the system, a framework for non conventional rough sets is provided. The main features of this framework, given in Figure 8.15, are as follows:

- The queries from the users are studied over a period of time to determine which attributes are most frequently used which are given higher priority to be included in the reducts. This requires the system to undergo a period of training.
- If prior knowledge of database is already known then the above process can be speeded up.

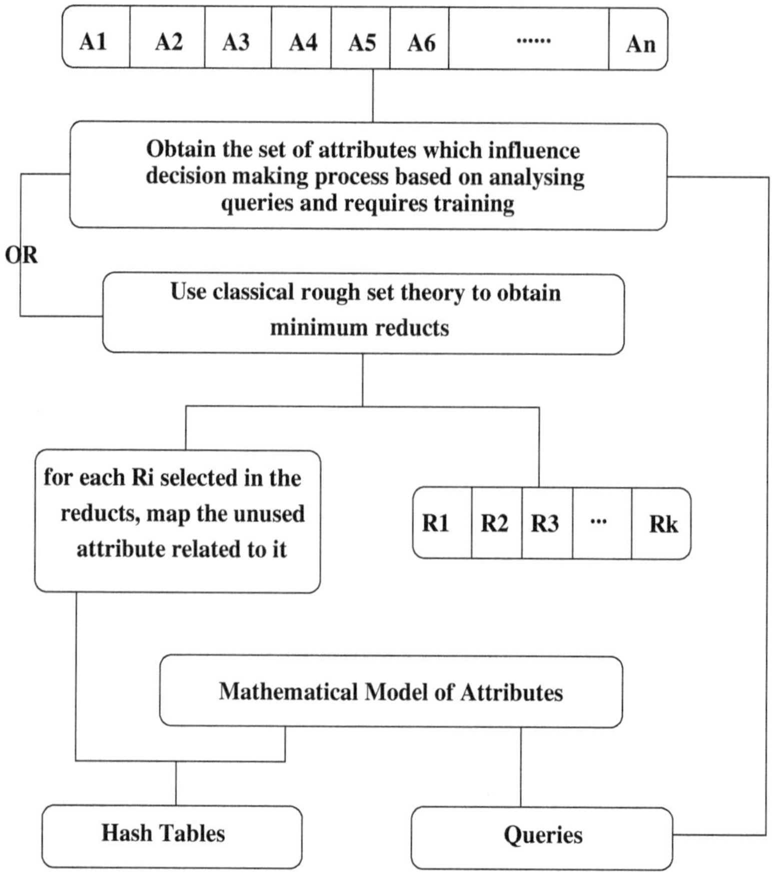

Fig. 8.15 Framework for adaptive Rough-Set model

- If the queries are inconsistent or substantial confidence is not obtained in classical training, classical rough set theory can be used to obtain the reducts.
- Once the reducts are obtained, for each R_i selected in the set, the dependency attributes (which have correlation with R_i) are mapped to corresponding reduct. This is done so that, should a relation be added to the database with missing tuples, approximate values of missing attributes can be obtained from other attributes.
- The subset of reducts obtained is used to store a representation of the entire dataset in the form of MMA.

8.8 Mathematical Model of Attributes (MMA)

The MMA takes advantage of the fact that all real data obtained usually have definite systematic patterns which are sometimes overlooked. Data sets obtained in real-scenarios usually fall into one of the many standard mathematical distributions. MMA is an attempt to store the data back in compressed form by representing the

Table 8.10 Illustrative Example

Marks	Hrs/Day	IQ
40	0.5	100
50	0.5	110
60	1.0	90
70	2.0	90
80	4.0	100
30	0.5	90
. .	. .	. .

attributes through their natural distribution functions, which can later be used for statistical reasoning without actually querying the dataset. This way the number of accesses to the disk is reduced and greatly improves the response time. Consider a student-database, given in Table 8.10, in which a relation which has 3 reducts: Marks (R_1), Average number of hours (R_2) and IQ (R_3).

If correlation between R_1 and R_2 are found and have same probability distributions, both of them can be represented using a single distribution but with different parameters. This would lose some of the inconsistent attributes and therefore such inconsistencies need to be separately stored when performing the actual query execution using GA. If R_3 can be represented using a different distribution then, MMA is designed such that the relation between R_1, R_2, and R3 are represented by multivariate distribution.

8.9 Two Phase Genetic Algorithms

The search bias during genetic search depends on the kind of problem solved and the structure of search space and the genetic operators. There are two possible ways to build meaningful blocks.

- Search through possible encodings for a good one while searching for a solution.
- Expand the number of good encodings by increasing the types of building blocks considered meaningful.

The general structure of GA is as follows:

```
t = 0;
P(t)=Initialize Population(no_of_attributes,
attribute_domains);
while(t < max_generations)
Evaluate fitness (P(t),dataset)
t=t+1
P(t) = select(P(t-1))
Crossover(P(t));
Mutate(P(t));
end while
```

The drawbacks of classical GA are as follows,

- Repetition of values from chromosomes.
- Clustering among best chromosomes.
- Inability to differentiate between chromosomes that are approaching the same maxima/minima.
- Initialization step is random and hence the prior knowledge obtained from the dataset is not put into use.
- If naïve GA is used to calculate fitness functions based on number of tuples belonging to a class in the database, then the cost of computation is high since each time the fitness function needs to be computed the actual data blocks have to be read into the memory.

As the basic problem with classical GA is that the offspring of two chromosomes are likely to be similar to its parents and hence scarcely has new information present carried through. Chromosomes of similar type survive through many populations and tend to produce redundant information. We seek to overcome this by using a two-phase GA that has spatial considerate chromosomal nets along with binary encoded GA. The fraction of population that has to be encoded using either method is decided by the success of that method. If classical GA is found to do better than the other, then the fraction of population that have to be binary encoded is dynamically increased, likewise for the other. Initially, the solution space is divided into a number of non-overlapping quarters with each quarter having a chromosomal net of its own, each net having identity C_i. Each chromosome has a field that contains the net to which it belongs to. The generation of the next population is evolved as follows:

1. The fitness values for the present chromosomes are calculated.
2. Let the fitness value of a chromosome at point (x, y) be v.
3. For each chromosome in a net, fitness values at points $(x \pm t, y \pm t)$ is calculated.
4. The collective fitness value of a chromosomal net is calculated based on all of its chromosomes.
5. A positive convergence is indicated by decrease in the area of the convex hull circumventing the chromosomal net for a particular sequence of $t_1, t_2, t_3, \ldots t_n$ for n chromosomes.
6. If the centroid of two chromosomal nets are within the threshold d then the two are merged together to form a new net.
7. If divergence of net is observed with decrease in collective fitness value then the net is dismantled.

The modified algorithm given below makes use of both binary GA along with a new type of chromosomes that have spatial consideration, but are more computationally intensive, with ability to exploit patterns in the solution space.

P_c: A parameter that denotes fraction of population that is used to produce the next generation
P_i: performance index
T_i: threshold for significant increase in performance

$G(t)$: binary encoded Generation a time t
$sG(t)$: spatial Generation a time t
Generate initial population using MMA and inconsistencies with s percent spatial chromosomes $sG(0)$ and (1-s) binary coded population $G(0)$.

$t = 1$
P_c=0.5
while(solution not found and t < maxgenerations)
{
Generate $G(t)$ using $G(t_1) * P_c$
Modify $sG(t)$ based on proximity sampling of $sG(t\text{-}1) * (1 \text{-}P_c)$
Calculate performance index P_i
Compare P_i with threshold value T_i
Increment/Decrement Pc proportional to $(T_i - P_i)$
$t = t$+1
}

The Experiments are performed on three real-life data sets taken from UCI Machine Learning Archive. The classification accuracy of all datasets is given in Table 8.11.

Classification: The classification accuracy of the proposed two phase genetic algorithm with Mathematical Model of Attributes is given in Table 8.11.

The size of summarized tables expressed as percentage of original dataset is shown in Table 8.12. $n_a$denotes the number of attributes that are a part of summarized tables.

Analysis of improvement with GA: The datasets are increased by 10% and experimented with GA. The emerging classification accuracy with two phase GA is given in Table 8.13. A slight improvement is seen in two data sets. There is no change observed in Iris dataset due to the absence of formation of new rules.

Table 8.11 Classification Accuracy of all datasets

Dataset	Classification Accuracy
KDD99	98.3
Cover-Type	64.2
Iris	97.6

Table 8.12 The sizes of summarized tables

Dataset	n_a	Size
KDD99	12	7.1%
Cover-Type	6	6.8%
Iris	4	2.6%

Table 8.13 Classification Accuracy with Two Phase GA

Dataset	Classification Accuracy
KDD99	98.9
Cover-Type	66.2
Iris	97.6

8.10 Summary

In this chapter we have proposed an intelligent query answering system using rough sets and genetic algorithms. The flexibility involved in building the summary tables of rough sets makes the system more scalable. The system performs well even with acceptable level of accuracy as justified in the experiments. Reinforcement of reducts with genetic algorithms leads to adaptive classification. The proposed framework can be used as alternative to system catalog. In future work, the system can be extended to other types of DMQL queries, since it is handling only a subset of queries.

References

1. Han, J., Kamber, M.: Data Mining: Concepts and Technique, 2nd edn. Morgan Kaufmann Publishers, San Francisco (2006)
2. Weiss, S.M., Kulikowski, C.A.: Computer Systems that Learn:Classification and Prediction Methods from Statistics, Neural Nets, Machine Learning, and Expert Systems. Morgan Kaufmann, San Francisco (1991)
3. Yamauchi, K., Yamaguchi, N., Ishii, N.: Incremental Learning Methods with Retrieving of Interfered Patterns. IEEE Transactions on Neural networks 10(6), 1351–1365 (1999)
4. Su, L., Gain, S.U., Yeo, Y.C.: Incremental Self-Growing Neural Networks with the Changing Environment. Journal of Intelligent Systems 11(1), 43–74 (2001)
5. DeJong, K.A., Spears, W.M.: Learning Concept Classification Rules using Genetic Algorithms. In: Proceedings of International Joint Conference on Artificial Intelligence, pp. 651–656 (1991)
6. Wojciechowski, M., Zakrzewicz, M.: Evaluation of the Mine Merge Method for Data Mining Query Processing. In: Advances in Databases and Information Systems, ADBIS (2004)
7. Zakrzewicz, M., Morzy, M., Wojciechowski, M.: A Study on Answering a Data Mining Query Using a Materialized View. In: Aykanat, C., Dayar, T., Körpeoğlu, İ. (eds.) ISCIS 2004. LNCS, vol. 3280, pp. 493–502. Springer, Heidelberg (2004)
8. Grosky, W.I., Tao, Y.: Multimedia Data Mining and Its Implications for Query Processing. In: DEXA Workshop (1998)
9. Pawlak, Z.: Rough Sets. International Journal of Computer and Information Sciences (1982)
10. Lingras, P.: Application of Rough Patterns. In: Rough Sets in Data Mining and Knowledge Discovery. Series Soft Computing Physics Verlag (1998)
11. Kolonder, J.L.: Case Based Reasoning. Morgan Kaufmann, San Francisco (1993)

12. Holland, J.H.: Adaption in Natural and Artificial Systems. Series Soft Computing Physics Verlag (1975)
13. Lingras, P., Davis, C.: Application of Rough Genetic Algorithms. Computational Intelligence (2000)
14. Skowron, A., Rauszer, C.: The Discernibility Matrices and Functions in Information Systems. In: Intelligent Decision Support, Handbook of Applications and Advances of the Rough Sets Theory, pp. 331–362. Kluwer Academic, Dordrecht (1992)
15. Shumway, R.H.: Applied Statistical Time Series Analysis. Prentice Hall, Englewood Cliffs (1988)
16. `http://kdd.ics.uci.edu/databases/kddcup99/kddcup99.html`
17. da Rocha, F.E.L., da Costa Jr., J.V., Favero, E.L.: A New Approach to Meaningful Assessment using Concept Maps Ontologies and Genetic Algorithms. In: Proceedings of the first International Conference on Concept Mapping, Spain (2004)
18. Costa Jr., V., Rocha, F.E.L., Fevero, E.L.: Linking Phrases in Concept Maps in Study on Nature of Inclusivity. In: Proceedings of First International Conference on Concept Mapping, Spain (2004)
19. Bentley, P.J., Wakefield, J.P.: Hierarchical Crossover in Genetic Algorithms. In: Proceedings of First Online Workshop on Soft Computing (1996)
20. Slezak, D., Wroblewski, J.: Order Based Genetic Algorithms for the Search of Approximate Entropy Reducts. Springer, Heidelberg (2003)
21. Guan, S.-U., Zhu, F.: An Incremental Approach to Genetic-Algorithms-Based Classification. IEEE Transactions on Systems, Man and Cybernetics - Part B 35(2), 227–239 (2005)
22. Blake, C.L., Merz, C.J.: UCI Repository of Machine Learning Databases, University of California, Irvine, Department of Information and Computer Sciences (1998), `http://www.ics.uci.edu/~mlearn/`

Chapter 9
Hashing the Web for Better Reorganization

Abstract. World Wide Web is a huge, distributed repository of global information. A large number of pages are added to the Web everyday and these pages contain countless hyperlinks and volumes of access and usage information. Web mining is a technique to discover knowledge from the Web documents and services. Web usage mining is a technique to discover useful access patterns from Web log data. The generated patterns are used to reorganize the Web structure for faster access and retrieval. In this paper, we propose an efficient hash based algorithm to analyse the Web log data and generate frequent access patterns for dynamic databases, which takes care of insertion and deletion of Web pages. The algorithm uses optimum memory and a single scan technique, which reduces access time enormously. This algorithm outperforms Web Access Pattern (WAP) tree and Generalized Sequential Patterns (GSP) both in time and space complexity. An algorithm to reorganize the existing Web structure from the frequent access patterns is proposed for improved navigation. Extensive simulations are performed on real and synthetic data and the performance of our algorithm is much superior to the existing algorithms.

9.1 Introduction

World Wide Web (WWW) is a huge, distributed repository of global information to service the enormous requirements of news, advertisement, consumer information, financial management, education, government, e-commerce etc.. The WWW contains huge dynamic collection of hyperlinks, Web page access and usage information, and providing useful sources for data mining. Data mining holds the key to uncover the authoritative links, traversal patterns and semantic structures that bring intelligence and direction to Web interactions. Web mining is a technique that automatically retrieves, extract and evaluate information for knowledge discovery from Web documents and services. The Web page complexity far exceeds the complexity of any traditional text document collection. Only a small portion of the Web pages contains truly relevant or useful information.

The Web mining tasks can be classified into following (i) Web structure mining, (ii) Web content mining and (iii) Web usage mining. The Web structure mining

K.R. Venugopal, K.G. Srinivasa, L.M. Patnaik: Soft Comput. for Data Min. Appl., SCI 190, pp. 197–215.
www.springerlink.com

generates structural summary about the organization of Web sites and Web pages. Web structure mining tries to discover the link structure of the hyperlinks at the inter-document level. Web content mining deals with the discovery of useful information from the Web data. The contents of the Web include a very broad range of data such as audio, video, symbolic, metadata and hyperlinked data in addition to text. Web content mining mainly focuses on the structure of inner-document, based on the topology of the hyperlinks, while Web structure mining categorizes the Web pages and generate the information, such as the similarity and relationship between different Web sites.

Web usage mining involves data from Web server access logs, proxy server logs, browser logs, user profiles, registration files, user sessions or transactions, user queries, book mark folders, mouse clicks, scrolls and any other data generated by the interaction of the users and the Web. Web usage mining provides a key to understand Web traffic behaviour, which can in turn be used for developing policies for Web caching, network transmission, load balancing and data distribution. Web content and structure mining utilize the real or primary data on the Web, while Web usage mining takes secondary data generated by the user's interaction with Web.

The Web usage mining basically extracts useful access information from the Web log data. The mined information can be used for analyzing the access patterns and concluding on general trends. An intelligent analysis helps in restructuring the Web. Web log analysis facilitates in building customized Web services for individual users. Since Web log data provides information about specific pages' popularity and the methods used to access them, this information is integrated with Web content and linkage structure mining for ranking the Web pages. They are used to classify Web documents and construct a multilayered Web information base. Web usage mining is used to improve and optimize the structure of a site, to improve the scalability and performance of Web-based recommender systems and to discover e-business intelligence for the purpose of online marketing. Web usage mining also provides patterns which are useful for detecting intrusion, fraud, attempted break-ins, etc.. The analysis of the mined data provides detailed feedback on user behaviour and providing the Web site designer with the information to redesign the Web organization.

9.1.1 Frequent Items and Association Rules

Frequent itemset discovery can be used to correlate pages that are most often referenced together in a single server session. Examples of frequent itemsets are as follows. (i) The *Home page* and *Food World page* are accessed together in 30% of the sessions. (ii) The *Car Race game* and *Soap Product pages* are accessed together in 1.5% of the sessions. Any set of n frequent items can further be broken into n separate association rules, where directionality is added to the rule. A frequent itemset of pages A and B leads to two association rules i.e., A $\rightarrow$ B and B $\rightarrow$ A. In the context of Web usage mining, frequent itemsets and association rules refer to set of pages that are accessed together with a *support* value exceeding some specified *threshold*. These pages may or may not be directly connected to one another via hyperlinks.

Sequential patterns: The technique of sequential pattern discovery finds inter-session patterns such that the presence of a set of items is followed by another item in a time-ordered set of sessions. For example, the *Snake and Ladder game* is accessed after the *Car Race* game page view 45% of the time. By using this approach, Web marketers can predict future visitor patterns, which is helpful in placing advertisements aimed at certain user groups. Trend analysis can be used to detect changes in the usage patterns of a site over a period of time, and change point detection identifies when specific changes take place. For example, (a) Page views for the *Car Race video* game have been decreasing over the last two quarters. (b) The *Car Race video* game page views increased from April through July, are steady until November, and then began to drop.

Frequent Itemsets: All the information contained in the server sessions is not necessary for frequent itemset discovery. The order and the number of occurrences of a page view or page file in a session are not required. Therefore, sessions must be reduced to a list of unique session ID/page pairs. A minimum support must be used in order to limit the number of discovered patterns. The support of an itemset is the fraction of the total sessions that appears in an ordered sequence. Let support S, for a set of n items, on a data-set D for an item l_i is defined as,

$$S = \text{Count}(\{l_1, \ldots, l_n\}\ D) \,/\, \text{Count}(D).$$

The frequent itemsets discovered can be used to rank the sites. Interest is defined as the support of a frequent itemset divided by the probability of all of the items appearing together in a set, if the items are randomly and independently distributed, and is represented by,

$$I = \text{S}(l_1, \ldots, l_n) \,/\, \Pi_{j=1}^{n}\ \text{S}(l_j).$$

The *confidence* of a rule is the fraction of sessions where the subsequence is present if the antecedent is also present. Confidence for a rule is defined as follows,

(i) $l_a = \{l_{s1}, \ldots, l_{sn}\}$,
(ii) $\text{C} = \text{S}(l_a, l_{s1}, \ldots, l_{sn}) \,/\, \text{S}(l_a)$.

Dynamic mining algorithms are proposed to handle the updation of associations when increments or decrements to database occur. It should be done in a cost effective manner, without involving the database already mined while permitting the reuse of the knowledge mined earlier. The two major operations involved are (i) Additions: Increase in the support of appropriate frequent access patterns and discovery of new patterns. (ii) Deletions: Decrease in the support of existing large frequent access patterns leading to the formation of new large patterns. The process of addition and deletion may result in the invalidation of certain existing associations. Dynamic Web usage mining helps in Web reorganization, so that the user can navigate the

Web with minimum efforts. The analysis of the usage of a Web site and the structure of the Web site facilitates the modifications to the existing Web structure for efficient access.

9.2 Related Work

The problem of mining association rules using an iterative and multi-scan method is introduced in [1]. An effective hash-based algorithm discussed in [2] is especially useful for the generation of candidate sets for large two-itemsets. The incremental updating technique presented in [3] is similar to Apriori algorithm and operates iteratively to make complete scan of the increment database in each iteration. This technique is better than direct application of Apriori algorithm on the entire database. Mining frequent itemsets and association rules on dynamic database using genetic algorithms for intra-transaction, inter-transaction and distributed databases are proposed in [4].

Sequential pattern mining and Generalized Sequential Patterns (GSP), which discovers frequent patterns in a sequence database, is introduced in [5, 6]. GSP is efficient when the sequences are not long and the number of transactions is less. The problem of mining access patterns from the Web logs using a data structure called Web Access Pattern (WAP) tree from pieces of logs is presented in [7]. WAP tree recursively reconstructs a large number of intermediate WAP trees during mining which makes it expensive. The solution for discovering structural association rules and other features of a semi-structured data is given in [8]. The problem of automatic knowledge classification from classified Internet documents using Intelligent Internet Document Organization and Retrieval is addressed in [9]. The intelligence and direction to Web interactions by uncovering the authoritative links, traversal patterns and semantic structures is surveyed in [10].

Sergey Brin [11] discusses the problem of extracting patterns and relations from the WWW and also deals with pattern relation duality. The Web site reorganization for faster retrieval and easy access based on mining user access patterns is proposed in [12]. Web personalization using clustering of URLs to prepare customized content of particular users is handled in [13] and Web personalization is improved using collaborative filtering based on clustering in [14].

9.3 Web Usage Mining and Web Reorganization Model

The Web mining model is presented in the Figure 9.1. Initially, the Web usage data generated by the user's interaction with the Web is taken (Information Selection). Information extraction is a process of selecting and pre-processing the log data to maximize the information content in order to facilitate efficient mining. Intelligent Web agents can be used to search the relevant information. A robust pre-processing system is required to extract the data as and when a user requests for a Web page. A variety of files like images, sound, video, etc., which are accessed are irrelevant or redundant (noise) to the mining task.

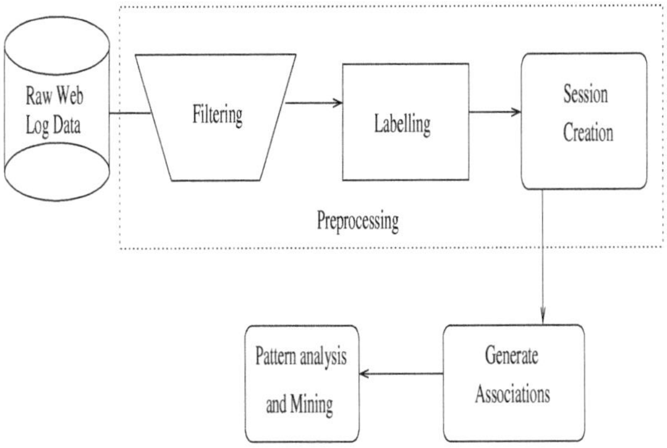

Fig. 9.1 Web Usage Mining Model

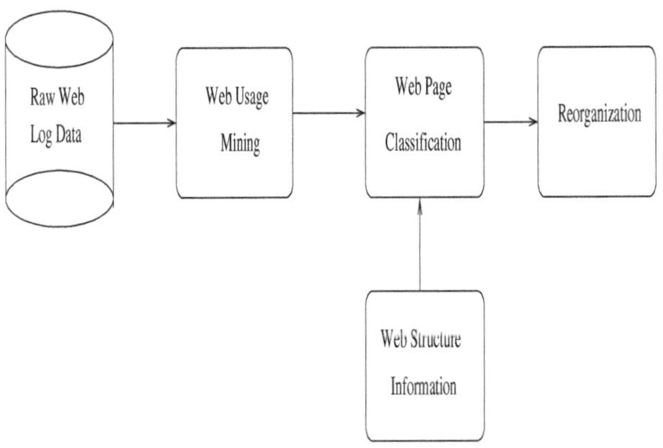

Fig. 9.2 Web Reorganization Model

Filtering process removes such noise from the raw Web log data. *Labelling* process assigns unique names (*labels*) to all the Web pages (*URI and Referrer*). New *sessions* are created for each login-session of every user, in addition new sessions are also created when the session times out (usually in 30 minutes). Web usage *Associations* are generated for each and every *URI-Referrer* pairs to obtain first level associations where no pruning is done. Later we proceed to generate n level associations in the pattern analysis phase based on minimum support criteria.

For reorganizing the Web Structure the frequent itemsets generated by mining the Web log data is used as input. Using the Web structure information, the Web pages are classified as either *content* page or *index* page. The result of reorganization generates a set of suggestions to the Webmaster to restructure the existing Web pages both in terms of content and structure. This results in the Web page being tailored to the needs of the users and hence improves the efficiency of Web page access. In addition, the filtered or the less frequent pages are also listed which can be removed from the Web sites either temporarily or permanently.

9.4 Problem Definition

Given a Web log access sequence database $D = \{T_1 T_2 ... T_m\}$, where each Web log access sequence $T_i = \{l_1, l_2 ... l_n\}$ is an ordered set of events and n is the length of the access sequence, the objectives are to

1. Explore and discover Web usage patterns.
2. Add/Remove Web records to/from the Web log database.
3. Generate frequently accessed Web pages with minimum support S
4. Reorganize the Web structure for efficient retrieval and faster access using a single scan technique.

Assumptions

- The Web log data contains the *URI*, *IP* address, Timestamp and Referrer.
- The events in the sequence need not be unique.
- Synthetic data generator is used for simulations.
- The real Web log data is taken from server sessions.

9.5 Algorithms

Input: The Web log data W
Output: Frequent Web access patterns F

1. *PreprocessData*( );
2. *LabelData*( );
{Generates unique labels to each *URI* and their corresponding Referrers}
3. *CreateSessions*( );
{Creates separate sessions for a required timestamp and for each new referrer}
4. *GenerateAssociations*( );
{Generates n level associations without pruning, access sequence database D}
5. *HashMine*( )
{Hash based method for mining associations}
6. *DynamicUsageMining*( )
{Handles insertions and deletions of data in batch mode}
7. *WebReorganization*( )
{Reorganization of existing Web structure based on the output of HashMine}

Initially, the n-level associations for each session are obtained after pre-processing to generate all the two level association pairs and their corresponding support and position, using a bucketing algorithm. At any instant, only two buckets are considered and the {Support, Position} pairs for every pair of labels is generated at each iteration. The {support, position} pairs are generated for all the labels in a file at the end of this phase. In the next phase, the contents of this file are read into a data structure(linked list) in the main memory that contains the position and supports.

Algorithm I: HashMine( )

```
Input: The n - level association database D
Output: Frequent Web access patterns F

Begin
  HashList ← Associations
  {A list of all the generated associations}
  for i = 1 to n - 1
  {n is the maximum length of the sequence T}
    for j = 1 to m
    {m is the number of access sequences in D}
     b_i   ← Merge(a_j(i), a_j(i+1)) // b_i is the i^th bucket.
     j   ← j + 1;
    endfor
    i   ← i + 1;
  endfor
  HashList ← b_{n-1} ;
  Index ← 0;
  for each HashEntry h_i in HashList
    for each association a_j in HashEntry
     HashIndex ← position(a_j[n]);
     SearchList ← HashList[HashIndex];
     for each Label in SearchList
      if (Label.Support > ξ)
      // ξ is the threshold support determine
      // MergePoints & TotalSupport
      // Add this new label to appropriate hash entry
      // along with position and support.
      endif
     endfor
    endfor
  endfor
End.
```

Using this data structure, we proceed to identify the frequent itemsets. In each iteration, usage information for patterns that have length (number of labels in the pattern) one more than the length of the patterns in the previous iteration is generated. The patterns not having sufficient support are eliminated and are not considered for the next iteration. This process comes to an end until either no further patterns can be generated or the maximum length of the association has been reached. While selecting labels for generating new combinations, a hashing technique is used to minimize the total time complexity. The merging and generation of new patterns occurs on the basis of the value of the last label. Two patterns P_1 and P_2 can only be merged if P_1 $.pos = (P_2.pos< \quad >1)$ and the last index of P_1 matches the first index of P_2. The hashing technique is a simple hashing with chaining. In case of conflicts,

Algorithm II: Dynamic Usage Mining

Input: The existing usage database D, Increment database Δd
Output: Updated database, Frequent Patterns

D - Existing usage database
Δd - New database, such that $\Delta d \geq 0.1D$
F- Frequent associations in D
S(x,y) - Support of x in database y
L - List of URI
ξ - Threshold support.
A_d - Associations in Δd

```
Begin
   for each association A_d in Δd
   if A_d ∉ D
     if S(A_d, Δd ) ≥ 2 ξ
       Add A_d
     for each URI U_a in A_d
       Add U_a to L
     endfor
   endif
  else if S(A_d, Δd ) ∨ S(A_d, Δd + D)≤ ξ
     Remove A_d
   endif
   endfor
End
```

the value is linked to previously stored value in that hash position in the form of a linked list. This kind of chaining does not affect the performance because all the feasible patterns are available at the same time for making the processing easier and simpler.

The Dynamic Usage mining algorithm is used whenever a new database with atleast 10% of the existing database is available. The new data undergoes pre-processing and associations are generated. New associations are added to the database only when their support is sufficiently large. Here we have heuristically taken it as twice the threshold. In such a scenario the Web pages involved are also added to URI list. Associations, which are already present in the existing database, is retained only if their combined support is $\geq \xi$ or 2ξ, the threshold support. For example consider a usage database D and the increment database Δd as shown. The threshold $\xi = 1/8$ which is 12.5%.

The supports of the URI's in D are $a = 10$, $b = 5$, $c = 3$, $d = 4$, $e = 6$, $f = 8$ out of 40 Web page access. Among these c and d fall below the threshold. The supports in Δd are a = 6, b = 1, c = 2, d = 2, e = 4, f = 3, g = 6 out of 24 Web page access. Now considering each page from Δd for addition we first have a with support 6/24 = 0.25 $> \xi$, similarly e and f also have threshold support. However as the support

Table 9.1 An Example Database D

acadef	abaef	aefe	bfead	bacdfe	abdfe	aefef	ace

$F_D = \{a, b, e, f\}$

Table 9.2 Increment Database δd

acefdgag	abdegfg	acagefeag

$F_{\Delta d} = \{a, e, f, g\}$

of b fall below threshold, b is removed from the frequent set. Also g which is a new occurrence has a support of 2ξ is included in the updated database and added to the URI list. The Frequent set in the resulting database is $F_{UPD} = \{a, e, f, g\}$. To reorganize the Web structure we start processing each Frequent Pattern and classify each page as an index page or a content page (Table 9.1 and 9.2).

Algorithm III: Web Reorganization

D – Web usage data
F - Frequent item sets, obtained from Usage Mining
W - Web structure data
$\mathbf{A}_f$ - An association in F
$\mathbf{U}_f$ - A Web URI or a Web Page

Begin
  for each association A_f in **F**
    for each URI U_f in A_f
    if *isContentPage*(U_f) in **D** $\wedge$ *isIndexPage*(U_f) in **W**
    U_f should appear as content page. {if the page consists
    of links rather than any textual content }
    else if *isIndexPage* (U_f) in **D** $\wedge$ *isContentPage* (U_f) in **W**
    $U_f$should appear as an index page in the Web site,
    link U_f to URI of all other pages. {if the page
    contains textual or other content }
    else if U_f *isContentPage*(U_f)
    Add a link to $U_f$in all the major index pages (Related index pages).
    else {if it is an index page }
    Add all the URI that links to U_f.
    endif
    Remove unused links.
  Endfor
  endfor
End

9.5.1 Classification of Pages

The two procedures *isIndexPage*( ) and *isContentPage*( ) are used to identify whether a Web page is a content page or index page. These two procedures when used with reference to usage information the decision is taken by looking into the access sequence in which the page is involved. If it is not possible to decide using the access sequence alone, then the average time spent on that page is computed from the Web log data. If it were to be an index page then the time spent on it is just a few seconds, but if the time spent is more, say a minute or a few minutes then it is a content page. The decision taken about the nature of the page based on the usage information rather than the structure information defines the actual role each page is playing; hence we make sure that each page's role and its corresponding place in the Web site do not contradict. For example if a page *P* is found to be a content page, but has too many other uninterested links, which are never referred to in the actual access patterns then those links ought to be removed from *P*, and at the same time all the major index pages in this context should have a link to *P*.

9.6 Pre-processing

The raw Web log data *W* is transformed into a set of unique labels for mining operations. It consists of following steps (i) Filtering (ii) Bit Representation (iii) Labelling (iv) Session Creation and (v) Association Generation.

(i) *Filtering*: The function *PreprocessData*( ) filters the raw Web log data *W* by removing all log entries concerned with image files. In addition, white spaces are removed and some special characters are translated for easier string processing. The standard Extended Common Log Format (ECLF) is as shown in Table 9.3, which consists of four fields <IP Address, Access Time, URI, Referrer> where URI and Referrer are the Web pages. In this format, each field is delimited by "||".

Table 9.3 The ECL Format

IP Address	Time of Access	URI	Referrer
20 chars	20 chars	Variable	Variable

(ii) *Bit Representation*: The bit representation shown in Table 9.4, is tailored to a 32 bit machine. The Hash Index field stores the index of all the associations stored in main memory. The association hence can be obtained as following; $A \leftarrow$ *HashList* [*HashIndex*].

The eight bit support field is the threshold count of the number of occurrences of particular associations (support S), where $0 \leq S \leq 100$ and is expressed as a percentage of the total database. For example in a database of size 3 million if a particular association occurs 3000 times then its support expressed in terms of percentage is 3000/3,000,000 which is 0.003 i.e., 0.3 percent. The first seven bits are used to store

Table 9.4 The bit representation

Pool	Support (S)	Hash Index
1 bit	8 bits (1+7)	23 bits

the percentage. MSB is 0 if the support is greater than one percent and the MSB is 1 if the support is less than one percent.

If the support of any associations falls below the threshold value (ξ), then it is discarded. Pool A corresponds to the content page and is represented by zero, whereas Pool B corresponds to an index page and is represented by one. The Bit representation helps in Web reorganization and the space and time complexity is significantly reduced.

(iii) *Labelling*: The Web log entry of a user's navigation pattern at *http://www.search.edu* site is given in Table 9.5. Each log entry is translated into a unique label. The corresponding labels for all URI's and Referrers of Table 9.5 are given in Table 9.6.

(iv) *Session Creation*: For mining access patterns, sessions are created depending on the time or the Referrer. New sessions are created based on two criteria. When the session is 30 minutes old, a new session is created. In addition, a new referrer also creates a new session and this is performed by *CreateSessions*( ) function.

(v) *Association Generation*: Initially, the function *GenerateAssociations*( ) generates a two level associations and the function can be used iteratively to generate n level associations. From the above table, we can derive the following associations, AA1 $\rightarrow$ AA2, AA2 $\rightarrow$ AA3, AA4 $\rightarrow$ AA5 and AA2 $\rightarrow$ AA6. In the next iteration, we generate three level associations AA1 $\rightarrow$ AA2 $\rightarrow$ AA3 etc. This procedure gives the Web access sequence database D.

Table 9.5 An Example for Web log data

Sl. No.	IP Address	Time of Access	URI	Referrer
1	123.456.78.9	2000/21/01-15:04:41	/index.html	http://www.search.edu/?key=search
2	123.456.78.9	2000/21/01-15:04:41	/1.html	http://www.search.edu/index.html
3	123.456.78.9	2000/21/01-15:04:41	/A.html	http://www.search.edu/10.html
4	123.456.78.9	2000/21/01-15:05:18	/C.html	http://www.search.edu/index.html

Table 9.6 Assignment of labels to each URI/Referrer

Sl. No.	Web Pages	Referrer
1	http://www.search.edu/?key=search	AA1
2	http://www.search.edu/index.html	AA2
3	http://www.search.edu/1.html	AA3
4	http://www.search.edu/10.html	AA4
5	http://www.search.edu/A.html	AA5
6	http://www.search.edu/C.html	AA6

The function *PreprocessData( )* takes the raw Web log data as input and filters it by removing all log entries concerned with image files. In addition, white spaces are removed and some special characters are translated for easier string processing. Then the data is sorted by time and the fields URI and Referrer are split and fed to the function *LabelData( )*.

LabelData( ) then merges both URI and Referrer and for each unique Web page a label is assigned, which is used while creating associations. The *CreateSessions( )* procedure now starts creating separate sessions for each user by populating the log entries in each session. A new session is also created whenever the session time-out (usually 30 minutes) occurs. The history of all the existing sessions are maintained in a singly linked list called *session list*, and is used when entries need to be added into existing sessions. The output of this procedure is a series of single level access sequence information per each session. Now the labelling information is read by the *GenerateAssociations( )* function and translates the access sequence output of the previous procedure into label representation, and generates upto n-level associations.

9.7 Example

Consider a database D = $\{T_1, T_2, T_3, T_4\}$ shown in Figure 9.3. It consists of a set of Web access sequences T_1, T_2, T_3 and T_4. Each column identifies a unique position (P_1, P_2, P_3, P_4 and P_5) for the construction of hash table. In the first iteration, two columns P_1 and P_2 are merged to form two level associations and placed in bucket b_1. For each pair of associations, positions and their corresponding support is indicated in Figure 9.4. The sequence b follows a is represented by $a \rightarrow b$. This pattern occurs thrice beginning with the first position, hence the support S of $a \rightarrow b$ is represented by $S_{a\rightarrow b} = 3$, in Figure 9.4. Similarly, the support of $b \rightarrow a$, i.e., $S_{b\rightarrow a} = 1$. The Frequent Pattern (FP) and the occurrences of the Web access sequences are illustrated in Figure 9.4.

The FP is placed in position P_2. In the next iteration (Figure 9.5) the patterns generated by the combined positions P_1 and P_2 are merged with position P_3 to form the set of three level associations and we derive $\{aba = \min(3, 2)\}$. This is obtained from the combination of *ab* and *ba*. The pattern *ab* occurs thrice beginning from position 1 and this is represented by (*ab* :: 3). Similarly, the pattern *ba* occurs twice beginning from position 2, i.e., (*ba* :: 2). So we have (i) (*ab* :: 3), (ii) (*ba* :: 2). From (i) and (ii) we can derive $\{aba = \min(3, 2)\}$, i.e., pattern *aba* occurs twice as shown in Figure 9.5. Similarly, the patterns $\{abc :: 1\}$ and $\{bab :: 1\}$ are generated. {In the following examples, the notation $\cap$ is used as a merge operator}.

Pos/Seq	P1	P2	P3	P4	P5
T1	a	b	a	c	
T2	a	b	c	a	c
T3	b	a	b	a	c
T4	a	b	a	c	b

Fig. 9.3 Initial Database

pos/Seq	P1	P2	P3	P4	P5
T1	a	b			
T2	a	b			
T3	b	a			
T4	a	b			

Fig. 9.4 The Merge of two positions P_1 and P_2 ($P_2 \leftarrow P_1 \cap P_2$)

FP: $a \rightarrow b$ ($S_{a\rightarrow b} = 3$), $b \rightarrow a$ ($S_{b\rightarrow a} = 1$)

pos/Seq	P1	P2	P3	P4	P5
T1	a	b	a		
T2	a	b	c		
T3	b	a	b		
T4	a	b	a		

Fig. 9.5 The Merge of two positions P_2 and P_3($P_3 \leftarrow P_2 \cap P_3$)

FP: $a\rightarrow b\rightarrow a$ ($S_{a\rightarrow b\rightarrow a}$= 2), $a\rightarrow b\rightarrow a$ ($S_{a\rightarrow b\rightarrow a}$= 2)

pos/Seq	P1	P2	P3	P4	P5
T1	a	b	a	c	
T2	a	b	c	a	
T3	b	a	b	a	
T4	a	b	a	c	

Fig. 9.6 The Merge of two positions P_3 and P_4($P_4 \leftarrow P_3 \cap P_4$)

FP: $a \rightarrow b \rightarrow a$ ($S_{a\rightarrow b\rightarrow a}$=3), $a \rightarrow b \rightarrow c$($S_{a\rightarrow b\rightarrow c}$=1)
$b \rightarrow a \rightarrow b$($S_{b\rightarrow a\rightarrow b}$=1), $b \rightarrow a \rightarrow c$($S_{b\rightarrow a\rightarrow c}$=2)
$b \rightarrow c \rightarrow a$($S_{b\rightarrow c\rightarrow a}$=1),$a \rightarrow b \rightarrow a \rightarrow c$($S_{a\rightarrow b\rightarrow a\rightarrow c}$=2)
$a \rightarrow b \rightarrow c \rightarrow a$($S_{a\rightarrow b\rightarrow c\rightarrow a}$=1),$a \rightarrow b \rightarrow c \rightarrow a$($S_{a\rightarrow b\rightarrow c\rightarrow a}$=1)

pos/Seq	P1	P2	P3	P4	P5
T1	a	b	a	c	
T2	a	b	c	a	c
T3	b	a	b	a	c
T4	a	b	a	c	b

Fig. 9.7 The Merge of two positions $P_4$and P_5($P_5 \leftarrow P_4 \cap P_5$)

FP: $a \rightarrow c$($S_{a\rightarrow c}$ = 2), $a \rightarrow c \rightarrow b$($S_{a\rightarrow c\rightarrow b}$ = 1) $b \rightarrow a \rightarrow c$(S $_{b\rightarrow a\rightarrow c}$=3), $a \rightarrow b \rightarrow a \rightarrow c$($S_{a\rightarrow b\rightarrow a\rightarrow c}$=3)

In Figure 9.6, the combined positions of P_1, P_2, and P_3 are merged with position P_4. The patterns generated are given in Figure 9.6. Finally, the positions P_1, P_2, P_3, and P_4 are merged with P_5 and patterns generated are shown in Figure 9.7.

Reorganization: An access sequence usually begins with an index page and ends with a content page. Moreover we see that in the above example that *a* is consistently the start page for a sequence and *c* is the end of web access sequence three out of four times, which implies that *c* is a content page while *a* is an index page, however it is not possible to take such an outright decision about *b* as *b* is once at the start of a sequence and once at the end. As a result we have to consider the average time spent on *b* by various users. This time spent on a particular page is one such metric used to infer whether a page is an index page or a content page, in addition inferences can also be drawn as in the above example where the type of pages *a* and *c* are known directly from the access pattern itself. Reorganization is necessary as the user's Web usage pattern keep changing with time. For example, Let *X* be an index page that has link to various content pages. Assuming a content page *Y* is the most accessed page which the Webmaster has highlighted in all the index pages including *X*. But as time progresses it may so happen that some other page (say *Z*) becomes more popular among users who visit *X* and in such a case it should be noted that page Z is highlighted. Our algorithm generates such decisions about the reorganization of existing pages based on the popularity of a page, while page *Y* which is no longer popular is deleted from the link *X*.

9.8 Performance Analysis

Simulations are performed on a system of physical memory capacity of 256 MB and Pentium III processor of speed 800 MHz. Experiments are conducted to compare the efficiency of *HashMine* algorithm with WAP tree and Generalized Sequence Patterns (GSP). A synthetic data generator is used which is based on Gaussian distribution. The sequence database is varied in the range of 100k to 500k. A random *n* level association is used as input to the algorithm.

Synthetic Data Generator: The Synthetic Data Generator uses a random number generator based on Gaussian distribution. It directly produces *n*- level associations and hence obviates the pre-processing and association generation steps. The number of associations and the maximum length of an association are taken as the input.

Figure 9.8 illustrates the performance of *HashMine* with GSP and WAP with increase in support from 1% to 3.0% for a database of 500k transactions. The algorithm GSP works on the principles of Apriori and makes multiple scans over the database, whereas WAP is better than the Apriori based algorithms but scans the database twice to construct the WAP tree and then recursively employs conditional search to find out the frequent patterns in a level wise manner. So the performance of HashMine is superior to those of GSP and WAP.

Figure 9.9 illustrates the comparison of the time complexity of the algorithm with increase in size of the database. The access sequence database is changed from 100k

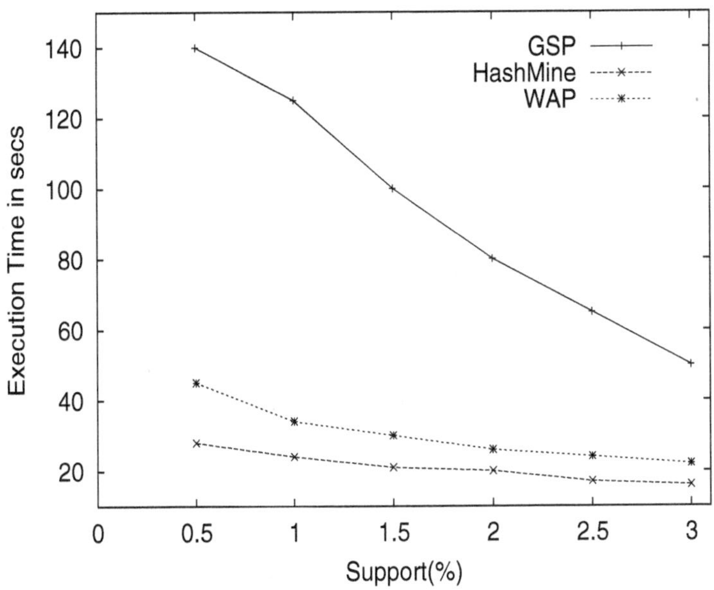

Fig. 9.8 The Execution Time versus varying support

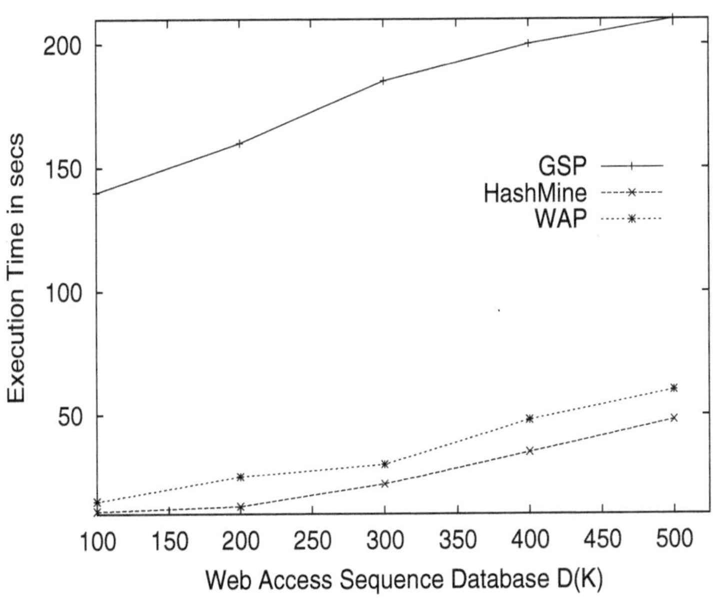

Fig. 9.9 The Execution Time versus varying Web access sequence database

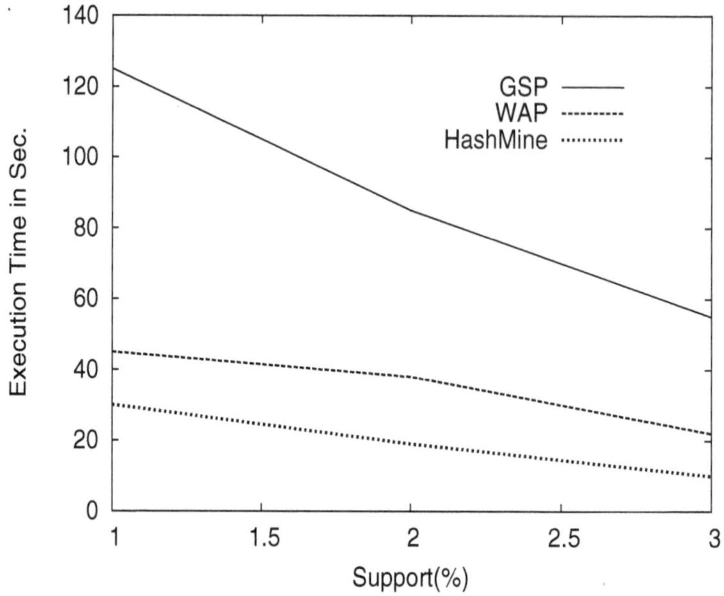

Fig. 9.10 The Execution Time versus varying Support

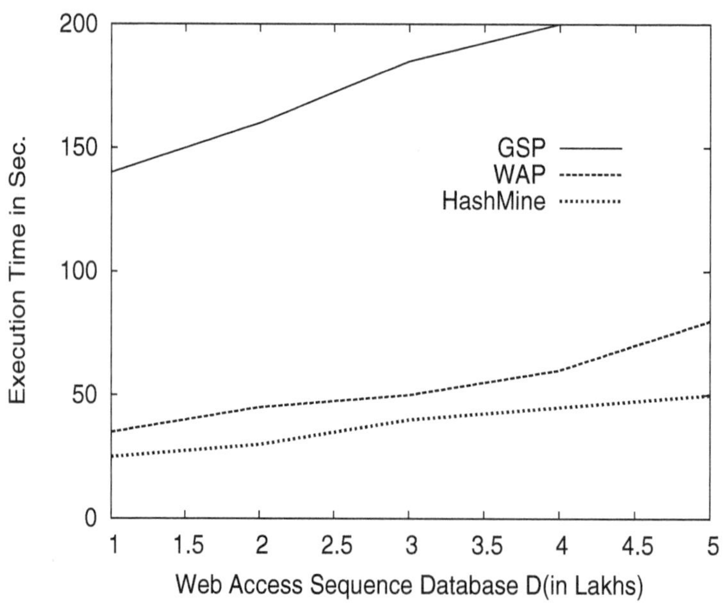

Fig. 9.11 The Support versus Speedup Factor

Sl. No	URI	LABEL
1	www.csee.usf.edu/music/machines/	AA8
2	hyperreal.com/music/machines/samples.html	AA6
3	hyperreal.com/music/machines/manufacturers/Roland/TR.909/samples/	AA5
4	guide.p.infoseek.com/machines/manufacturers/Roland/TB.303/	AA1
5	www.hyperreal.com/machines/manufacturers/	AA9
6	www.hyperreal.com/machines/samples.html	AA11
7	hyperreal.com/music/machines/manufacturers/Sequential/Tom/samples/	AA12
8	hyperreal.com/music/machines/categories/drum.machines/samples/	AA13
9	hyperreal.com/music/machines/images.html	AA4
10	www.hyperreal.com/machines/samples.html	AA11

Fig. 9.12 The URI Database

Sl. No	REFERRER	LABEL
1	www.csee.usf.edu/ gould/synth.html	AA7
2	hyperreal.com/music/machines/	AA3
3	hyperreal.com/music/machines/samples.html	AA11
4	guide.p.infoseek.com/Titles?qt=devilfish+sample	AA2
5	www.hyperreal.com/machines/manufacturers/Roland/TB.303/	AA10
6	www.hyperreal.com/machines/manufacturers/	AA9
7	hyperreal.com/music/machines/samples.html	AA6
8	hyperreal.com/music/machines/samples.html	AA6
9	hyperreal.com/music/machines/	AA3
10	hyperreal.com/music/machines/images.html	AA4

Fig. 9.13 The Referrer Database

Fig. 9.14 The Generated Associations

AA3 ⟶ AA4 ⟶ AA6
AA10 ⟶ AA9 ⟶ AA11
AA6 ⟶ AA12
AA7 ⟶ AA8
AA6 ⟶ AA13
AA2 ⟶ AA1
AA3 ⟶ AA6 ⟶ AA5

to 500k with a support of 0.5%. *HashMine* is much faster than other algorithms on account of single scan method and low time complexity of the hashing technique. The time complexity increases with the size of the database linearly and the algorithm is scalable. Figure 9.10 shows the plot of the execution time of our algorithm versus the support. Increment of 20% is added to the existing database of size 500K and the support is varied from 1% to 5%. The graph clearly illustrates that the time taken for the dynamic update (DUPD) is much lower than that for the static update (SUPD) as dynamic update (DUPD) procedure makes use of the associations already available from mining the original database (D) whereas the static update (SUPD) makes a complete scan of the entire updated database (D + Δd).

A good speedup factor of 2 to 4 is achieved for various supports for a database of size 500K (Figure 9.11). Experiments are conducted on real data sets from UCI data repository. The sample database of the URI is shown in Figure 9.12 and the corresponding Referrer database is shown in Figure 9.13. The generated associations are shown in Figure 9.14.

9.9 Summary

In this chapter, we examine the issue of Web usage mining to discover frequent usage patterns. The *HashMine* is efficient in generating frequent access patterns on a dynamic database. The frequent access patterns generated are used to reorganize the Web structure for faster access and retrieval. The algorithm uses a single scan technique resulting in low time complexity. The use of bit vector to represent all the necessary information of the log entry in a single word reduces the memory requirements significantly and helps in the reorganization of the Web structure. The algorithm *HashMine* outperforms WAP and GSP in both time and space complexity.

References

1. Agrawal, R., Srikant, R.: Fast Algorithms for Mining Association Rules. In: Proc. Very Large Data Bases (VLDB), pp. 487–499 (1994)
2. Park, J.S., Chen, M.-S., Yu, P.S.: An Effective Hash-Based Algorithm for Mining Association Rules. In: Proc. ACM-SIGMOD, pp. 175–186 (1995)
3. Cheung, D.W., Han, J., Vincent, T., Wong, C.Y.: Maintenance of Discovered Association rules in Large Databases. In: Proc. On Data Engineering, pp. 106–114 (1996)
4. Shenoy, P.D., Srinivasa, K.G., Venugopal, K.R., Patnaik, L.M.: Evolutionary Approach for Mining Association Rules on Dynamic Databases. In: Proc. Of Int'l Conf on Pacific Asia Conference on Knowledge Discovery and Data Mining. LNCS (2003)
5. Agrawal, Srikant, R.: Mining Sequential Patterns. In: Proc. on Data Engineering, pp. 3–14 (March 1995)
6. Srikant, R., Agrawal, R.: Mining Quantitative Association Rules in Large Relational Tables. In: Proc. ACM-SIGMOD on Management of Data, June 1996, pp. 1–12 (1996)
7. Pei, J., Han, J., Mortazavi-asl, B., Zhu, H.: Mining Access Patterns Efficiently from Web Logs. In: Terano, T., Chen, A.L.P. (eds.) PAKDD 2000. LNCS, vol. 1805. Springer, Heidelberg (2000)
8. Wang, K., Liu, H.: Discovering Structural Association of Semi structured Data. IEEE Transactions on Knowledge and Data Engineering 12(3) (May/June 2000)
9. Lin, S.-h., Chen, M.C., Ho, J.-M., Huang, Y.-M.: ACIRD: Intelligent Internet Document Organization and Retrieval. IEEE transaction on knowledge and data engineering 14(3), 599–614 (2002)
10. Srivastava, J., Cooley, R., Deshpande, M., Tan, P.N.: Web Usage Mining: Discovery and Applications of Usage Patterns from Web Data. In: SIGKDD Explorations, pp. 12–23 (2000)

11. Brin, S.: Extracting Patterns and Relations from the World Wide Web. In: WebDB Workshop at 6th Intl. Conf. on Extending Database Technology, EDBT (1998)
12. Fu, Y., Shis, M.-Y., Creado, M., Ju, C.: Recognizing Web Sites Based on User Access Patterns. In: Proc. on Information and Knowledge Management (2001)
13. Mobasher, B., Cooley, R., Srivastava, J.: Creating Adaptive Web Sites through Usage-Based Clustering of URLs. In: IEEE Knowledge and Data Engineering Workshop, KDEX 1999 (November 1999)
14. Mobasher, B., Dai, H., Luo, T., Nakagawa, M.: Discovery and Evaluation of Aggregate Usage Profiles for Web Personalization. Kluwer Academic Publishers, Netherlands (2001)

Chapter 10
Algorithms for Web Personalization

Abstract. Personalization is a process of gathering, storing and analyzing information about site visitors and delivering the right information to each visitor at the right time. A personalization technique can enable a website to target advertisement, promote products, personalize news feeds, recommend documents, make appropriate advice and target e-mail. In this chapter, an adaptive weight update algorithm based on the standard search engine, using the combination of query refinement, recommender system and user profiles is explored. The algorithm proposed captures and records the user interest with respect to the indicated keywords to the search engine. The user profile is built dynamically on the recorded interests of the user and used as a recommender system to refine the query on further hits by adding additional keywords depending upon the users interest. A dynamic weight update equation is used to adapt to the changing user interests. The algorithm is tested on varying keywords and the results obtained are better when compared with the Googles page rank system.

10.1 Introduction

Web Personalization is a set of actions that can tailor the Web experience to a particular user or a group of users. The experience can be like, users browsing patterns, sequence of pages visited by a user, usual query patterns, etc.. The actions can range from simply making the presentation more pleasing to anticipating the needs of a user and providing customized and relevant information to the user. To achieve effective personalization, organizations must rely on all available data, user behavior, the site content, the site structure, domain knowledge, as well as user demographics and profiles [1].

One of the techniques to achieve effective personalization is by having visitors of a site fill out the forms with information fields that populate a database. The Web site then uses the database to match a user's needs to the products or information provided at the site, with middleware facilitating the process by passing data between the database and the Web site. Consider an example of Amazon.com, which is used for online book purchase/selling. One of the facilities provided by Amazon.com for

K.R. Venugopal, K.G. Srinivasa, L.M. Patnaik: Soft Comput. for Data Min. Appl., SCI 190, pp. 217–230.
springerlink.com

its registered users is the suggestion of books/CDs depending on their previous purchase history or his interests captured while browsing the Web. Customers tend to buy more when they know exactly what's available at the site and they do not have to hunt around for it. Cookies may be the most recognizable personalization tools. Cookies are bits of code that sit in a user's Internet browser memory and inform Web sites about a person. That is how a Web site is able to greet its users by name [2].

Web personalization is also achieved by collaborating filtering software that resides on a web site and tracks the path of the users interest and viewing habits depending upon various parameters like types of pages they visit, the amount of time they spend on a page, etc., Collaborative-filtering software compares the information it gains about one user's behavior against data about other customers with similar interests. In this way, users get recommendations like Amazon's "Customers who bought the book x also bought other books y and z". Domain-specific search engines are growing in popularity because of their increased accuracy and extra functionality that is not possible with the general search engines. For example, *scholar.com* of Google allows complex queries over Internet by type of file, prioritization of institutions, size, and location. *Yahoo* is another example for domain-specific search engines. Unfortunately, these search engines are difficult to use for naive users and time-consuming to maintain.

The key challenge in personalization is to display to the visitor a sub graph that contains both relevant content items, and also organizes them in a coherent and meaningful manner. The problem of automatically constructing Personalized Site Maps is discussed in [3]. The approach is based on the assumption that the best way to indicate the relationship between a given pair of pages is to show the path between them that has been most popular with past visitors. An ontology based personalized retrieval of audio information is addressed in [4]. The crux of the work is the development of an ontology-based model for the generation of metadata for audio and the selection of audio information in a user customized manner.

The data mining methods are very much exploited to build the customer profiles. The system constructs personal profiles based on customers transactional histories and then uses data mining techniques to discover a set of rules describing customers behavior and supports human experts in validating the rules [5]. The study of personalized recommendation to build improved algorithm for learning Bayesian networks efficiently and effectively is proposed in [6]. The algorithm reduces the number of independence tests and database passes, while effectively restricting the search space and by means of the heuristic knowledge of mutual information. In [7], a web personalization system based on web usage mining and automatic customization is proposed. It is an effective technique for capturing common user profiles based on association rule discovery and usage-based clustering. The techniques for combining knowledge with the current status of an ongoing Web activity to perform real-time personalization are also discussed.

The use of Agent technology for personalized recommendation, which is fuzzy cognitive in nature, to give personalized suggestions based on the current users preferences, general users common preferences, and the experts knowledge is described in [8]. Fuzzy cognitive agents are able to represent knowledge via extended fuzzy cognitive maps, to learn users common preferences from most recent cases and to help customers to make inference/decisions through numeric computation instead of symbolic and logic deduction. The idea of Web reconfiguration based on page personalization rules is discussed in [9]. A page personalization rule for providing reconfigurable Web content including hyperlinks is described in a declarative language. The concept of geo-cookie for reconfiguring Web pages based on a user's physical and logical location is also used. An algorithm that can generate and display helpful links while users navigate a site can increase a Web sites usability and help Web designers and the user achieve their goals has been presented in [10]. Partial evaluation is a technique popular in the programming languages community. It is applied in [11] as a methodology for personalizing Web content. Two different implementations of the methodology demonstrate their effectiveness. Domain Specific Search engines [12] are developed to search for information regarding only a specific field. Such systems maintain a user profile of the histories of web pages that a user has visited. The problem here is that lack of narrowing of the search field by supplying more keywords. Manual work has to be done apriori, to categorize a collection of pages according to domains. They have the problem of scalability and do not adapt to users changing interests.

10.2 Overview

In this chapter, we propose a hybrid system of Web Search Personalization using a combination of Query Refinement, Recommender Systems and User Profiles. Consider a Search Engine *S*, which accepts queries in the form of keywords and returns a list of near relevant web pages. The queries are of the form $kw_1[(op)kw_i], for\ i=2$ to n, where,kw_i is the ith keyword supplied to *S*, (*op*) is a Boolean operator like OR, AND, n is the number of keywords supplied to *S* and [...] indicates the parameters are optional.

The search query supplied to S contains keywords. The presence of these keywords indicate that the user is interested in them. This interest has to be captured and recorded for future use by the system. By recording this interest, the system can modify the search query supplied to S in subsequent searches to narrow the search field and concentrate on those topics that are of interest to the user. Once S returns its results, those web pages of interest to the user have to be scanned for keywords that are already present in the User Profile. This indicates the relation of the returned keywords to the one supplied to S for searching. We address these issues in the following sections.

10.3 Data Structures

The data structure used in this approach is discussed below.

```
Integer Relative_Search for global;
Integer Domain_Search[Number of Domains];
Structure Domain  {
```

```
        String Keyword;
        String Type;
        Real Weight;
        Integer Keyword_Search;
        Integer Rel_Keyword_Search;
}
```

Each domain is associated with one or more keywords. Each search instance in the system is specified by a set of keywords. When a search is performed, the keywords are matched to their corresponding domains. A keyword may be associated with more than one domain. In such a case, in the first domain it is assigned to its Type as primary. In any subsequent assignments to other domains, the Type in those domains is auxiliary.

Each keyword has an associated weight. The weight decides the keywords that have to be chosen to add as keyword spices to the user specified keyword(s) while performing a search. It also indicates the users interest in a particular domain. Each time a new keyword is created, it is assigned an initial weight zero. Hence all the keywords start with the same weight or same opportunity in the User Profile. The weights for a keyword are dynamic in nature. This is done to reflect the users interest in certain domains. A higher weight indicates a greater interest in that domain. For each search instance, the weights for all the keywords in that domain are varied as defined by the functions weight_increment() and weight_decrement(). When the weight for a keyword crosses a pre-defined lower threshold, it is deleted from the User Profile. This reflects the fact that the user has lost interest in that area in the domain.

A global variable Relative_Search is maintained by the system, which keeps track of the total number of searches conducted by the user. It is incremented by one after every search. It is required to deduce the relative importance of the domains in the User Profile. Each domain has an associated counter called Domain_Search that keeps track of the User Profile. Each time a new domain is created, the Domain_Search variable is assigned an initial value of one. For each subsequent search in any domain, the value of this variable is incremented by one.

Each keyword has an associated counter that indicates the absolute number of times the keyword has been used in a search. Each time a new keyword is created, the Keyword_Search variable is assigned an initial value of one. For each subsequent search on that keyword, the value of this variable is incremented by one. Each keyword has an associated counter that shows the absolute number of times, the domain which contains the keyword has been used in a search, since the keyword has been introduced into the User Profile. Each time a new keyword is created, the Rel_Keyword_Search variable is assigned an initial value of one. For each subsequent search the user performs in the domain, the value of this variable is incremented by one. At no point can the value of Keyword_Search exceed the value of Rel_Keyword_Search. It can only be less than or equal to it.

Organization of the Data Structures for retrieval: The data structure Domain has to be stored for use by the system. It has to be stored in a permanent storage medium for use after the system has been powered down and re-started. Each domain keyword creates a new instance of the Domain structure. Storing all the domains in a single file is not an efficient method. This is because, each time a domain has to be accessed, the entire file has to be searched to find the location of the desired domain.

Another approach is to store each domain in its own file. The files are named according to the name of the domain. All these files are stored under the same directory tree. This transfers the search for a domain location to the OS. Since the OS has to anyway set up its File Allocation Tables every time it is run, it does so for the domain files as well and reduces the overhead of having to search for the domain locations. Now, the basic file operations and only the name of file need to be specified to the OS to manipulate a domain.

A RDBMS system is used for the storage of the User Profile. Each Domain is represented by a table in the database. Each Keyword is represented by a tuple in the domain. This facilitates the search for a particular keyword in a domain using any query language and the retrieval of top-k weighted keywords is more efficient.

10.4 Algorithm

The function weight_increment() is used to decide the amount, the value of the Weight variable has to be incremented to reflect the users changing interests. This reflects the user's growing interest in an area of a domain. The following terms are used in function weight_increment(). The variable increment is the value of the weight for a keyword that has to be incremented, the variable factor is used to represent keyword's Weight that has to increase after it is used in a search. The following equation is used for denoting the factor to increment the weight for a keyword,

$increment = (1/Weight) * log(factor)$. Hence we have
$Weightnew = Weightold + increment$ and
$factor = (Tf * INL)/(log(Pr) * ROL)$.

We use Term Frequency (*Tf*) as an indicative factor to the information represented in a page. It is directly proportional to the information with respect to an area of interest. The Number of IN Links (INL) indicates the 'authoritativeness' of a page on a topic. Higher the value of INL, the more likely it is to be an Information page. We assume that the user is interested more in information than links to information. The rank of the page (*Pr*) represents an independent ranking of the page by an external source. Since here we use the Pagerank provided by Google, the increment is inversely proportional to *Pr*. Higher value of *Pr* represent lower ranks. The Related Out Links (ROL) indicates the status of the page as a 'hub'. Higher the value of ROL, the more likely it is to be a hub. This term is used to balance the *Tf* as in a hub page; higher *Tf* indicates more links to the information rather than information itself.

The *weight_increment_new()* uses same equation as defined in *weight_increment()*. However since there is no *Weightold* for a new keyword, we use the modified form, $increment = log(factor)$ and $Weightnew = increment$.

The function *weight_decrement*() is used to decide the amount the value of the Weight variable has to be decremented to reflect the users changing interests. This makes it possible to prevent the User Profile from growing indefinitely, since every new keyword encountered in a search instance creates its own domain. As mentioned earlier a domain can be deleted from the User Profile when its weight crosses a pre-defined lower threshold . The following terms are used in function *weight_decrement*(). The variable decrement is the value of the weight for a keyword that has to be decremented, the variable counter is used to represent the factor by which the keyword has not been involved in the searches performed by the user and the variable life is used to represent the lifetime of the keyword. The lifetime of the keyword indicates the usage of the keyword from the instant it has been introduced into the User Profile. We use the following equation for denoting the factor to decrement the weight for a keyword,

$decrement = Weight * (counter/life)$,
$Weightnew = WeightoldWeightold * (counter/life)$
$counter = Rel\_Keyword\_Search - Keyword\_Search$
$life = Rel\_Keyword\_Search$

A dynamic Weight Update Algorithm for Web Personalization (WUAWP) is given as follows.

1. The user profile U is initially empty. The profile is divided into domains and each domain has a collection of keywords. U = ϕ initially,
2. Let QS be the query string supplied to S. $QS = kw_1[(op)kw_i], for\ i = 2\ to\ n.$ Parse QS to obtain the individual keywords and check if any kw_i is a primary keyword in any domain. Let m be the number of kw_i that are present in U.
3. If $(m = 1)$ then add the top-k weighted keywords to QS from that domain. Let this be QS_{mod}.
4. If $(m > 1)$ then group the keywords into p groups by the domains in which they exist
   a. Case 1: If any p groups of keywords have the same number of keywords then;
      i. for all $p\ sum_j = \Sigma_{i=1}^{x} weight(kw_i)$;
      ii. calculate sum_j as the sum of weights of all keywords in j^{th} group,
      iii. where x is number of keywords in j^{th} group
      iv. kw_i are the keywords of jth group
      v. Add the top-k weighted keywords to QS from the domain(s) to which the jth group(s) belongs.
      vi. Let this be QS_{mod} and $sum_j > sum_i$; where $i = 1\ \ to\ \ p$ and $i \neq j$
   b. Case 2:If no p groups of keywords have the same number of keywords then Add the top-k weighted keywords to QS from the domain to which the group with the highest number of keywords belong. Let this be QS_{mod}
5. Submit QS_{mod} to S as the query string.
6. Collect user response to the pages returned by QS_{mod}.
7. For all pages wp_i found interesting, parse wp_i to find keywords fk_w.

8. If $m = 0$ and the fk_w do not belong to any existing domain, then create a new domain with these keywords. Set $WEIGHT(fk_w) = 0$ and $WEIGHT(kw_i) = 0$.
9. If $m = 0$ and fk_w belong to any p domain(s) in U add kw_i as primary keywords to MSD and as auxiliary keywords in the other domains; where $MSD = Domain$ *having* $mdsr$ and $msdr = max(DOMAIN_SEARCH/RELATIVE_SEARCH)$ over p domains
10. for each keyword kw_i in all domain(s) on which search was applied do
    - If keyword kw_i is included in the search and $WEIGHT(kw_i) \neq 0$ then change weight by weight_increment()
    - If keyword kw_i included in the search and $WEIGHT(kw_i) = 0$ then change weight by weight_increment_new()
    - If keyword kw_i not included in the search then change weight by the function weight_decrement()
    - If $WEIGHT(kw_i) < \delta$ then delete kw_i from the domain, where, δ is the user-defined lower threshold for pruning the User Profile

Since counter represents the factor that the keyword has not been involved in any searches the user has performed, the value of decrement is directly proportional to counter. This reflects the fact that the weight for a keyword decreases by a higher factor if it is not being used often. Since life represents the lifetime of the keyword from the instant it was introduced into the User Profile, the value of decrement has to be inversely related to it. This reflects the fact that keywords with a larger lifetime indicate sustained user interest in them and hence must be given more importance. The following points can be observed from the equation,

- The value of decrement is always between zero and one.
- If counter = 0 then no decrement is performed. This shows *Rel_Keyword_Search* = *Keyword_Search* and reflects the fact that the keyword has been involved in searches every time from the time it was introduced into the User Profile

At any instant the counter cannot be equal to life. If it is true then Keyword_Search would have to be zero. This is impossible because if an entry exists for a keyword then it would have to be searched at least once and therefore Keyword_Search would be minimally one. Hence at any instant the value of decrement cannot be equal to Weight for a keyword and the weight after decrement would never be zero. Therefore, a lower threshold value has to be selected for pruning the user profiles.

10.5 Performance Analysis

The test results are obtained by running sample queries on the commercial search engine, Google [www.google.com]. The training phase of the system, i.e., the construction of the User Profile up to an initial state, is done, by asking a single user to enter queries and provide feedback on the search results. The feedback involved identifying the web pages that the user found interesting (relevant) and then parsing the pages to obtain the keywords. These keywords are added to the User Profile

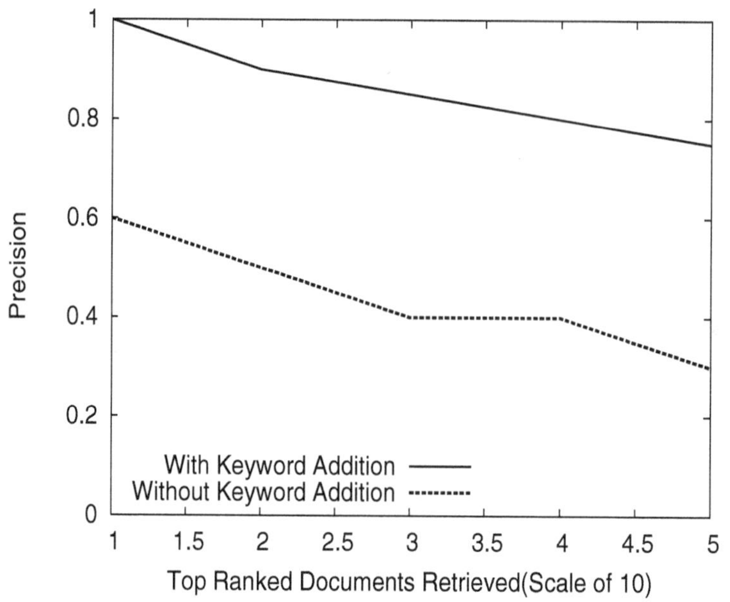

Fig. 10.1 Precision of Requests with and without keyword addition

Table 10.1 Domain: Music

Keyword	Type	Weight	Keyword Search	Rel. Keyword Search
music	p	3.98	14	15
guitar	a	1.15	4	9
sheetmusic	p	2.47	10	13
tablature	p	2.47	10	13
lyrics	a	0.33	1	5
piano	a	0.62	2	6

according to our algorithm. In the testing phase, the same user is asked to provide queries. The user should also provide relevance judgments regarding the web pages he expects from the search engine to return. These queries are run on the search engine twice. The first time, the queries are run without the addition of any keywords and the second time, the system added keywords to the query according to our algorithm. For different top-k web pages returned, the precision is calculated for both the test runs. The precision is defined as the ratio between the number of selected relevant pages and the total number of selected pages. The results are as shown in Figure 10.1.

From Figure 10.1 it is obvious that the precision of the search engine increases with the addition of relevant keywords, otherwise, the precision suffers because of

Table 10.2 Domain Artificial Intelligence

Keyword	Type	Weight	Keyword Search	Rel. Keyword Search
Artificial Intelligence	p	3.42	12	20
Neural Networks	p	2.38	10	20
Computer	a	0.47	3	20
Pattern recognition	p	1.62	7	14
Genetic algorithms	p	2.25	8	12
Expert systems	p	1.81	5	8

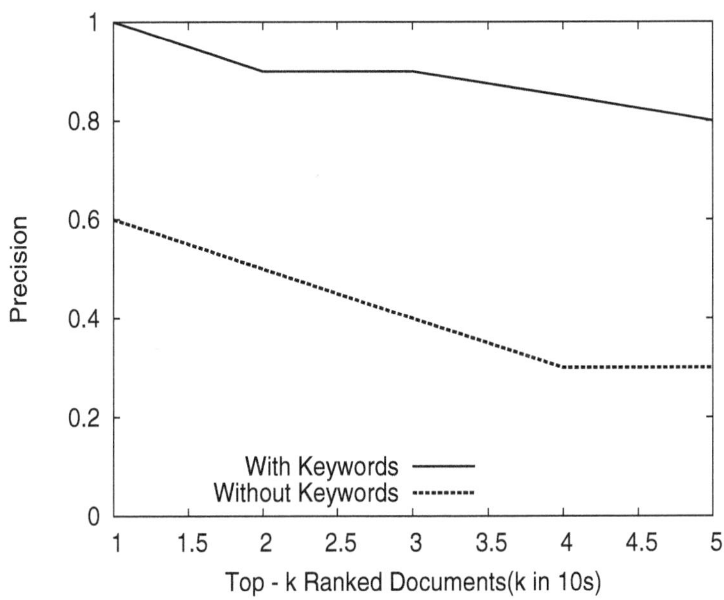

Fig. 10.2 Precision for Query: Expert Systems Original Query: Expert Systems; Keywords Added: Artificial Intelligence, Neural Networks

broader search. A snapshot of the domains for Artificial Intelligence and Music is as shown in Table 10.1 and Table 10.2. This state of the domain is obtained after a period of time the user builds his profile. The results for the queries run by the user with and without the addition of keywords are as shown in Figure 10.1. The precision is calculated for the top-k pages.

Experiments are conducted on varying keywords and adding keywords to the original query. Figure 10.2, demonstrates the precision of top-k ranked documents retrieved with or without additional keywords. In Figure 10.2, since the original query is Expert Systems and it belongs to the Artificial Intelligence domain the top ranked keywords in that domain are Artificial Intelligence and Neural Networks.

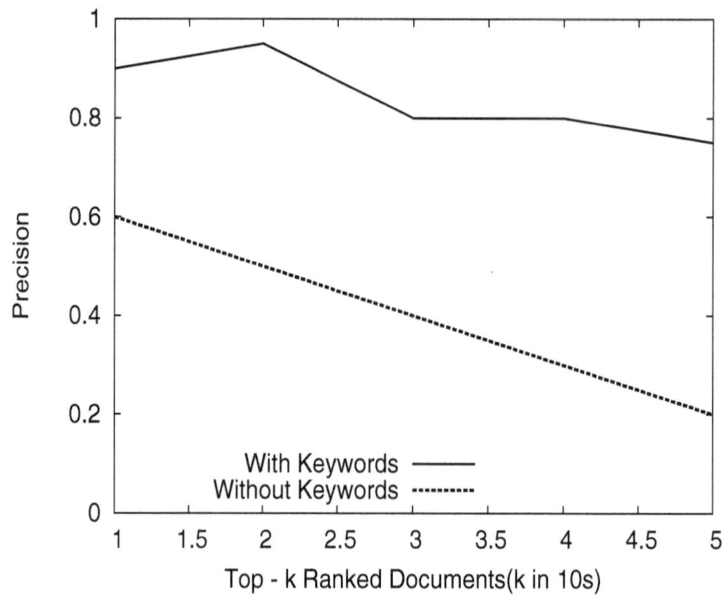

Fig. 10.3 Precision for Query: Pattern Recognition Original Query: Pattern Recognition; Keywords Added: Artificial Intelligence, Neural Networks

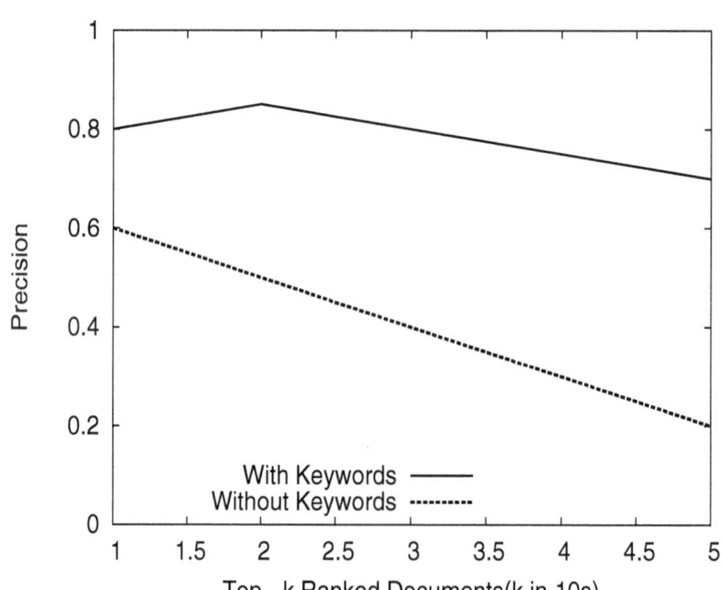

Fig. 10.4 Precision for Query: Guitar Original Query: Guitar; Keywords Added: Music, Talature

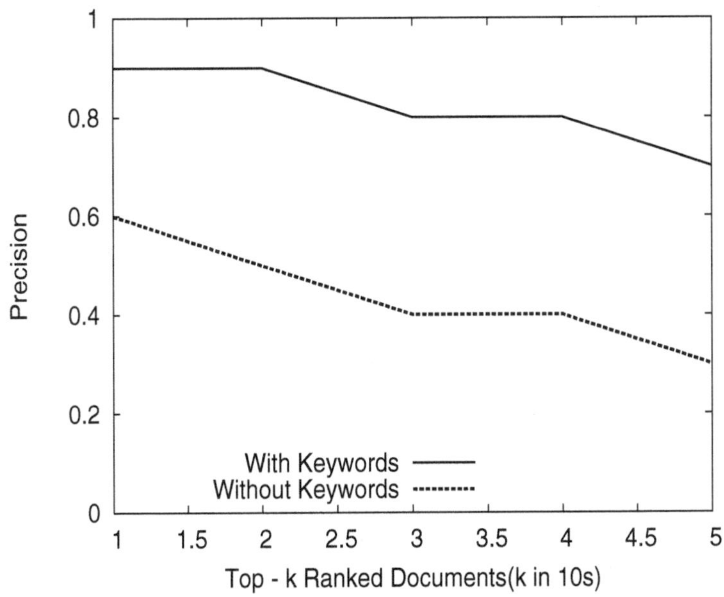

Fig. 10.5 Precision for Query: Piano Original Query: Piano; Keywords Added: Music, Sheet Music

Table 10.3 Efficiency of the Proposed Algorithm

Original Query	Keywords added	Top-k pages	
		Without Keywords	With Keywords
Expert Systems	Artificial Intelligence Neural Networks	20	11
guitar	music, tablature	38	17
piano	music, sheetmusic	28	14
Pattern Recognition	Artificial Intelligence Neural Networks	30	14

Therefore, these two keywords are added in addition to the given query. The precision of the retrieved documents with addition of two keywords is compared with the precision of retrieved documents without any keyword addition. Similarly, the experiments are conducted on original queries Pattern Recognition, Guitar and Piano in Figure 10.3, 10.4 and 10.5 respectively. In all the cases, the top-2 ranked keywords of that particular domain are added and the precision of the retrieved documents with keyword addition is better as compared to retrieval without any addition of keywords.

Table 10.3 represents the top-*k* pages that need to be examined by the user to obtain 10 pages, which are relevant to him. For example, in the case of the query on Expert systems, the user is expected to search the top 20 pages to find 10 relevant pages when the keywords are not added to the query but the number drops to 11 when the query is run by adding relevant keywords to the original query. The listing of retrieved URLs with and without the addition of keywords for different queries is given below.. It is observed from the results that when the algorithm adds the keywords, more relevant pages appear at higher ranks in the results obtained.

Original Query: Expert Systems
Top 10 Webpages Retrieved without Addition of Keywords:
-www.aaai.org/AITopics/html/expert.html
-www.ghg.net/clips/CLIPS.html
-www.sciencedirect.com/science/journal/09574174
-herzberg.ca.sandia.gov/jess/
-www.computer.privateweb.at/judith/
-www.cee.hw.ac.uk/ alison/ai3notes/chapter2_5.html
-www.blackwell-synergy.com/servlet/useragent?func=showIssues&code=exsy
-www-2.cs.cmu.edu/Groups/AI/html/faqs/ai/fuzzy/part1/faq.html
-www.aiinc.ca/demos/whale.html
-www.claes.sci.eg

Top 10 Webpages Retrieved with Addition of Keywords:
Artificial Intelligence, Neural Networks
-www.pcai.com/web/ai_info/expert_systems.html
-www.pcai.com/
-www.compinfo-center.com/tpai-t.htm
-www.sigma-research.com/bookshelf/default.htm
-ai-depot.com/Applications/1092.html
-www.webopedia.com/TERM/A/artificial_intelligence.html
-www.blackwellpublishing.com/journals/exsy
-www.wargaming.net/Programming/113/Expert_Systems_index.htm
-www.decision-team.com/knowledge_base/ knowledge_base_showarticle.asp?
-www.nasatech.com/software/ai.html

Original Query: Guitar
Top 10 Webpages Retrieved without Addition of Keywords:
- www.guitar.com
- www.guitarnotes.com
- www.guitarworld.com
- www.olga.net
- www.gibson.com/
- www.guitarplayer.com
- www.guitarcenter.com/
- www.ultimate-guitar.com
- www.acousticguitar.com

Top 10 Webpages Retrieved with Addition of Keywords: Music, Tablature
- www.olga.net
- www.harmony-central.com/Guitar/tab.html
- www.1christian.net/guitar/
- www.guitarnotes.com/tabs
- www.guitarnotes.com/guitar/chords.shtml
- www.mxtabs.net
- alt.venus.co.uk/weed/music/classtab
- adamschneider.net/music
- dir.yahoo.com/Entertainment/Music/Tablature/
- www.countrytabs.com

10.6 Summary

In this chapter, a weight update algorithm for consolidating the user interests, domain specific search and addition of keyword for a better web personalization is proposed. The algorithm is tested on Google results and Google PageRank to obtain the top 10 results and the results obtained are much better compared to the output of Googleś PageRank system.

References

1. Kolari, P., Joshi, A.: Personalized Web Search for Improving Retrieval Effectiveness. In: Computing in Science and Engineering, pp. 49–53. IEEE, Los Alamitos (2004)
2. Pokorny, J.: Web Searching and Information Retrieval. In: Computing in Science and Engineering, pp. 43–48. IEEE, Los Alamitos (2004)
3. Toolan, F., Kusmerick, N.: Mining Web Logs for Personalized Site Maps. In: Proceedings of the Third International Conference on Web Information Systems Engineering (Workshops), WISE 2004, pp. 232–237 (2004)
4. Khan, L., McLeod, D.: Audio Structuring and Personalized Retrieval Using Ontologies. IEEE Intelligence Systems (2000)
5. Gediminas, A., Alexander, T.: Using Data Mining Methods to Build Customer Profiles. IEEE Computer, 74–82 (2001)
6. Ji, J., Liu, C., Yan, J.: Bayesian Networks Structure Learning and Its Application to Personalized Recommendation in a B2C Portal. In: Proceedings of IEEE/WIC/ACM International Conference on Web Intelligence (WI 2004) (2004)
7. Mobasher, B., Cooley, R., Srivastava, J.: Creating Adaptive Web Sites Through Usage-Based Clustering of URLs. In: IEEE Knowledge and Data Engineering Exchange Workshop (KDEX 1999) (1999)
8. Miao, C., Yang, Q., Fang, H., Goh, A.: Fuzzy Cognitive Agents for Personalized Recommendation. In: Proceedings of 3rd International Conference on Web Information Systems Engineering (WISE 2002) (2002)

9. Kiyomitsu, H., Takeuchi, A., Tanaka, K.: Web Reconfiguration by Spatio-Temporal Page Personalization Rules Based on Access Histories. In: Proceedings of 2001 Symposium on Applications and the Internet (SAINT 2001) (2001)
10. Jenamani, M., Mohapatra, P.K.J., Ghose, S.: Online Customized Index Systhesis in Commercial Web Sites. IEEE Intelligent Systems, 20–26 (2002)
11. Ramakrishnan, N.: PIPE: Web Personalization by Partial Evaluation. IEEE Internet Computing, 21–31 (2000)
12. Oyama, S., Kokubo, T., Isida, T.: Domain-Specific Web Search with Keyword Spices. IEEE Trans. On Knowledge and Data Engg. 16(1), 17–26 (2004)

Chapter 11
Classifying Clustered Webpages for Effective Personalization

Abstract. Personalization is the process of presenting the right information to the right user at the right moment, by storing browsing history about the user and analysing that information. Personalization methods enables the site to target advertising, promote products, personalize news feeds, recommend documents, make appropriate advice, and target e-mail. In this chapter, we introduce a clustering algorithm, which creates clusters of similar sites, Naive Bayesian probabilistic model, which classifies the sites into different categories and a hybrid model, which is a combination of both. The user profile is built dynamically on the recorded interests of the user, which are categories of the site in which the user browses. The algorithms are tested on varying keywords and the results obtained are compared with the Google's page rank system.

11.1 Introduction

Internet is one development, which fuelled the growth of IT industry to a large extent. It has single handedly changed the way we communicate, collaborate and cooperate with each other. It has been responsible for breaking the barriers of time and geographical limitations and helped to set a new business order. From being an information repository it has today grown to be a strategic tool in business. Internet by its very nature does not support systematic storage of information. The units of storing information in Internet servers are websites which are uniquely identified by their URL. The problem which everyone had to face aftermath the explosion of information in Internet is to get relevant information on time. Remembering URLs proved cumbersome and difficulty in mining for correct information was proving as hindrance for the further growth of Internet.

Search engines proved to the right medicine for this ailing problem of the industry. But search engines faced a typical problem since they relied on keywords for conducting search. The present day search engines have failed to understand the specific requirements of the user. Users experience increasing difficulty finding

K.R. Venugopal, K.G. Srinivasa, L.M. Patnaik: Soft Comput. for Data Min. Appl., SCI 190, pp. 231–247.
springerlink.com

documents relevant to their interests as search engines throw same result to everyone irrespective of their expectation. Google's personalized search drives to make search experience more relevant to the user. Using personalized search, user can get the results most relevant to user, based on what user has searched in the past, view and manage user's past searches, including the web pages, images, Google results which he has clicked on and create bookmarks. Personalized search orders search results based on user's past searches, as well as the search results and news headlines the user has clicked on. User can view all these items in the search history and remove any items. Early on, user may not notice a huge impact on his search results, but as he builds up the search history, his personalized search results continues to improve.

This work tries to bridge this gap; a gap created between the search engine's capabilities and user expectations. By trying to learn from the previous behaviour of the user, it tries to readjust the generic results thrown by the search engine. The behaviour of the user is identified by the categories the user browses. Using snippets as the input, the proposed algorithms identify the category to which the site belongs. The algorithms make use of clustering concepts and Naïve Bayesian with a new flavor as a tool for categorization.

With Google becoming the default search engine and it giving away the Application Programming Interface (API) for development freely, it's natural that the present system takes Google results as its source. Alternatively, we can even parse the results retrieved for getting snippets, which actually form the source for identifying category. This chapter is a demonstration of effort that can be carried on to increase the usability and usefulness of a common Internet service like search engine.

11.2 Related Work

Conceptual search can be done by explicitly providing the meaning of the content in a Web page. So one way to address this problem is by having the authors of the content explicitly specify the meaning associated with a page using a knowledge representation language. One of the knowledge representation languages is Ontobroker and is discussed in [1]. Domain-specific Web search engines are effective tools for reducing the difficulty experienced when acquiring information from the Web. Building of a domain-specific search engine simply by adding domain-specific keyword, called "keyword spices," to the user's input query and forwarding it to a general-purpose Web search engine is presented in [2]. A tool that assists an end-user or application to search and process information from Web pages automatically by separating the primary content sections from the other content sections is presented in [3].

Ontology is a specification of a conceptualization. Sophisticated ontologies incorporate logical relationships and membership rules. However, concept hierarchies can also be used as simple ontologies. Use of Yahoo! categories as a concept hierarchy and classifying documents into it using an n-gram classifier is discussed in [4]. The user profiles are a representation of the user's interests. Systems such as Wisconsin Adaptive Web Assistant (WAWA) builds profiles non-invasively by observing the patterns of web usage of its users over a period of time [5]. They generally use the profile to suggest related Web pages to the users as they browse. The study of personalized recommendation in a B2C portal to build improved algorithm, EI-B&B-MDL, for learning Bayesian networks effectively and efficiently is proposed in [6]. The algorithm reduces the number of independence tests and database passes while effectively restricting the search space.

A personalized recommendation agent which is fuzzy cognitive in nature to give personalized suggestions based on the current users' preferences, general users' common preferences, and the experts' knowledge is given in [7]. Fuzzy cognitive agents are able to represent knowledge via extended fuzzy cognitive maps, to learn user's common preferences from most recent cases and to help customers to make inference/decisions through numeric computation instead of symbolic and logic deduction. An algorithm that can generate and display helpful links while users navigate a site can increase a Web site's usability and help Web designers and the user achieve their goals [8]. In the literature, data mining methods are very much exploited to build the customer profiles [9]. The 1:1 Pro system constructs profiles based on customers' transactional history. The system uses data mining techniques to discover a set of rules describing customers' behaviour and supports human experts in validating the rules.

The vision of ontology learning including a number of complementary disciplines that feed on different types of unstructured, semi structured, and fully structured data to support semiautomatic, cooperative ontology engineering is presented in [10]. A new method for tracking the dynamics of user interests from a minimal number of relevance judgments is proposed in [11].

11.3 Proposed System

Problem Formulation: Search engines are affected by problems such as ambiguity [12] and the results are ordered by web site popularity rather than the user interests. Natural language queries are inherently ambiguous. For example, consider a user query "Pluto". Due to ambiguity in the query terms, the user gets the results that are either related to astronomy or cartoon. Most users enter queries in just a word or two, without providing enough information. These short queries are often ambiguous. Providing little more information to a search engine on which to base its selection of the most relevant Web pages among millions is always preferred. But explicit detailed unambiguous query from the user cannot be expected always. A user profile that represents the interests of a specific user can be used to supplement queries and thus narrowing down the number of topics considered when retrieving the results. For the user in our example, if we know that he has a strong interest in Astronomy but little or none in cartoon, the Astronomy – related results of Pluto can be presented to the user preferentially. Therefore, user profile creation is an important step for personalization.

Our approach of building user profiles is based on the user's interactions with a particular search engine. For this purpose the tool GoogleWrapper, a wrapper around the Google search engine [13], which logs the queries, search results and clicks on a per user basis is used. The snippets, which is extracted using wrapper, forms the input for our algorithms to identify the category of that site. This information is then used to create user profiles and these profiles are used in a controlled study to determine their effectiveness for providing the personalized search results.

Information can be collected from users in two ways. First using explicit way, for example asking the user to fill up the forms when he logs in for the first time and ask him to give preferences or ratings. Secondly using implicit way, for example observing user behaviors such as the categories of the URLs visited, time spent and number of inner clicks made in a particular site. Explicit construction of user profiles has several drawbacks. The user provide inconsistent or incorrect information. The profile built is static whereas the user's interests may change over time, and the construction of the profile places a burden on the user that they may not wish to accept. User browsing history i.e., the categories the user has browsed so far are the most frequently used source of information about user interests. We use this information to create user profiles represented as user interest. Classifying the collected Web pages with respect to a category in which the user is interested creates the user profile. The fact that a user has visited a page and spent some time is an indication of user interest in the content of that page i.e., in that category.

Problem Definition: Consider a search engine E, which accepts queries in the form of keywords and returns a list of near relevant web pages. The queries are of the form k_1 [**(op)** k_i] for i= 2 to n, Where k_i is the ith keyword supplied to E, **(op)** is a Boolean operator like OR, AND..., n is the number of keywords supplied to E, and [...] indicate the parameters are optional.

Search engine E returns its results when an user clicks on some site and the corresponding snippet S is used as an input to the proposed algorithm to identify the category of the site. This category actually represents the users' interest and is added to the user profile. Each category has an associated weight. The weight decides the category of the results to be shown to the user. Each time a new category is created, it is assigned an initial weight $C_{Weight} = 0$. Hence all categories start with the same weight or same opportunity in the user profile. The weights for a category are dynamically updated. The dynamic updation is carried out to reflect the users' interest in a certain domain. The weight of the category also depends on the amount of time the user spends in a particular site and therefore in a particular category. A higher weight indicates a greater interest in that domain. When the weight for a category crosses a pre-defined lower threshold, it is deleted from the user profile. This reflects the fact that the user has lost interest in that area in the domain. The steps involved in the personalization system are given below.

Input: Query
Output: Personalized results

1. GoogleWrapper ( );
   {Gets the snippets of the retrieved results.}
2. PreprocessData ( );
   {Stop words are removed from Snippets.}
3. Choose (Classifier / Clustering)
   If (Cluster Analysis is chosen)
   {Create cluster ( ) : Clusters of similar sites are formed and sites within the cluster are ranked.}
   If (Classifier is chosen)
   {Naïve Bayesian Probabilistic model: Classify sites into their respective categories.}

System Architecture: The system consists of four modules, (i) Google Wrapper, (ii)Pre-processing, (iii) Identification of the category and (iv) Clustering/Classifying the URLs with snippets. The overall system architecture is shown in Figure 11.1. Once the query is passed to the system, the following steps are executed.

GoogleWrapper( ): Google wrapper takes query string as an input. Each time the user issues a search request to the Google service, a response is returned back to the user. This response contains many result elements, two of which are URL and Snippet, where,

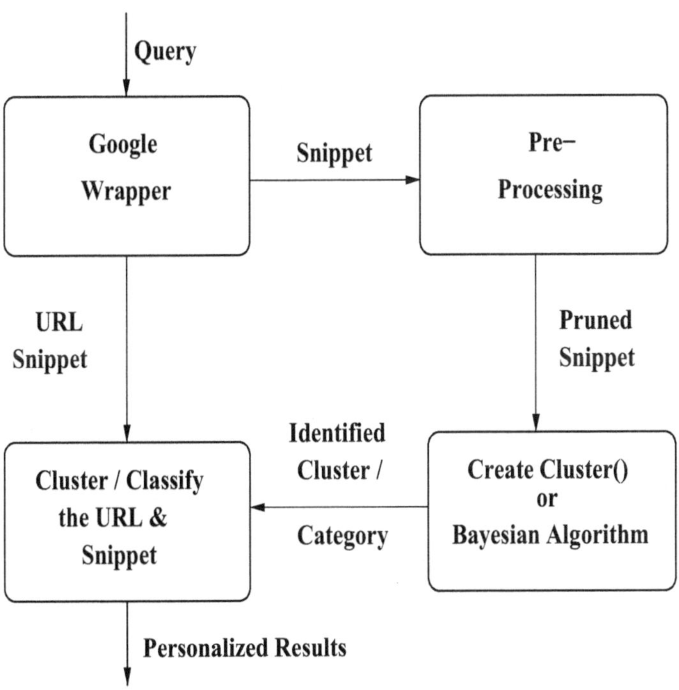

Fig. 11.1 Proposed System

<URL> - the Uniform Resource Locator(A Web Link) of the search result is returned as text, with an absolute URL path.

<Snippet> - A text excerpt from the results page that shows the query in context as it appears on the matching results page. This is formatted HTML and usually includes <B> tags within it. Query terms are highlighted in bold in the results and line breaks are included for proper text wrapping. If Google has searched for stemmed variants of the query terms using its proprietary technology, those terms are also highlighted in the snippet. This snippet is then preprocessed.

PreprocessData ( ): Here the snippet obtained from the Google wrapper is pruned. The stop words such as {a, is, the, are, in, as etc.,} stored in stop words list are removed from the snippet and forwarded to the algorithm.

Algorithm I: Create cluster( ): Find the similar sites using snippets and rank the sites within a cluster.
Input: Snippets of the results retrieved from the Search Engine.
Output: Clustered results.
Method:
Begin
$n \leftarrow$ *Number of Clusters;* $S \leftarrow$ *Snippet of the URL;*
$S_w \leftarrow$ *Stop words;* $S_p \leftarrow$ *Pruned Snippet;*
$W_p \leftarrow$ *Words in* S_p; $W_c \leftarrow$ *Words in the Cluster;*
$U_c \leftarrow$ *URLs in the Cluster;*
count $\leftarrow$ *0;*
If URL_is_clicked **then**
$S_p \leftarrow S - S_w$;// Prune step: remove stop words
If (n == 0) **then** // No clusters
Create cluster with W_p and Label it;
else
for *i*= 1 to *n*
if check_similarity (W_p, W_c) **then**
$U_c \leftarrow U_c$**U** URL; // add new URL
Rank = *count* / number of W_c;
$W_c \leftarrow W_c$**U** S_p; // add new words
endfor

Procedure check_similarity(W_p: Words in S_p; W_c : Words in the cluster)
for each W_p
for each W_c
if $(W_p \bigcap W_c)$ then // words match
count $\longleftarrow$ *count* + 1; // number of words matched
return TRUE (1);
return FALSE (0);
endfor
endfor
End

11.4 Example

Let us assume that a new user logs in for the first time and submits a query 'Mickey Mouse', many cartoon related sites of Mickey Mouse will be retrieved. Now if a user clicks on some URL, immediately the corresponding snippet is preprocessed by removing stop words (the, in, a, of, etc.,). With the remaining words (mickey, mouse, Disney, land, animation, cartoon) a cluster, say A_1 is created as shown in Table 11.1.

Now, when the same user submits a query 'Pluto' (which is ambiguous having two meaning, cartoon and planet), the retrieved sites are checked for similarity with the cluster A_1. Similarity is found considering the word as a binary variable i.e., if a word is found then it is 1 or else 0 as shown in Table 11.1. If the snippet of the site contains a single word or more that is available in cluster A_1, then it is said to be similar and shown to the user. Now the sites 1, 2, 5, 7, 8 and 9 are said to be similar and are clustered with A_1 and displayed to the user. The ranking (Rank = sort in decreasing order of weight) of these sites is based on the number of words matched to the total number of words in the cluster. Here, three cases are discussed:

Case 1: Yes, the results retrieved are as needed by the user.

- Increment the weight of that particular cluster,
  $C_{Weight} = C_{Weight} + 1$;
- Add additional words to the cluster.

Case 2: No, the results retrieved are not as per the users' need.

- Show the remaining results.
- Create a cluster using the snippet on which user clicks.

Case 3: Yes, the retrieved results are correct but the user also wants other results.

- Increment Case 1: Yes, the results retrieved are as needed by the user.
- Increment the weight of that particular cluster, $C_{Weight} = C_{Weight} + 1$;
- Add additional words to the cluster.

Table 11.1 A Relational table containing mostly binary attributes and site's weight

CLUSTER A1 Query: Pluto	Mickey	Mouse	Disney	Land	Animation	Cartoon	Similar? Y/N	Weight x
Site 1	0	0	1	0	1	1	Y	3/6=0.500
Site 2	0	0	1	1	0	1	Y	3/6=0.500
Site 3	0	0	0	0	0	0	N	0/6=0.000
Site 4	0	0	0	0	0	0	N	0/6=0.000
Site 5	0	0	1	1	1	1	Y	4/6=0.670
Site 6	0	0	0	0	0	0	N	0/6=0.000
Site 7	0	0	1	0	0	1	Y	2/6=0.340
Site 8	0	0	1	1	1	1	Y	4/6=0.670
Site 9	0	0	0	0	0	1	Y	1/6=0.167
Site 10	0	0	0	0	0	0	N	0/6=0.000

Case 2: No, the results retrieved are not as per the users' need.

- Show the remaining results.
- Create a cluster using the snippet based on the weight of that particular cluster, $C_{Weight} = C_{Weight} + 1$;
- Show the remaining results.
- Create a cluster using the snippet on which user clicks.

Case 1: User is interested in this cluster, so the weight of the cluster is incremented i.e., weight of cluster A_1 now becomes 2 which was initially 1 when created. Now let us say the user clicks on Site 2 containing the words Disney, Land, Cartoon and in addition to that Drawing, friend etc., and these additional words are also added to the cluster.

Case 2: User is not interested in this cluster, so there is no need of incrementing the weight of this cluster. Show the remaining sites retrieved for the same query 'pluto', i.e., sites 3, 4, 6, and 10. Now if the user clicks on some site, a cluster is created with the words from the corresponding snippet after pruning (say, Pluto, Planet, Fastest Sun, Smallest) and assign a new label say A_2 and assign weight 1.

Case 3: User shows interest in this cluster but also wants other results. In this case, increment the weight of cluster A_1, as in Case 1, then show the other remaining results and continue as in Case 2.

These weights of the cluster show the user's interest. But users may click erroneously on the sites although they are not interested in that site, providing some error weight to the cluster. To overcome this error, the weight of the cluster also depends on the *amount of time spent* by the user in a site falling in a particular cluster. If the user spends more time on a particular site it implies that he is more interested

Table 11.2 Training Data

Category (2 levels)	Keywords	Total
Arts > Entertainment	actors, animation, movies, games, literature, museum, radio, television, cartoons, characters.	10
Science > Astronomy	astrophysics, cosmology, sky, eclipses, planets.	5
Science > Biology	anatomy, biochemistry, biomechanics, bio-physics, biotech, ecology, evolution, genetics, taxonomy, zoology.	10
Science > Computer	artificial intelligence, genetic algorithms, graphics, software engineering, database	8
...	...	...

in that site. Assuming that the user does not spend more time in the sites in which he has entered erroneously, the *time-spent* attribute compensates the error.

For identifying the categories using naïve Bayesian probabilistic model, the training data used is as shown in the Table 11.2. The training data is obtained from yahoo directory search [14, 15] or Open Directory Project(ODP) [16]. ODP is readily available for download from their web site in a compact format. In addition, it is becoming a widely used as an informal standard. Only the top 2-level and 3rd level categories along with other related words are used as keywords for classification.

11.5 Algorithm II: Naïve Bayesian Probabilistic Model

Step 1: The data sample is Snippet $S = (w_1, w_2, ..., w_n)$, where, w_i for i = 1 ...n are the words in the snippet after removing stop words.

Step 2: Suppose that there are m categories $C_1, C_2, ..., C_m$. Given an unknown Snippet S (i.e., with no label), the classifier predicts that S belongs to the category of having highest probability. By Bayes Theorem, it is given that,

$$P(C_i/S) = P(S/C_i) * P(C_i)/P(S)$$

Step 3: $P(S)$ is constant for all categories, therefore only $P(S/C_i) * P(C_i)$ need to be maximized. If the prior probability of a category is not known, then it is assumed that categories are equally likely i.e., $P(C_1) = P(C_2) = ... = P(C_m)$. Therefore maximize only $P(S/C_i)$. Otherwise, maximize $P(S/C_i) * P(C_i)$. This even identifies the users interest in a particular category and the prior probability of the Category is estimated as,

$P(C_i)$ = (Number of times the user visited the category C_i / Total number of visits)

Step 4: Now, Naïve assumption of class conditional independence is made. That is, $P(S/C_i) = \Sigma_{k=1}^{n} P(S_k/C_i)$, Thus $P(w_1/C_i), P(w_2/C_i), ..., P(w_n/C_i)$ can be estimated from training samples, where $P(S_k/C_i)$ = (Number of words in C_i having value S_k / number of words in C_i).

Step 5: In order to classify an unknown snippet S, $P(S/C_i) * P(C_i)$ is evaluated for each category C_i, Sample snippet S is then assigned to the category C_i, if and only if $P(S/C_i) * P(C_i) > P(S/C_j) * P(C_j)$ for $1 \leq j \leq m$, $j \neq i$. In other words, it is assigned to the category C_i for which $P(S/C_i) * P(C_i)$ is maximum.

Observation: Our version of naïve Bayesian probabilistic model differs in computation of prior probability of a category $P(C_i)$ (Step 3). Traditional model estimates the prior probability of a category by $P(C_i)$ = (number of training samples in the category C_i/total number of training samples), instead we make use of user profile for computation of $P(C_i)$. We also classify the sites into the categories found in the *user profile* (Table 11.3) and not rest, since user would be most interested in those categories found in his profile. This reduces the cost of the model to a large extent because we are not classifying the sites into all the available categories. It

Table 11.3 User Profile

Category (C_i)	Weight (No. of visits)	P (C_i)
Science > Astronomy	2	2/10 = 0.2
Science > Computer	5	5/10 = 0.5
Science > Biology	3	3/10 = 0.3

only tries to classify into other categories when user is no more interested in these categories, which seldom happens. In that case $P(C_i)$ is not considered, i.e., only $P(S/C_i)$ is maximised. Here three cases exist:

Case 1: Yes, the results retrieved are as needed by the user.

- Increment the weight of that particular category, $C_{Weight} = C_{Weight} + 1$.

Case 2: No, the results retrieved are not as per users need.

- Classify the remaining results and display.
- Append the category of the site on which the user clicks and assign weight 1.

Case 3: Yes, the retrieved results are correct but user also wants other results.

- Increment the weight of that particular category, $C_{Weight} = C_{Weight} + 1$.
- Classify the remaining results and display.
- Append the category of the site on which user clicks and assign weight 1.

Example: Lets assume that a user has logged in and browsed 10 sites, 5 of which fall under Science > Computer, 3 of which come under Science > Biology and remaining 2 Science > Astronomy as shown in his profile (Table 11.3). Now if he submits a query say 'Genetics' (Ambiguous because genetics can be related to both genetic algorithms and genetics of biology), all the possible results of genetics is retrieved initially. Now we apply Naïve Bayesian algorithm to classify these results into categories found in the profile. Let us say there is an unknown snippet, which is to be classified, the following scenarios are possible.

- It has words w_i={genetics, powerful} after getting pruned.
- Number of training samples belonging to category Science>Astronomy(SA)=5
- Neither of the words w_i are found in training set of SA (so the query is not in this category).
- Number of training samples belonging to category Science>Biology(SB)=10
- This has only one word, genetics.
- P(genetics/SB) * P(SB) = (1/10) * 0.3 = 0.03
- Number of training samples belonging to category Science>Computer(SC)=8
- This has only two words, genetics and algorithm.
- P(genetics/SC) * P(SC) = (1/8) * 0.5 = 0.0625
- P(genetics/SC) * P(SC) > P(genetics/SB) * P(SB) therefore the snippet S belongs to SC i.e., (Science>Computer) category.

Hybrid Model: Algorithm II suffers with the overhead of storing training data and in Algorithm I we find the similarity of sites taking binary variables into consideration and it is not a probabilistic model. To combine the advantages of both the algorithms, we present a hybrid model which uses the clustering concepts as in algorithm I and instead of computing the similarity of the sites with cluster using binary variables, it uses Naïve Bayesian probabilistic model as in algorithm II.

11.6 Performance Analysis

Experiment 1: Average rank of the Google search results remain the same irrespective of user interest, where as the rank of the search results of the proposed technique is better because the system knows users' interest better. When user interest is zero i.e., the system does not know anything about the user, then the rank is same as the Google rank, say 5. As user interest approaches to one, i.e., the system is learning the user interest gradually, then the ranking improves to 3 as shown in Figure 11.2. Google pagerank works on the basis that if a website *abc.com* has been linked from an another website *xyz.com*, *abc.com* must have some good content and therefore Google counts the link from *xyz.com* as a vote for *abc.com*.

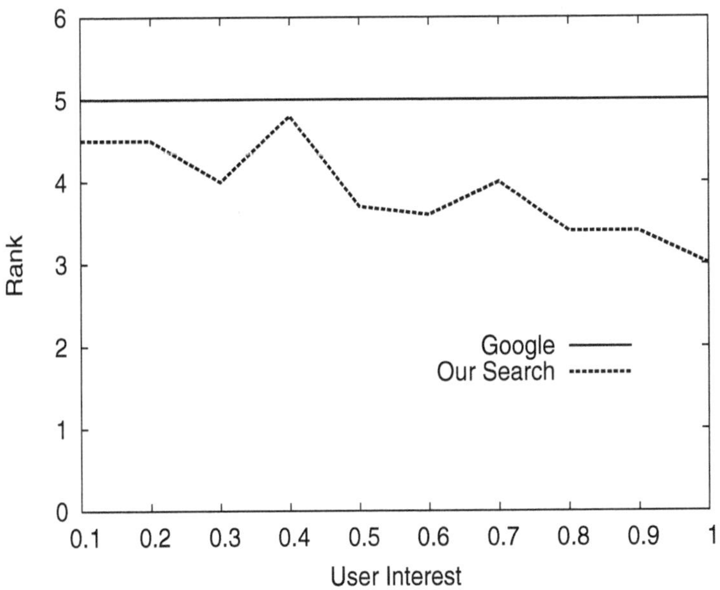

Fig. 11.2 Variation in ranking of Google results and results of the proposed technique as the user interest is varied.

Experiment 2: Precision is the percentage of retrieved documents that are in fact relevant to the query (i.e., "correct" responses). Precision of the system depends on the number of times the users use the system and the sequence in which the user proceeds.

$$\text{Precision} = |\{\text{Relevant}\} \cap \{\text{Retrieved}\}| / |\{\text{Retrieved}\}|$$

Figure 11.3 shows a comparison of the precision of the search results obtained from the proposed technique with that of Google. In Google the precision does not vary much, where as the precision of the proposed technique grows linearly as shown in Figure 11.3, if the flow of logins is as assumed below. Let us say the system retrieves 100 results every time and the query is Pluto.

Login 1: 20 relevant documents (user interested in cartoon) are found and user clicks on cartoons site. So precision is 0.2.

Login 2: Query is Pluto. Only cartoon sites are retrieved as user showed interest in cartoons last time. Now user does not want cartoons site, but wants planets. Therefore there are no relevant sites and the user opts for planets sites. So precision is 0.

Login 3: Query is Pluto again. This time the system retrieves both cartoons and planets because both categories have equal weight, and user is interested in any one of them, say user shows interest in cartoons, so cartoon site gets more weight. Precision is 0.5.

Login 4: Query is Pluto. System retrieves 75 cartoon results and 25 planet results based on weight. Users wants cartoon. Therefore precision is 0.75.

Login 5: Query is Pluto. System shows 80 results of cartoon and 20 of planets and the user wants cartoon sites. Hence precision is 0.8.

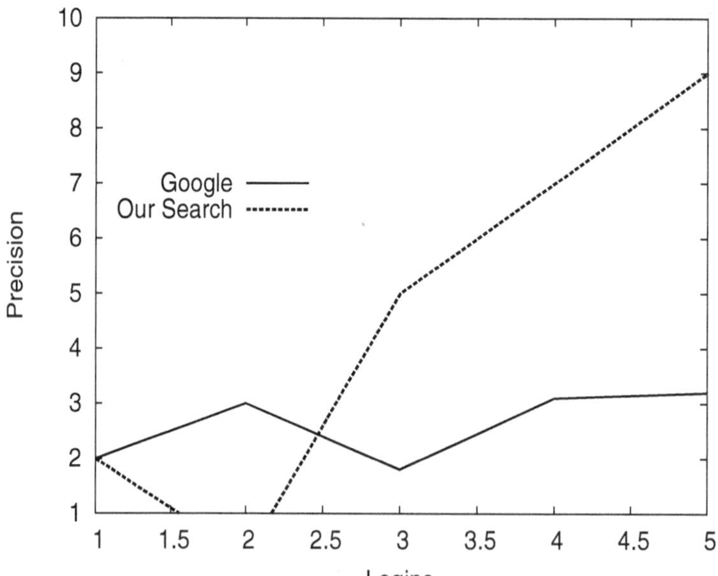

Fig. 11.3 Variation in precision of Google results and the search results of the proposed technique as the number of times user logs in i.e., as the system is used more

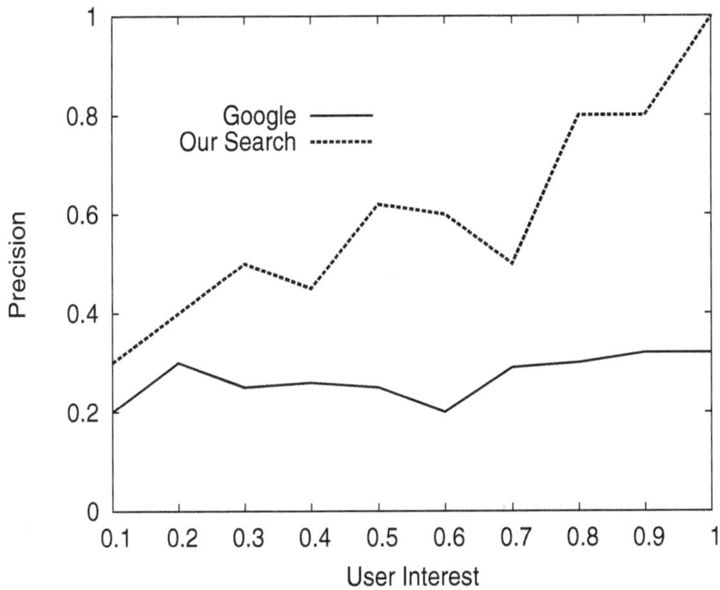

Fig. 11.4 Variation in precision of Google results and the proposed technique as the user interest is varied

Therefore, the precision of the results is function that is dependent on training the system.

Experiment 3: In this experiment one can see that the precision of the Google search results and the search results of the proposed technique vary when user interest changes. Figure 11.4 shows that google search results' precision does not change much whereas the precision of the search results of the proposed technique varies from 0.2 to 1 as user interest approaches one. Therefore, the precision increases as the system learns the user interest.

Experiment 4: Recall is the percentage of documents that are relevant to the query and are, in fact, retrieved. It is formally defined as

$$\text{Recall} = |\{\text{Relevant}\} \cap \{\text{Retrieved}\}| / |\{\text{Relevant}\}|$$

Figure 11.5 shows a variation in recall of Google results and of the proposed technique as the user interest is varied. Recall in case of Google does not change much as it does not depend on the user interest, whereas recall varies from 0.2 to 1 as the user interest reaches one in our case, which demonstrates the effeciency proposed technique.

Experiment 5: Figure 11.6 shows a comparison of recall between the two techniques. Recall in case of Google remains the same i.e., one, because every time the user logs in and submits a query, say pluto there is at least one result relevant to the

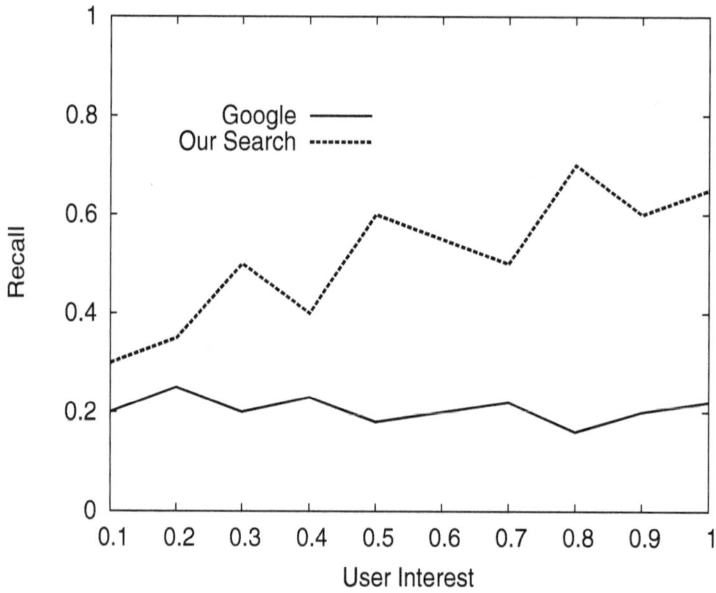

Fig. 11.5 Variation in recall of Google results and the proposed technique as the user interest is varied

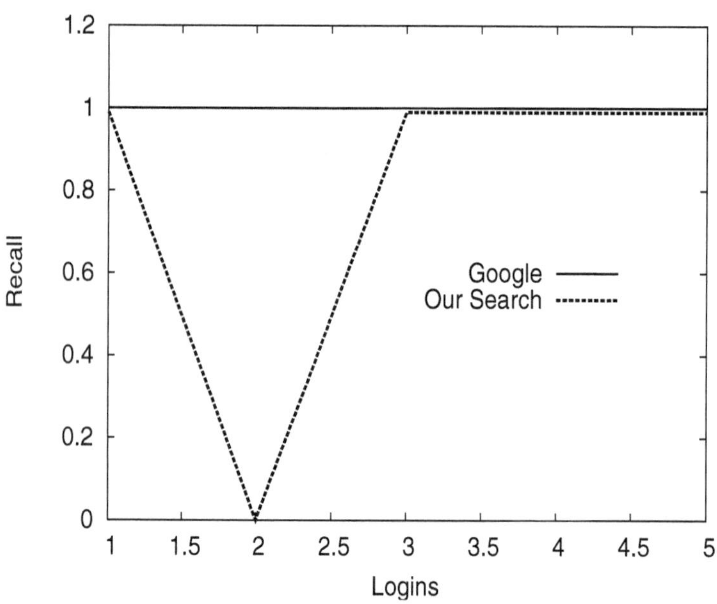

Fig. 11.6 Variation in recall of Google results and the proposed technique as the number of times user logs in i.e., as the system is used more

user among the retrieved. If it does not retrieve any result relevant to users interest then the recall becomes zero according to its definition. Where as the recall in our case is described with respect to the flow of logins as given below. Let us say the system retrieves 100 results every time and the query is Pluto.

Login 1: 20 relevant documents (user interested in cartoon) are found and user clicks on cartoons site. So recall is 1 because for recall to be 1, it is sufficient to have one relevant site of user's interest.

Login 2: Query is Pluto. Only cartoon sites are retrieved as user showed interest in cartoons last time. But now user is does not want cartoons site, he wants planets. Therefore no relevant sites and the user goes for planets sites. So precision is 0.

Login 3: Query is Pluto again. This time the system retrieves both cartoons and planets because both categories have equal weight, and user is interested in any one of them; say user shows interest in cartoons, so cartoon site gets more weight. Recall again becomes 1 as user gets what he wants.

Table 11.4 Comparison of Google search results with proposed technique for the query Pluto

Query: Pluto	
Top 10 results of Google search	
www.nineplanets.org	
www.solarviews.com	
dosxx.colorado.edu	
en.wikipedia.org	
www.plutobooks.com	
portals.apache.org	
pluto.jhuapl.edu	
www.pluto.net.nz	
www.plutorecords.com	
solarsystem.nasa.gov	
Top 10 results of our search	
User interest: Cartoons	User interest: Planets
disney.go.com	www.nineplanets.org
en.wikipedia.org	www.kidsastronomy.com
www.bcdb.com	www.nasm.si.edu
www.toonzone.net	www.brightsurf.com
www.melaman2.com	www.kidskonnect.com
www.barksbase.de	www.enchantedlearning.com
www.animationshow.com	www.studyworksonline.com
disneyshorts.toonzone.net	www.netmoon.com
www.imdb.com	www.solarviews.com
www.ultimatedisney.com	solarsystem.nasa.gov

Table 11.5 Comparison of Google search results with proposed technique for the query Genetic

Query: Genetic	
Top 10 results of Google search	
www.genetic-programming.org	
gslc.genetics.utah.edu	
www.geneticalliance.org	
www.aic.nrl.navy.mil	
www.savingsandclone.com	
www.genengnews.com	
www.genome.gov	
www.nsgc.org	
www.greenpeace.org	
www.genetichealth.com	
Top 10 results of our search	
User interest: Computers	User interest: Biotech
www.genetic-programming.org	www.icgeb.org
www.babicgroup.com	www.vivo.colostate.edu
www.rennard.org	www.i-biotechnology.co.uk
www.doc.ic.ac.uk	www.medbioworld.com
www.mathworks.com	www.cln.org
ai-depot.com	www.geneinfo.net
www.jaga.org	www.carolina.com/biotech/
www.aridolan.com	www.genetics.wisc.edu
www.nd.com	www.biotech.wisc.edu
www.codeproject.com	www.genengnews.com

Login 4: Query is Pluto. System retrieves 75 cartoon results and 25 planet results based on weight. Users wants cartoon. Recall is again 1.

Login 5: Query is Pluto. System shows 80 results of cartoon and 2 of planets and user wants cartoon sites. Recall is again 1.

11.7 Summary

In this chapter, a clustering algorithm is proposed, which creates clusters of similar sites, our version of Naive Bayesian probabilistic model, which classifies the sites into different categories and a hybrid model, which is a combination of both is used for classification of web pages. These algorithms are tested on Google's results and Google PageRank to obtain the top 10 results and the results obtained are good. Table 11.4 shows the Comparison of Google search results against the proposed technique for the query *Pluto*. Table 11.5 shows the comparison of Google search results against the proposed technique for the query *Genetic*.

References

1. Decker, S., Erdmann, M., Fensel, D., Studer, R.: Ontobroker: Ontology based Access to Distributed and Semi-Structured Information. In: Proceedings of W3C Query Language Workshop QL 1998 (1998)
2. Oyama, S., Kokubo, T., Ishida, T.: Domain-Specific Web Search with Keyword Spices, 1041-4347/04/2004 IEEE
3. Debnath, S., Mitra, P., Pal, N., Giles, C.L.: Automatic Identification of Informative Section of Web Pages, 1041-4347/05/2005 IEEE
4. Labrou, F.: Yahoo! as an Ontology: using Yahoo! Categories to Describe Documents. In: Proceedings of Eighth International Conference on Information and Knowledge Management, Kansas City, Missouri (1999)
5. Shavlik, J., Calcari, S., Eliassi-Rad, T., Solock, J.: An Instructable, Adaptive Interface for Discovering and Monitoring Information on the World Wide Web. In: Proceedings of International Conference on Intelligent User Interfaces, Redondo Beach, CA, pp. 157–160 (1999)
6. Ji, J., Liu, C., Yan, J.: Bayesian Networks Structure Learning and Its Application to Personalized Recommendation in a B2C Portal. In: Proceedings of the IEEE/WIC/ACM International Conference on Web Intelligence (WI 2004) (2004)
7. Miao, C., Yana, Q., Fang, H., Goh, A.: Fuzzy Cognitive Agents for Personalized Recommendation. In: Proceedings of the 3rd International Conference on Web Information Systems Engineering (WISE 2002) (2002)
8. Jenamani, M., Mojapatra, P.K.J., Ghose, S.: Online Customized Index Synthesis in Commercial Web Sites. IEEE Intelligent Systems, 20–26 (2002)
9. Gediminas, A., Alexander, T.: Using Data Mining Methods to Build Customer Profiles. IEEE Computer, 74–82 (2001)
10. Maedche, A., Staab, S.: Ontology Learning for the Semantic Web. IEEE Intelligent Systems (2001)
11. Widyantoro, D.H., Ioerger, T.R., Yen, J.: Tracking Changes in User Interests with a Few Relevance Judgments. In: CIKM 2003 (2003)
12. Krovetz, R., Croft, B.W.: Lexical Ambiguity and Information Retrieval. Proceedings of ACM Transactions on Information Systems 10(2), 115–141 (1992)
13. `http://www.google.com/api`
14. Labrou, Y., Finin, T.: Yahoo! As An Ontology – Using Yahoo! Categories To Describe Documents. In: Proceedings of the 8th International Conference On Information Knowledge Management (CIKM), pp. 180–187 (1999)
15. Yahoo!, `http://www.yahoo.com`
16. Open Directory Project, `http://dmoz.org`

Chapter 12
Mining Top - k Ranked Webpages Using SA and GA

Abstract. Searching on the Internet has grown in importance over the last few years, as huge information is invariably accumulated on the Web. The problem involves in locating the desired information and corresponding URLs on WWW. With billions of webpages in existence today, it is important to develop efficient means of locating the relevant webpages on a given topic. A single topic may have thousands of relevant pages and of varying popularity. Top - k ranked webpages pertaining to a given topic are of interest to the Web user. In this chapter, we propose an efficient top-k document retrieval method (*TkRSAGA*), that works on the existing search engines using the combination of Simulated Annealing and Genetic Algorithms. The Simulated Annealing is used as an optimized search technique in locating the top-k relevant webpages, while Genetic Algorithms helps in faster convergence via parallelism. Simulations are conducted on real datasets and the results indicate that *TkRSAGA* outperforms the existing algorithms.

12.1 Introduction

Data mining and web mining are emerging areas of immense interest for the research community. These two fields deal with knowledge discovery on the Internet. Extensive work is being carried out to improve the efficiency of existing algorithms and to devise new and innovative methods of mining the Web. Such efforts have direct consequences on e - commerce and Internet business models.

The Internet can be considered as a huge database of documents, which is dynamic in nature and results in ever-changing chaotic structure. Search engines are the only available interface between the user and web. It allows the user to locate the relevant documents in WWW. A huge number of webpages may exist on any given topic in the order of 10^4 to 10^6. It becomes tedious for the user to sift through all the web pages found by the search engine to locate the documents of interest to the user.

The problem of page ranking is common to many web-related activities. The basic goal of ranking is, providing relevant documents on a given search topic.

K.R. Venugopal, K.G. Srinivasa, L.M. Patnaik: Soft Comput. for Data Min. Appl., SCI 190, pp. 249–258.
springerlink.com

Top -k selection queries are being increasingly used for ranking. In top - k querying, the user specifies target values for certain attributes and does not expect exact matches to these values in return. Instead a ranked list of top - k objects that best match the attribute values are returned.

Simulated Annealing (SA) is powerful stochastic search method applicable to problems for which little prior knowledge is available. It can produce high quality solutions for hard optimization problems. The basic concept of SA comes from condensed matter of physics. In this technique, the system (solid) is first heated to a high temperature and then cooled slowly. The system settles in a minimum energy state if the cooling point of the system is sufficiently slow. This process can be simulated on a computer. At each step of the simulation, a new state of the system is generated from the current state giving a random displacement to a randomly selected particle. The new generated state is accepted as the current state, if the energy of the new state is not greater than that of the current state. If not, it is accepted with the probability, $e^{(-(E_{new-state}-E_{current-state})/T)}$, where E is the energy of the system and T is the temperature. This step can be repeated with a slow decrease of temperature to find a minimum energy state [1,2].

Another tested soft computing approach is Genetic Algorithms (GA), which works on the concept of evolution. Every species evolves in a direction suited for its environment. The knowledge they gain in this evolution is embedding in their chromosomal structure. The changes in chromosomes causes changes in the next generation. The changes occur due to mutation and crossover. Crossover means the exchange of parts of genetic information between parents to produce the new generation. Mutation makes it possible for chromosomes to get a structure that is suitable for the environment [3].

A combination of SA and GA is appropriate to the problems that place a premium on efficiency of execution, i.e., faster runtimes. This is an important consideration in any web-based problem as speed is of the utmost importance. The SA and GA techniques can be combined in various forms. GA can be applied before or after or even during the annealing process of the system under consideration [4].

Any page ranking algorithm has to be applied online and should be fast and accurate. The existing page ranking algorithms, though they give complete results, they produce enormous number of webpages resulting in low efficiency. The use of soft computing approaches can give near optimal solutions, which are better than existing algorithms. In this chapter, we combine Simulated Annealing with Genetic Algorithms to devise an efficient search technique. The Simulated Annealing is used because of its ability to handle complex functions and Genetic Algorithms is used to choose between the set of points in the intermediate states of Simulated Annealing, so as to eliminate the points that do not satisfy the fitness function. We thus achieve more accurate results with fewer runs of SA [5, 6].

Problem Definition: Given a query to a search engine, results in a large number of relevant web documents in terms of URLs (Uniform Resource Locators). Each

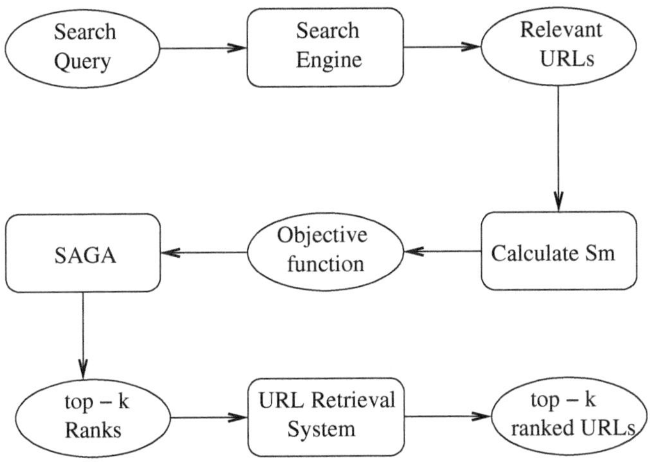

Fig. 12.1 The System Architecture

webpage is characterized by the number of hits(the number of times a URL has been accessed by past users), number of referrer pages(incoming links), number of referred pages(out going links) and the number of occurences of the specified keywords of the given query. Let E be the dataset containing the set of URLs and their corresponding characterstics, i.e., $E = \{U_m, S_m\}$, where $1 \leq m \leq n$ and n is the total number of URLs returned. The function $S_m = N_m + I_m + O_m + D_m$, where, N_m is the number of hits, I_m is the number of incoming links, O_m is the out going links and D_m is the number of occurences of query keywords for the corresponding m^{th} URL. Our objective is to find the top - k relevant web documents from the dataset E using combination of Simulated Annealing and Genetic Algorithms.

System Architecture: This section deals with the various modules involved in the system. The first step is to submit a query to a commonly used search engine. The query is a string or collection of strings that represent a set of keywords for a particular topic in which the search is being performed. Each string in the query is separated by a space or a special symbol. The query is represented as a set, $S = \{s_1, s_2, s_3, ...s_n\}$, s_k is the kth string in the query. The query is submitted to the search engine. Once the search engine completes the search process, it returns a set of n relevant unique web documents (URLs). It can be represented as the set, $E = \{U_m, S_m\}$ where $1 \leq m \leq n$. U_m actual address of mth URL in the result and S_m is the function on URL U_m.

The resulting URLs are categorized by their characterstic function S_m to ease the retrieval process. Once the search engine returns n relevant URLs, an objective function over S is generated using harmonic analysis. The algorithm TkRSAGA is executed on the objective function *f(x)* and outputs the top - k ranked URLs.

12.2 Algorithm *TkRSAGA*

Top - k Document Retrieval using Simulated Annealing and Genetic Algorithms:

Step 1: *Preprocessing:* Submit a query to an existing search engine like Google. The search engine returns a list of *n* URLs (webpages) of relevance to the topic. Each entry *E*, in the list of returned URLs must be composed of two entries {U, S}. Thus E = {U, S}, where *U* is the actual URL and *S* is the function over the corresponding to URL U and is denoted as $\{(U_1,S_1),(U_2,S_2),...(U_n,S_n)\}$.

Step 2: *Harmonic Analysis:* Let the output of Step 1 be denoted as $\{(n_1,s_1),(n_2,s_2),...(n_n,s_n)\}$, where n_m is the mth URL and s_m is the function over m^{th} URL and the objective function over these *n* points can be generated using the formula $f(x) = a_0 + a_k cos(k\pi) + b_k sin(k\pi)$, where $1 \leq k \leq n$.

Step 3: *Performing Search:* The combination of Simulated Annealing and Genetic Algorithms can be applied over the objective function *f(x)* as given below,

Generate initial states $\alpha_0, \alpha_1, ...\alpha_{n-1}$ at random.
Generate initial temperature T_0.
loop
  for each α_i in $\{\alpha_0, \alpha_1, ...\alpha_{n-1}\}$
  loop
    β_i = generate_state(α_i, T_j);
  until point α_i satisfies $curve \pm \varepsilon$, where ε is the error,
  if accept_state$(\alpha_i, \beta_i, T_j), then \alpha_i = \beta_i$,
  next α_i,
    for each i $0 \leq i \leq n-2$
      crossover_pairs(α_i, α_{i+1})
      α_i = calculatefitness (α_i, α_{i+1})
  next i,
  T_{j+1} = update_state(T_j),
  $j = j+1$;
until *k* states remain.

Let the initial states $\alpha_0, \alpha_1, ...\alpha_{n-1}$ be randomly chosen set of points from the objective function $f(\alpha)$, where $0 \leq \alpha_i \leq 2\pi$. The points α_i are chosen on the *x* - axis at evenly spaced intervals. However, the actual initial states are computed using the objective function f(x). The Simulated Annealing technique cools the system uniformly and slowly from a higher initial temperature T_0 to a lower final temperature $T_k (T_0 > T_k)$. In the next iteration, a random state is generated by the function generate_state(α_i, T_j) and is determined by the probability $G_{\alpha\beta}(T_j)$ of generating a new state β_i from an existing state α_i at temperature T_j. The generation function is defined as $g_i(Z) = 2 * (| +1/ln(1/T_j)) * ln(1 + ln(1/T_j))$. The generation probability is given by $G_j(Z) = \frac{1}{2} + (\Sigma z * ln(1+ | z | ln(1/T_j)))/2 * ln(1 + ln(1/T_j))$.

The newly generated state β_i is checked for acceptance by the function accept_state(α_i, β_i, T_j) and is determined by the probability $A_{\alpha\beta}(T_j)$ of accepting state β_i after it has been generated at temperature T_j. The acceptance probability $A_{\alpha\beta}(T_j)$ is given by, $A_{\alpha\beta}(T_j) = min1, exp(-(f(\beta) - f(\alpha))/T_j)$, where $f(\alpha)$ is the objective function considered for optimization. The new state β_i is accepted only if it has lower energy state than the previous state α_i.

The rate of cooling of the Simulated Annealing Technique(Annealing Schedule) is represented by ρ. It is a control parameter used to change the system temperature as the time progresses. The annealing schedule used in the algorithm is of the form, $T_k = T_0/e^{e^k}$, where k represents the kth iteration. For practical considerations, the annealing schedule is set to $T_{n+1} = \rho T_n$. The function update_state(T_j) updates the temperature with respect to the annealing schedule. The function crossover_pairs(α_i, α_{i+1}) performs the genetic crossover operation on states α_i and α_{i+1}. The random one-point crossover is carried on two states i and j.

Finally, the function calculate_fitness(α_i, α_{i+1}) performs the fitness calculation that is used to select the two states which are allowed to propagate to the next generation. The fitness function calculates the Euclidean distances of points α_i and α_{i+1} to the objective function *f(x)* and returns the closer point. Thus, the algorithm starts with few initial number of states and terminates with k final states.

Step 4: Once the algorithm returns k final states, they represent the points on the global minima over the objective function *f(x)*. These points can be mapped to the corresponding URLs and these URLs represent the top - k ranked URLs.

12.3 Performance Analysis

The algorithm TkRSAGA works in two basic phases. The first phase involves the generation of the Fourier coefficients to determine the objective function *f(x)* and is linear with respect to the number of URLs supplied. The second phase is the application of combined SA and GA on the objective function *f(x)* to obtain the top-k ranked URLs. The convergence of the second phase depends on the number of initial states, the annealing schedule and the initial temperature. Keeping these parameters constant for the test runs, we see that the performance curve for TkRSAGA tends to be linear. The execution time is higher for smaller number of URLs and relatively lower for larger URLs. The graph of execution time versus the number of URLs for the algorithms TkRSAGA and HITS is shown in Figure 12.2. It shows that the algorithm TkRSAGA works better for larger databases.

The Figure 12.3, shows the graph of execution time versus the number of initial states and the performance curve is roughly logarithmic. As the number of initial number of states increases by a factor x, the execution time increases by a factor of $log(2x)$. This is obvious since, the initial states only influence the number of iterations made by GAs. After every crossover operation, exactly half the new generation is retained for future propagation. The graph in Figure 12.4, shows the execution time versus the desired top -k ranks. The graph is plotted for varying number of

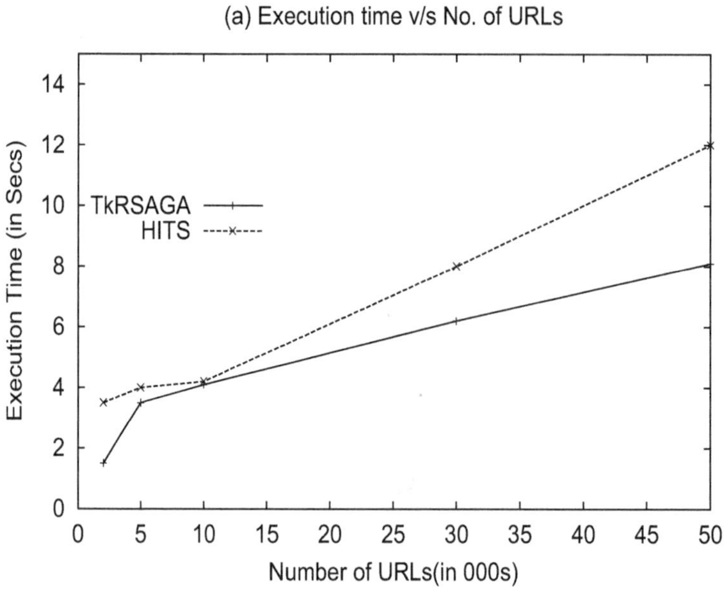

Fig. 12.2 The graph of execution time versus number of URLs (Number of initial states = 128)

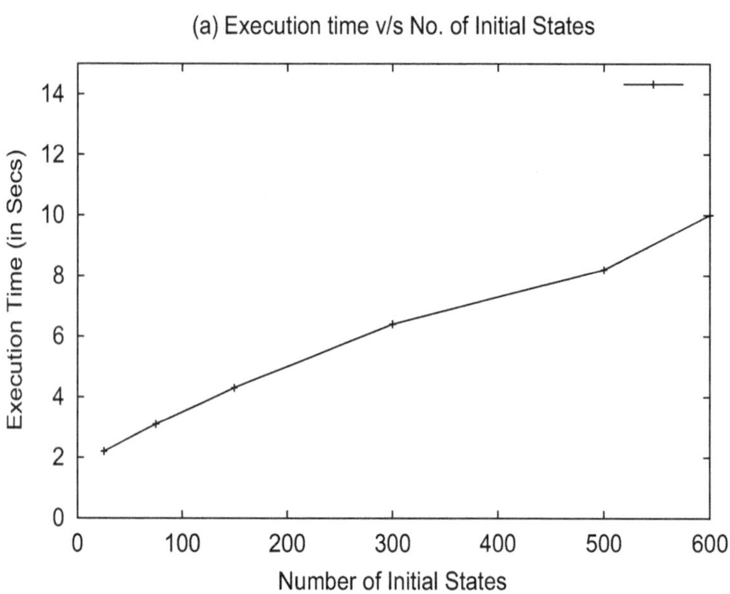

Fig. 12.3 The graph of execution time versus number of initial states (Number of URLs = 10,000); for, Annealing schedule $(\rho) = 0.95$, k = 4

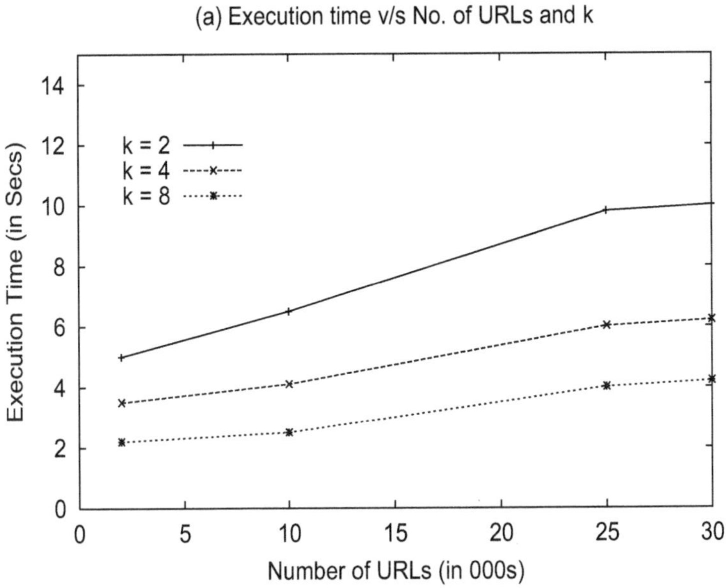

Fig. 12.4 The graph of execution time versus varying number of URLs and k (Number of initial states = 128 and Annealing schedule $(\rho) = 0.95$)

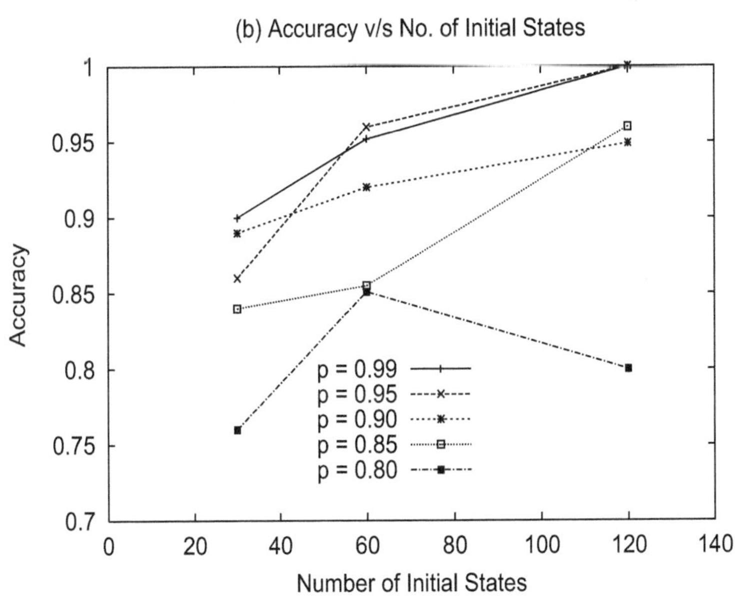

Fig. 12.5 The graph of Accuracy versus number of initial states (Number of URLs = 10,000 and k = 4);

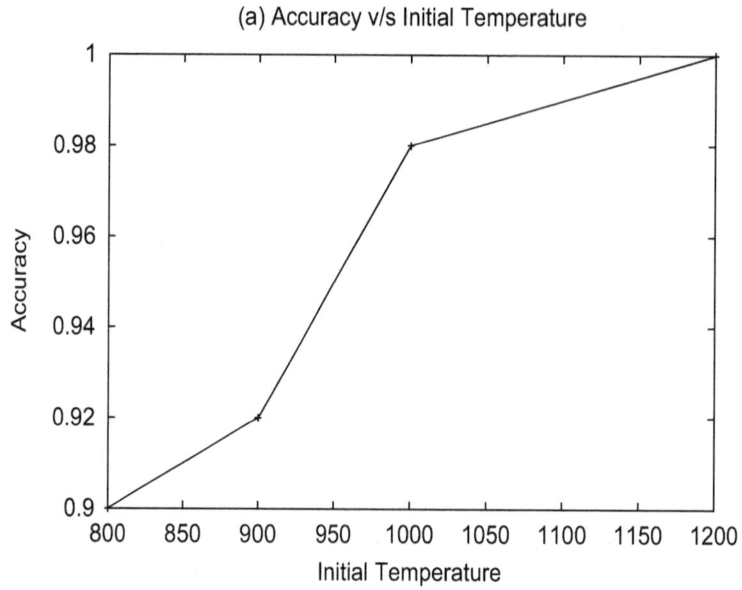

Fig. 12.6 The graph of accuracy versus initial temperature

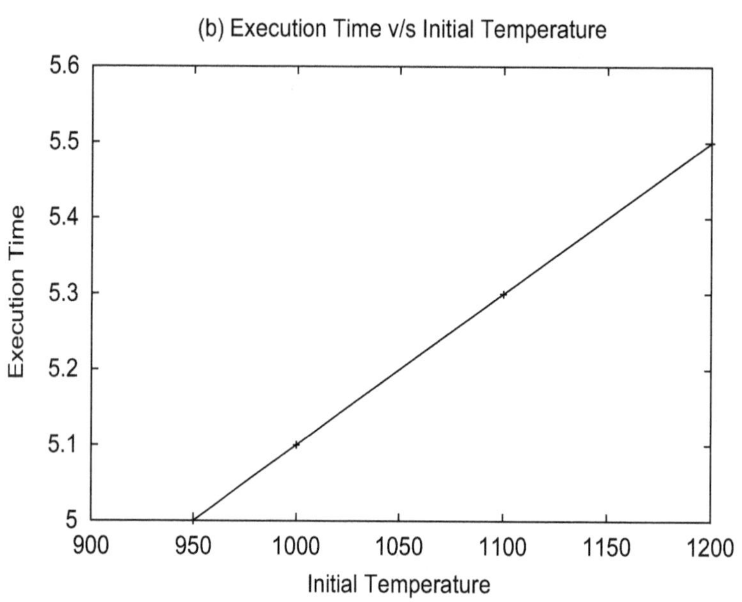

Fig. 12.7 The graph of execution time versus initial temperature; for, (Number of initial states = 128, Annealing schedule $(\rho) = 0.95$, Number of URLs = 10,000 and k = 4)

URLs and varying k. Since the number of iterations increases for lower values of k, the curve is logarithmic.

Figure 12.5, shows the graph of accuracy of the retrieved top-k documents versus varying annealing scheduling (ρ) and the initial number of states. The accuracy parameter defines the ratio of the number of top-k ranks returned by the TkRSAGA to the desired top-k. The accuracy increases with the number of iterations. For higher values of initial states, better results are obtained. This is because the GAs produce the generations satisfying the fitness function. Similarly, for higher annealing schedules, the accuracy increases as SA performs more number of iterations in search of global optima.

The initial temperature T_0 determines the temperature of the system as it starts cooling. The higher the temperature, the more time it takes the system to reach the lower equilibrium state, i.e., the algorithm performs more number of iterations and takes longer time to reach the final k states. However, the number of iterations is directly proportional to the number of intermediate states being generated. Therefore, more the number of intermediate states, higher the accuracy and hence generates accurate k final states. Thus, there exists a tradeoff between execution time and accuracy of results obtained, based on the initial temperature T_0. Figure 12.6, depicts the graph of initial temperature versus accuracy. Therefore, as the initial temperature increases accuracy increases, inturn increasing the execution time. Figure 12.7, shows the linear relationship between the initial temperature and the execution time.

Experiments on real datasets: The datasets of university link files from *cs.wlv.ac.uk* are used for our experiments. A set of n webpages and corresponding number of hits are available. The number of hits is used to compute the harmonics for the objective function $f(x)$. The output of TkRSAGA is a set of k values representing the

Table 12.1 Sample URLs taken from *cs.wlv.ac.uk*

URL(U_m)	No. of hits(N_m)
www.canberra.edu.au/UCsite.html	25482
www.canberra.edu.au/secretariat/council/minutes.html	1501
www.canberra.edu.au/Staff.html	199950
www.canberra.edu.au/Student.html	218511
www.canberra.edu.au/crs/index.html	178822
www.canberra.edu.au/uc/privacy.html	15446
www.canberra.edu.au	258862
www.canberra.edu.au/uc/convocation/index.html	16702
www.canberra.edu.au/uc/staffnotes/search.html	38475
www.canberra.edu.au/uc/search/top.html	190852
www.canberra.edu.au/uc/help/index.html	156008
www.canberra.edu.au/uc/directories/index.html	6547
www.canberra.edu.au/uc/future/body.html	25006
www.canberra.edu.au/uc/timetable/timetables.html	257899
www.canberra.edu.au/uc/hb/handbook/search.html	54962

Table 12.2 The output of *TkRSAGA* and HITS for (T_0 = 1200, No. of Initial States = 256, $(\rho) = 0.95$, $k = 4$)

RANK	*TkRSAGA*	HITS
1	www.canberra.edu.au	www.canberra.edu.au/uc/timetable/timetables.html
2	www.canberra.edu.au/uc/timetable/timetables.html	www.canberra.edu.au
3	www.canberra.edu.au/Student.html	www.canberra.edu.au/Student.html
4	www.canberra.edu.au/Staff.html	www.canberra.edu.au/Staff.html

top-k relevant webpages. These values are mapped to the URLs to obtain the actual addresses. The HITS [6] algorithm is executed on the same database and the results of TkRSAGA and HITS algorithm are compared. The Table 12.1 shows the list of URLs and their corresponding number of hits. Table 12.2 shows the outputs of both TkRSAGA and HITS. The outputs of both the algorithms are same and our algorithm TkRSAGA executes much faster than HITS algorithm. From Table 12.2, we can conclude that TkRSAGA outperforms the HITS in execution time without compromising with the accuracy of the results obtained.

12.4 Summary

In this chapter, we have proposed an efficient algorithm TkRSAGA, for mining top-k ranked web documents using the combination of Simulated Annealing and Genetic Algorithms. The ability of SA to solve harder problems and the combination of GA to reduce the number of iterations of SA and the inherent parallelism has made the algorithm efficient and effective.

References

1. Yao, X.: Simulated Annealing with Extended Neighbourhood. International Journal of Computer Mathematics 40, 169–189 (1991)
2. Yao, X.: Optimization by Genetic Annealing. In: Proc. Second Australian Conference on Neural Networks, Sydney, pp. 94–97 (1991)
3. Srinivas, M., Patnaik, L.M.: Genetic Algorithms: A Survey. IEEE Computer, 17–26 (1994)
4. Szu, H.H., Hartley, R.L.: Fast Simulated Annealing. Physics Letters A 122, 157–162 (1982)
5. Ingber, L.: Very Fast Simulated Re-Annealing. Mathl. Comput. Modelling 12(8), 967–973 (1989)
6. Kleinberg, J.M.: Authoritative Sources in a Hyperlinked Environment. In: Proceedings of ACM-SIAM Symposium on Discrete Algorithms (1998)

Chapter 13
A Semantic Approach for Mining Biological Databases

Abstract. A variety of biological databases are currently available to researchers in the XML format. Homology-related querying on such databases presents several challenges, as most available exhaustive mining techniques do not incorporate the semantic relationships inherent to these data collections. This chapter identifies an index-based approach to mining such data and explores the improvement achieved in the quality of query results by the application of genetic algorithms. Our experiments confirm the widely accepted advantages of index and vector-space based model for biological data and specifically, show that the application of genetic algorithms optimizes the search and achieves higher levels of precision and accuracy in heterogeneous databases and faster query execution across all data collections.

13.1 Introduction

Research in molecular biology and molecular medicine has accumulated enormous amount of data. This includes genomic sequences gathered by the human genome project, gene expression data from micro array experiments, protein identification and quantification data from proteomics experiments, and SNP data from high-throughput SNP arrays, etc.. There have been perseverant attempts at generating knowledge discovery and data mining models that are suitable to biological systems. But very few such models take into account the underlying complexity of biological data. Exhaustive search techniques tend to treat biological data as any other collection of terms or abstracts, while semantically inclined techniques apply heuristic rules, tested on ordinary data. Mining biological data needs a more comprehensive model that aims at making such data relevant and quickly accessible to researchers [1].

Biological databases refer to structured or non-structured collection of specimens (DNA, cells, and tissues) or information of a genetic or proteomic nature from a variety of sources - including medical and other health files, and genealogical, socioeconomic, and environmental information- which is stored electronically as a single entity or as part of a larger database. Biological databases are created and managed

K.R. Venugopal, K.G. Srinivasa, L.M. Patnaik: Soft Comput. for Data Min. Appl., SCI 190, pp. 259–278.
springerlink.com

in a constantly evolving, research-focused setting. Most large biological databases are published and available on IP networks and on the World Wide Web in the Extended Markup Language (XML) Format. Examples include RCSB Protein Data Bank and the InterPro PROSITE Database of the European Bioinformatics Institute.

This chapter deals with the storage of vast amounts of information related to three example biological databases- Protein Structure, Medical Abstracts and Carbohydrate Sequences and the issues related to mining necessary information from such databases. The database of Medical Abstracts has been previously tested for semantic index-based mining techniques. However, in our work, after having established the semantic approach to mining, we test for the optimization achieved by the application of genetic algorithms.

13.2 Understanding the Nature of Biological Data

XML that complements HTML promises to increase the benefits that can be derived from the wealth of information found today on IP networks around the world. This is because XML provides a uniform method for describing and exchanging structured data [2]. The ability to describe structured data in an open text-based format and deliver this data using standard HTTP protocol is significant for two reasons:

- XML facilitates more precise declarations of content and more meaningful search results across multiple platforms.
- Secondly, once the data is located, XML enables a new generation of viewing and manipulating the data [3].

With respect to biological databases available in electronic format, specifically in XML, a large number of stakeholders are involved in various capacities and on many levels: the human subjects affected by genetic research, the researchers and research community, healthcare providers, research sponsors, regulating bodies, users of the results, special interest groups, and the media. It becomes imperative then to allow these diverse and possibly naïve end users to search available databases in an efficient manner.

Consider for example a database containing listings and details of proteins. Proteins are found in every cell and are essential to every biological process, protein structure is very complex: determining a protein's structure involves first protein sequencing - determining the amino acid sequences of its constituent peptides; and also determining what conformation it adopts and whether it is complexed with any non-peptide molecules. Discovering the structures and functions of proteins in living organisms is an important tool for understanding cellular processes, and allows drugs that target specific metabolic pathways to be invented more easily. A tremendous amount of research has already mapped, and continues to map millions of protein sequences. This information has a wide range of applications, both academic and commercial. A snapshot of such an XML collection is shown in Figure 13.1. The most important issue in the interpretation of such data is the understanding the functional associations between data fields. Previously, several

```
<protein>
  <name> 104 kDa microneme–rhoptry antigen precursor </name>
  <name> p104 </name>
</protein>
<gene> <name type = "ORF"> TAO8425 </name>   </gene>
<organism key = "1">
        <name type = "scientific"> Theileria annulata </name>
<dbReference type = "NCBI Taxonomy" id = "5874" key = "2" />
<lineage>
          <taxon> Eukaryota </taxon>
          <taxon> Alveolata </taxon>
          <taxon> Apicomplexa </taxon>
          <taxon> Piroplasmida </taxon>
          <taxon> Theileriidae </taxon>
          <taxon> Theileria </taxon>
</lineage>
</organism>
```


Fig. 13.1 Snippet of Protein Sequence Database

approaches have been described to extract relationships from various biological databases using term-matching methods. However, more flexible automated methods are needed to identify functional relationships, both explicit and implicit.

Previous works have explored the utility of Latent Semantic Indexing (LSI), a vector space model for information retrieval, to automatically identify conceptual gene relationships from titles and abstracts [4]. Latent Semantic Analysis (LSA) is a theory and method for extracting and representing the contextual-usage meaning of words by statistical computations applied to a large corpus of text [5]. The underlying idea is that the aggregate of all the word contexts in which the existence of a given word provides a set of mutual constraints that largely determines the similarity of meaning of words and sets of words to each other. The adequacy of LSA's reflection of human knowledge has been established in a variety of ways. It has been found that LSI identifies gene-to-gene and keyword-to-gene relationships with high average precision. In addition, LSI identified implicit gene relationships based on word usage patterns in the gene abstract documents. Also, it has been shown that pair-wise distances derived from the vector angles of gene abstract documents can be effectively used to functionally group genes by hierarchical clustering. These findings provide proof-of-principle that LSI is a robust automated method to elucidate both explicit and implicit gene relationships from the biomedical literature. These features make LSI particularly useful for the analysis of novel associations discovered in genomic experiments.

The use of Genetic Algorithms(GA) helps in improving the precision and the query execution time. A GA is a search technique used in computer science to find approximate solutions to optimization and search problems. Genetic algorithms are a particular class of evolutionary algorithms that use techniques inspired by evolutionary biology such as inheritance, mutation, natural selection, and recombination. Genetic algorithms are typically implemented as a computer simulation in which a population of abstract representations of candidate solutions to an optimization problem evolves toward better solutions. Traditionally, solutions are represented in binary as strings of 0s and 1s, but different encodings are also possible. The evolution starts from a population of completely random individuals and happens in generations. In each generation, the fitness of the whole population is evaluated, multiple individuals are stochastically selected from the current population, modified to form a new population, which becomes current in the next iteration of the algorithm.

13.3 Related Work

Ongoing research in data mining of biological databases encompass a range of activities along a continuum, from the formative, theoretical development of new algorithms, data structures and tools specific to the management of biological information, through the development of new information resources to the enhancement of established resources needed by whole communities of biological researchers. Examples include theoretical research on data structures; new database architectures more tuned to the complexity of biology; planning and prototype development of new types of biological data- or knowledge-bases; and design of easy-to-use interfaces and tools for data input, manipulation, analysis and extraction.

Some of the commercial search applications available like BIOSIS and EMBASE use unique indexing systems that allow flexible searching over biological databases. Other related approaches being continually explored and applied include:

- Development of new I/O scheduling techniques for accessing disk-resident biological data, using the developed tools for genome comparisons (e.g., human vs. mouse) and shotgun assembly, and extending the above pair-wise comparison techniques to multiple alignment of genomes.
- Building of new data models for providing seamless access to heterogeneous data sources such as strings, structures, and pathways.
- Supporting interactive queries and efficient access to distributed datasets through the use of prediction and statistics to identify meaningful data.
- Development of new algorithms for comparison of pathways. Use these algorithms for construction of phylogenetic trees based on pathways. Develop techniques for combining information about pathways with sequences and structures. Realize the information about pathways and use these models to make predictions.

Analysing data from biological databases often requires the consideration of data from multiple relations rather than from one single table. Recently, approaches are being studied that utilize multi-relational data and yet meet the efficiency requirements of large-scale data mining problems. It is difficult and requires profound understanding of both knowledge discovery and computational biology to identify problems and optimization criteria which, when maximized by knowledge discovery algorithms, actually contribute to a better understanding of biological systems. Identification of appropriate knowledge discovery problems and development of evaluation methods for knowledge discovery results are ongoing efforts.

13.4 Problem Definition

Given a substantially large XML file, containing genomic or other biological data, it is required to propose a methodology for mining information from the file in a fast and accurate manner. Specifically, the search methodology must incorporate semantic knowledge related to the nature of the data stored in order to produce the required results. The problem at hand can be expressed in three steps as follows:

1. To identify a suitable method to index the data contained in the XML database - The indexing must take into consideration the underlying relationships between terms in the biological data.
2. To identify a search technique to query the index.
3. To test the performance of the search technique and to improve accuracy and precision, possibly with the use of GAs.

Biological data is currently stored in the electronic medium in a plethora of different data formats. Several technologies have been proposed to integrate data stored in different formats, or spread across different databases. This chapter restricts itself to biological data available in the XML format. Further, rather than the integration of data, the performance studies are concentrated on mining for search terms within a well-defined but difficult-to-query database.

13.5 Identifying Indexing Technique

Genomic and biological databases are most frequently queried for homology searching. The size of these databases is growing exponentially, given the intensive ongoing research in molecular biology. In searching such data collections, it is desirable to use inherent semantic relations between data fields. It has been experimentally shown that this requirement is best fulfilled by index-based approach. Moreover, index-based searching has been proven to be computationally less expensive and more, accurate than existing exhaustive search schemes [6].

Given that we are dealing with XML documents, we must also explore the possibility of using standard term-search techniques over such data collections. However, such techniques can at best return links to documents and not to specific fragments thereof. This is problematic, since large XML documents may contain thousands of elements storing many pieces of information that are not necessarily related to each other. For example, in a pattern entry in a protein database, a date entry is associated with each of the pattern IDs registered into the database on that date but not with other pattern IDs. Actually, if a search engine simply matches the search terms against the documents, it may return documents that do not answer the user's query. This occurs when distinct search terms are matched to unrelated parts of an XML document as is likely in a database that contains such information as protein patterns. The problem arises since the document is treated as an integral unit. Since a reference to a whole XML document is usually not a useful answer, the granularity of the search should be refined. Instead of returning entire documents, an XML search engine should return fragments of XML documents. It is not possible to present queries that explicitly refer to XML tags. Hence, it is difficult, and sometimes even impossible to formulate a search query that incorporates semantic knowledge in a clear and precise way [7,8].

It becomes imperative therefore, to identify an indexing technique that not only overcomes the drawback of simplistic XML indexing schemes but also satisfies the current requirement of a scheme that recognizes heuristic rules in biological data collections. Essentially, various search engines vary with regard to the indexing methodology they use to create and maintain a repository of documents and the search algorithm they employ to generate a response to a query. The repository maintained is implicitly an estimation of the size of the solution space itself. Thus the index of a large database can itself be large enough to adversely affect the performance of the search algorithm, given the main memory constraints [9].

It is widely accepted that vector-space approaches to data mining enables the user to search for concepts rather than specific terms and rank the results of the search according to their relative match to the original query. There are several vector-space approaches currently in use in various commercial and non-commercial search engines. LSI is one such vector-space approach. Experiments have shown that this technique can achieve up to 30% better retrieval performance than lexical searching techniques by employing a reduced-rank model of the term-document space. The vector-space approach represents the information items and the query as vectors, and it commonly uses the angle between the information item vectors and the query vector to determine their similarity [10].

LSI has been widely documented as an advanced vector-space based approach. According to Letsche and Berry, the LSI information retrieval model builds upon the prior research in information retrieval and, using the Singular Value Decomposition (SVD) to reduce the dimensions of the term-document space. LSI explicitly represents terms and documents in a rich, high-dimensional space, allowing the underlying semantic relationships between terms and documents to be exploited during searching. LSI relies on the constituent terms of a document to suggest the document's semantic content. However, the LSI model views the terms in a document as somewhat unreliable indicators of the concepts contained in the document. It assumes that the variability of word choice partially obscures the semantic structure of the document. By reducing the dimensionality of the term-document space, the underlying, semantic relationships between documents are revealed, and much of the noise like differences in word usage, terms that do not help distinguish documents, etc., is eliminated. LSI statistically analyses the patterns of word usage across the entire document collection, placing documents with similar word usage patterns near each other in the term-document space, and allowing semantically related documents to be near each other even though they may not share terms [11].

The application of this approach to biological data sets appears intuitively apposite. Given the nature of querying on such data sets, an approach that statistically analyses usage patterns and relies on semantic relations is likely to aid the search process.

13.6 LSI Model

In the LSI model, terms and documents are represented by an *m by n* incidence matrix ***A***. Each of the *m* unique terms in the document collection are assigned a row in the matrix, while each of the *n* documents in the collection is assigned a column in the matrix [4]. A non-zero element a_{ij}, where $A = [a_{ij}]$, indicates term *i* occurs in document *j*, and also the number of times the term appears in that document. Since the number of terms in a given document is typically far less than the number of terms in the entire document collection, *A* is usually very sparse.

LSI typically uses both a local and global weighting scheme to increase or decrease the relative importance of terms within documents and across the entire document collection, respectively. The product of the local and global weighting functions is applied to each non-zero element of *A* is given by a_{ij} = L(ij) * C(i), where L_{ij} is the local weighting function for term *i* in document *j* and C(i) is the global weighting function for term *i*.

Once the *m by n* matrix *A* has been created and properly weighted, a rank-*k* approximation to *A*, where, $k << \min(m,n)$, is computed using SVD. The SVD of the matrix A is defined as the product of three matrices, A=U$\sum V^T$, where the columns of *U* and *V* are the left and right singular vectors, respectively, corresponding to the monotonically decreasing in value diagonal elements of $\sum$ which are called the singular values of the matrix *A*. The first *k* columns of the *U* and *V* matrices and the first (largest) *k* singular values of *A* are used to construct a rank-*k* approximation to *A* via the above product definition. The columns of *U* and *V* are orthogonal, such that, $U^T U = V^T V = I_r$, where *r* is the rank of the matrix *A*. A theorem due to Eckart and Young suggests that A_k, constructed from the *k*-largest singular triplets of *A* is the closest rank-*k* approximation to *A*.

In the LSI model, queries are formed into pseudo-documents that specify the location of the query in the reduced term-document space. Given *q*, a vector whose non-zero elements contain the weighted term-frequency counts of the terms that appear in the query, the pseudo-document, $q*$ can be represented by $q* = q^T \mathrm{U}_k \Sigma_k^{-1}$. Thus, the pseudo-document consists of the sum of the term vectors $(q^T U_k)$ corresponding to the terms specified in the query scaled by the inverse of the singular values (Σ_k^{-1}). The singular values are used to individually weight each dimension of the term-document space.

Once the query is projected into the term-document space, one of several similarity measures can be applied to compare the position of the pseudo-document to the positions of the terms or documents in the reduced term-document space. One popular similarity measure, the cosine similarity measure, is often used because, by only finding the angle between the pseudo-document and the terms or documents in the reduced space, the lengths of the documents, which can affect the distance between the pseudo-document and the documents in the space, are normalized. Once the similarities between the pseudo-document and all the terms and documents in the space have been computed, the terms or documents are ranked according to the results of the similarity measure, and the highest-ranking terms or documents, or all the terms and documents exceeding some threshold value, are returned to the user [12,13].

13.7 Search Optimization Using GAs

The genetic algorithm involves three basic types of operators: selection, crossover and mutation. *Selection* operator selects chromosomes in the population for reproduction. The better the chromosomes, more often are they likely to be selected to reproduce. *Crossover* operator randomly chooses a locus and exchanges the subsequences before and after that locus between two chromosomes to create two offspring. For example, the strings 10000100 and 11111111 could be crossed over after the third locus in each to produce the two offspring 10011111 and 11100100. The crossover operator roughly mimics biological recombination between two single chromosome (haploid) organisms. Finally, *Mutation* operator randomly flips some of the bits in a chromosome. For example, the string 00000100 might be mutated in its second position to yield 01000100. Mutation can occur at each bit position in a string with some probability, usually very small (e.g., 0.001). The general process of GAs is given below.

1. Randomly select initial population of term documents in the term-document space.
2. Evaluate the fitness function for each of the term documents in the population.
3. Generate a *mating pool* from the population where the probability that a term-document is selected is proportional to its fitness.
4. Randomly select a number of points for mating (crossover) and mutation.
5. Carry out these operations deleting the original term documents on which they are performed to form a new generation.
6. Stop if the termination criterion has been satisfied and adopt the fittest term-document as the solution. Otherwise goto step (2).

Problem Formulation: There are two basic steps to representing a problem so a GA can be applied.

(*i*) *Encoding Solutions:* Encode the solution space as strings of a fixed length over some fixed alphabet.
(*ii*) *Fitness Function:* Determine a suitable function that takes a string as input and returns a number that is a measure of that solution's performance in solving the problem. The definition of a chromosome is represented as:

$$J = (j_1, j_2, ..., j_i, ..., j_L)$$

Where j_i denotes the weight of the tag i and L is the number of tags to be considered. Each gene represents a tag weight. The genes of initial chromosomes are generated randomly and the range of weight values is from 0.0 to 8.0 for experiments. The fitness function used is given by:

$$\text{Fitness} = \alpha\ (p) + (1\text{-}\ \alpha)R$$

where precision p is defined as the proportion of retrieved documents that are relevant. Recall R is defined as the proportion of relevant documents that are retrieved. The selection operator used is stochastic universal sampling. Intermediate recombination is used in producing new phenotypes around and between the values of the parents' phenotypes. Offspring are produced according to the rule

$$O_1 = P_1 * \alpha\ (P_2 - P_1)$$

where P_1 and P_2 represent parent populations. A real valued mutation is also applied. A mutation rate of $1/n$ (n :number of individuals per individual) means that per mutation only one variable per individual is changed, making the mutation independent of the size of the population.

13.8 Proposed Algorithm

The proposed methodology for mining over the test databases can now be summarized as follows. Given a large biological data collection, stored in the XML format, we,

- Apply LSI to the data to obtain a term-document incidence matrix A and consequently assign rank-k approximation to the matrix A [14, 15].
- We then feed the query vector q to the term-document space, to retrieve the pseudo-document that consists of the sum of the term vectors corresponding to the terms specified in the query scaled by the inverse of the singular value.
- Next, a similarity measure is applied to compare the position of the pseudo-document to the positions of the terms or documents in the reduced term-document space. Once the similarities between the pseudo-document and all the terms and documents in the space have been computed, the terms or documents are ranked according to the results of the similarity measure.
- We then execute the GA over the reduced term-document space and the highest-ranking terms or documents, or all the terms and documents exceeding some threshold value, are returned to the user.

The overview of the system is given below.

1. Let $\tau_1, \tau_2, \ldots \tau_n$ represent the XML document trees of the documents with identification numbers $(1, 2, 3, \ldots n)$, where n is the number of documents.
2. Applying LSI, terms and documents are represented by an m by n incidence matrix A. Each of the m unique terms in the document collection are assigned a row in the matrix, while each of the n documents in the collection is assigned a column in the matrix.

3. Once the mXn matrix A has been created and properly weighed, a rank - k approximation to A, where, $k << \min(m,n)$, is computed using SVD.
4. The query vector q is projected into the term-document space, and cosine similarity measure is applied to compare the position of the pseudo-document to the positions of the terms or documents in the reduced term-document space.
5. Next, to apply modified GA, each document vector is encoded as a chromosome:

$$J = (j_1, j_w, \dots j_i, \dots j_L)$$

where j_i denotes the weight of tag i and L is the number of tags to be considered. The simplified fitness function used is given by

$$Fitness = \alpha(p) + (1 - \alpha)R$$

where precision P is defined as the proportion of retrieved documents that are relevant.
6. GA is then applied and the gain in accuracy and query execution time is computed.

13.9 Performance Analysis

The experiments are carried out on a Pentium V, running the Windows XP Operating System. Matlab6.5 is used as the programming environment. The Algorithm is executed on three sample databases:

1. The UniProt Swiss Protein XML Database, a manually annotated protein knowledgebase established in 1986 and maintained since 2003 by the UniProt Consortium, a collaboration between the Swiss Institute of Bioinformatics (SIB) and the Department of Bioinformatics and Structural Biology of the Geneva University, the European Bioinformatics Institute (EBI) and the Georgetown University Medical Center's Protein Information Resource (PIR). The database contains function(s) of the protein, post-translational modifications, secondary structure, quaternary structure etc.. The terms contained in the database are manually processed into term matrices, assigned with singular values.
2. Similarly, the Medical Abstracts Database of MEDLARS, used in several previous works related to LSI. These abstracts are related to journal entries on neonatal medicine.
3. The Cabos Carbohydrate Sequence Database is defined in the Carbohydrate Markup Language, similar to XML. It contains XML descriptions of carbohydrate types and structures[16].

The term list for each is created with each term assigned with an SVD value. Each Matrix file is based on the Harwell-Boeing (compressed column) sparse matrix format. Each record in a Term List file contains the term, its *id*, and its global weight based on a Log-Entropy weighting scheme.

Figure 13.2 is a snapshot of a two-dimensional plot of the nonzero elements of the upper left corner of the term-by-document matrix of the non-zero elements of the term-by-document matrix UniProt. A non-zero element represented by a dot

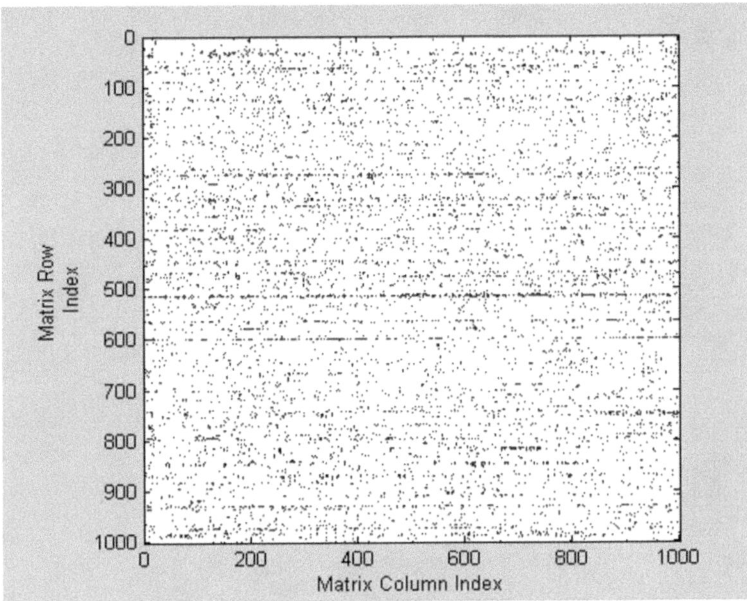

Fig. 13.2 Term by Term Document Matrix for UniProt

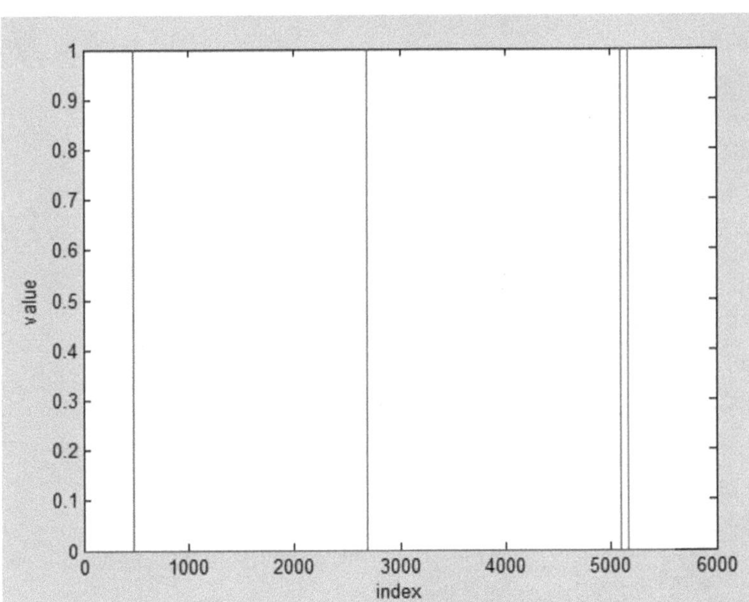

Fig. 13.3 A Plot of Value of a 3 word query q against its index

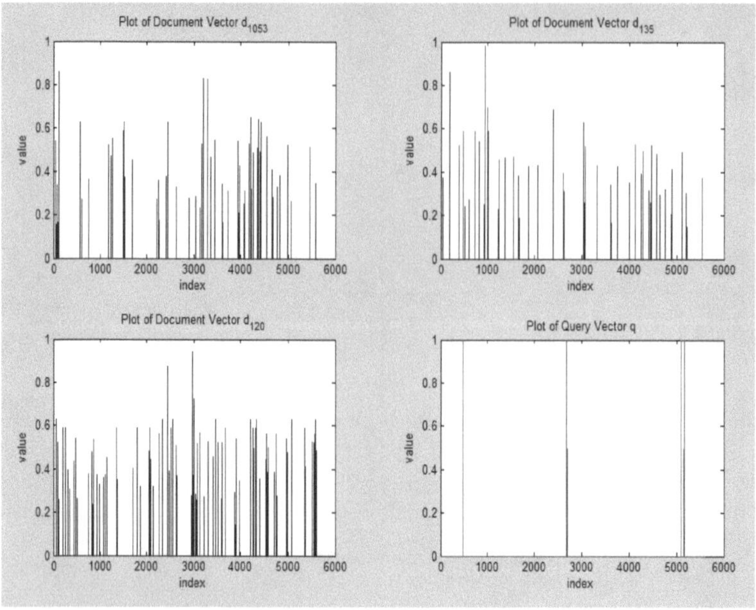

Fig. 13.4 Plot to Analyse Relevance of Results against Query

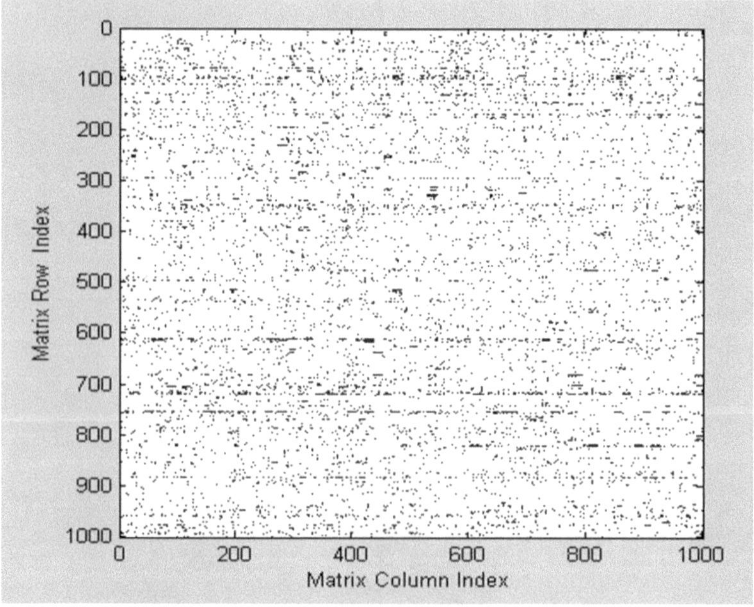

Fig. 13.5 Term by Document matrix MED

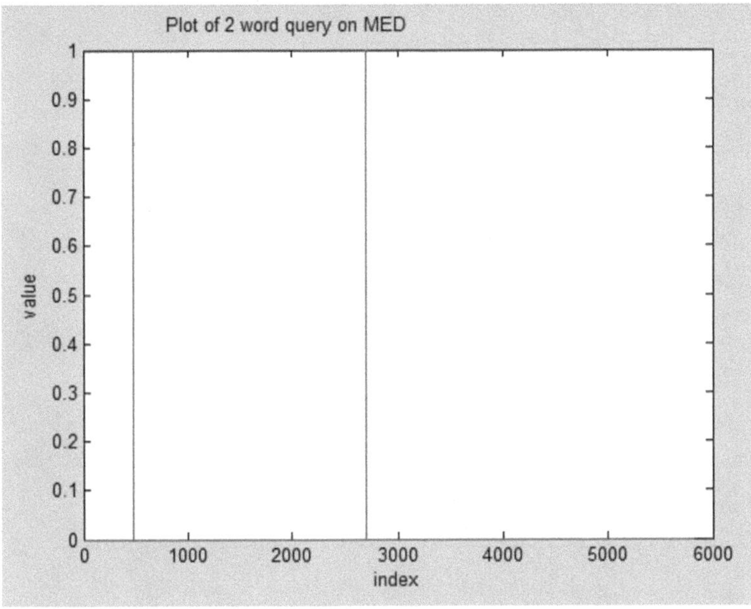

Fig. 13.6 A Plot of value of a 2 word query q against its index

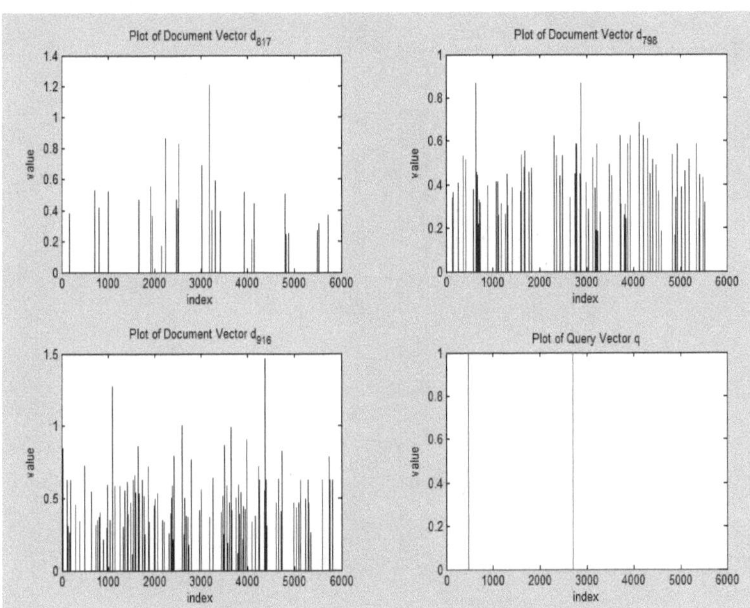

Fig. 13.7 Plot to Analyse Relevance of Results against Query

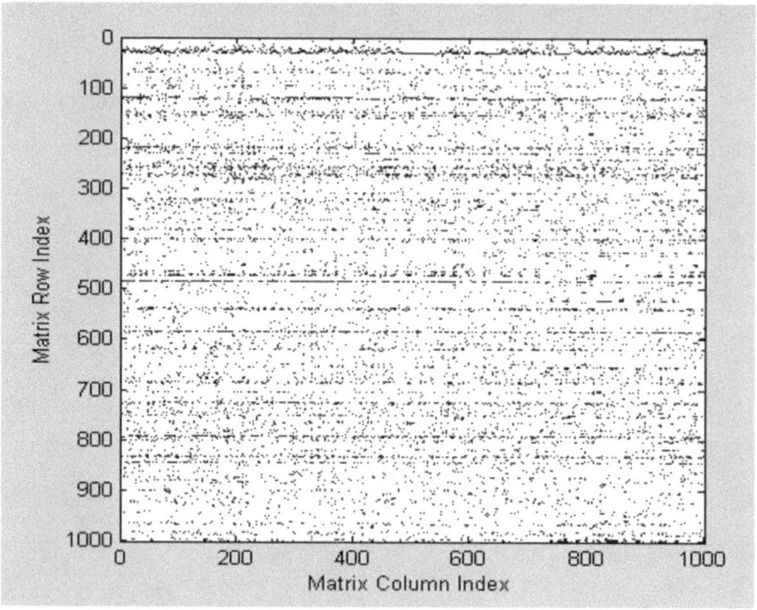

Fig. 13.8 Non Zero elements of the matrix Cabos

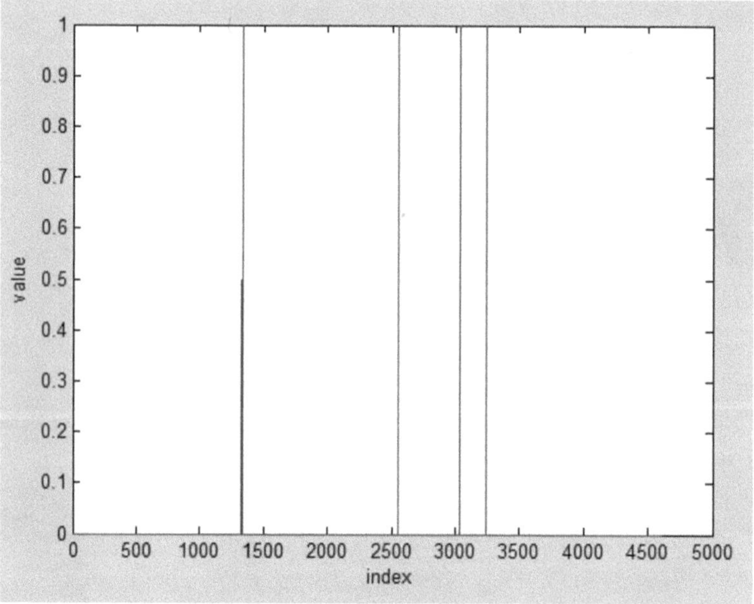

Fig. 13.9 A Plot of value of a 4 word query q against its index

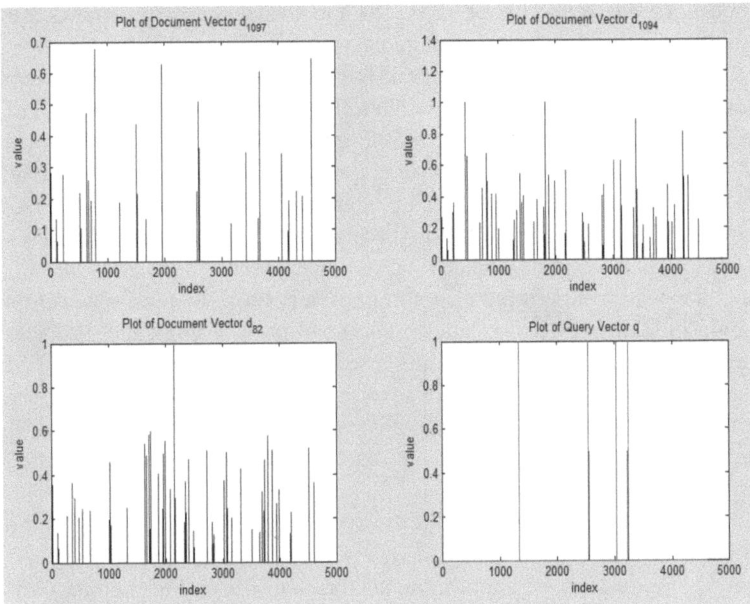

Fig. 13.10 Plot to analyse Relevance of Results against Query

indicates not only that term i occurs in document j, but also the number of times the term appears in that document. While such plot of a very large term-by-document matrix would be sparse *i.e.*, contain many zero elements, the same is not true of any of the three relatively small matrices under observation.

A three word query q is posed on the Matrix, say "constituents of glycine". The plot of the query q is shown below, displaying the nonzero elements of the query vector q. The x-axis represents the index of the document vector while the y-axis represents the value of the vector at that index. For example, a spike at x =500 with $y = 1$ means that the 500th element of the document vector is 1. Figure 13.3 depicts the similar situation.

Using a truncation point for the SVD, we mine n most relevant document vectors for query q. This involves the application of the cosine similarity measure. The retrieved document vectors are then sorted by their angle measure. The plot for the most relevant document vector, along with the plot for query q is as shown in Figure 13.4. Each spike represents the location of a member of the document vector. It may be observed from comparison with the query plot that the plot of the most relevant result has the closest similarity to the plot of the original query.

In Figure 13.5, similarly, is a snapshot of the non-zero elements of the term-by-document matrix MED. A non-zero element represented by a dot indicates that term i occurs in document j, and also the number of times the term appears in that document. A 2 word query "muscular atrophy" is posed on the Matrix. The plot of the query vector is shown in Figure 13.6. The most relevant document vectors retrieved are plotted along with the original query vector in Figure 13.7.

Figure 13.8 shows the non-zero elements of the matrix Cabos. A 4 word query *multiple glucose monosaccharide polymer* is fed into the term-document space. The plot of the weighted query vector is shown in Figure 13.9. The plot of the most relevant results is shown along with the original query vector in Figure 13.10. The significance of the LSI technique is that the document vectors retrieved in each case, are not exact term matches but rather, those vectors which are similar to the query vector.

The results of the search experiments comprehensively prove that the application of latent semantic indexing to term-relation identification in biological data paves the way to the construction of a suitable mining system for such datasets. The LSI engine provides accurate results in terms of term relevance. More importantly, the results show that the LSI model does away with the exhaustive approach of finding approximation between a database and a query and instead eliminates mismatched queries by improved indexing.

The second part of our performance studies is to compare the performance of the system on the basis of selected queries with and without the application of the genetic algorithm. In this section we capture the performance results of the proposed search technique. The experiments are conducted wholly in the Matlab Environment. The functions of the GA Toolbox are used to apply the GA and indexing and storage is carried out with the help of functions defined by the Geodise Lab XML Toolbox.

Query Execution Time: The query execution time is the single major factor, which controls the efficiency of a search engine [5]. In the context of XML documents it is more relevant due to large corpus size. Single term queries are faster than queries involving more than one term. This is an aspect of relevance since the nature of searches over such databases usually constitute single or two term queries as opposed to searches over non-biological databases. Also the query execution time for the first search tends to be high. For the subsequent searches, the query execution time shows considerable improvements. This is again relevant as for the first search; the index is not yet loaded into the main memory. This finding is encouraging to the proposal to apply GAs to partition stored indices in order to negate the constraint that main memory imposes on the use of the more straightforward, large single indices. The graphs in Figures 13.11 to 13.13 shows the estimated query execution time for each of the three databases before and after the application of GA.

Thus we find that the application of GA facilitates a significant gain in execution time. The results are plotted in the graphs below. In each graph, the execution time is plotted in *ms* for a query of a particular length before and after the application of GA. The x-axis denoted the 'number of terms' or the number of keywords in the query. Tests are carried out for single-word queries, 2, 3 and 4 and 5 word queries.

Accuracy of the Search Results: A good search technique should retrieve the results that are relevant to its users. At the same time, the number of irrelevant results

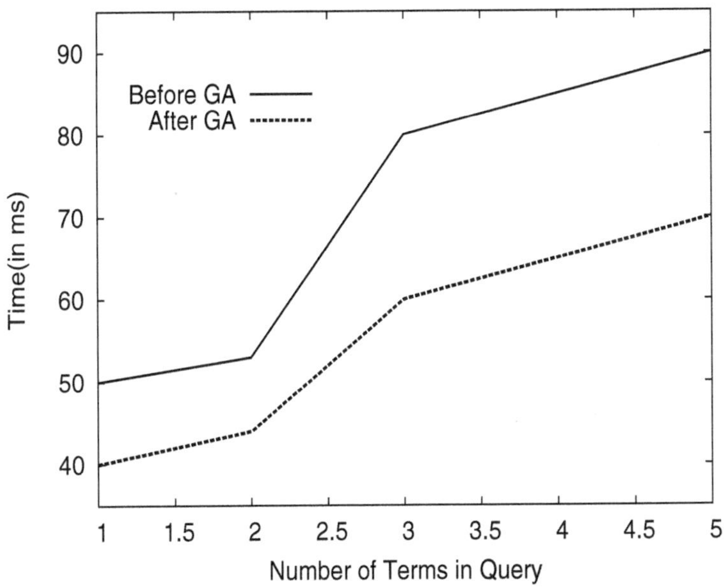

Fig. 13.11 Execution Time versus Number of Terms in Query for UniProt Database

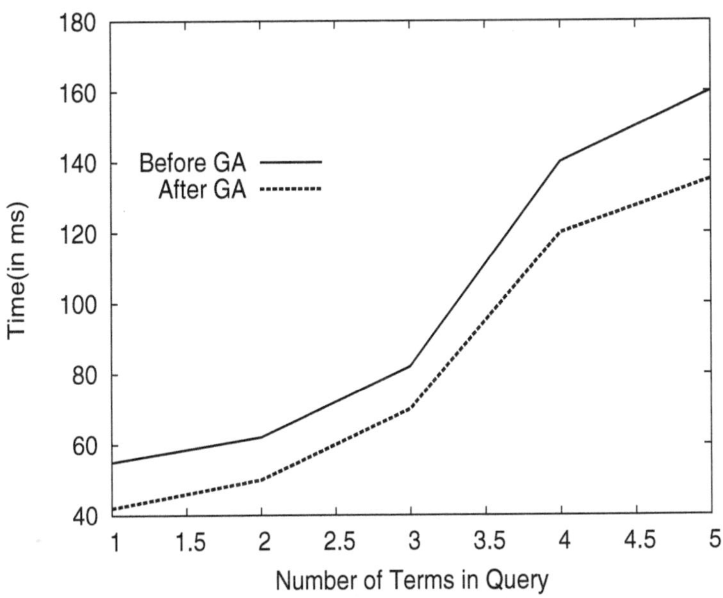

Fig. 13.12 Execution Time versus Number of Terms in Query for MED Database

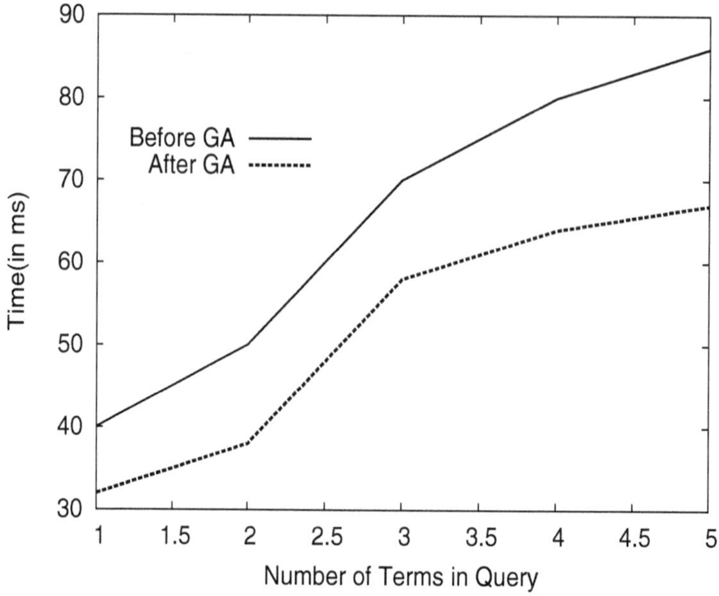

Fig. 13.13 Execution Time versus Number of Terms in Query for Cabos Database

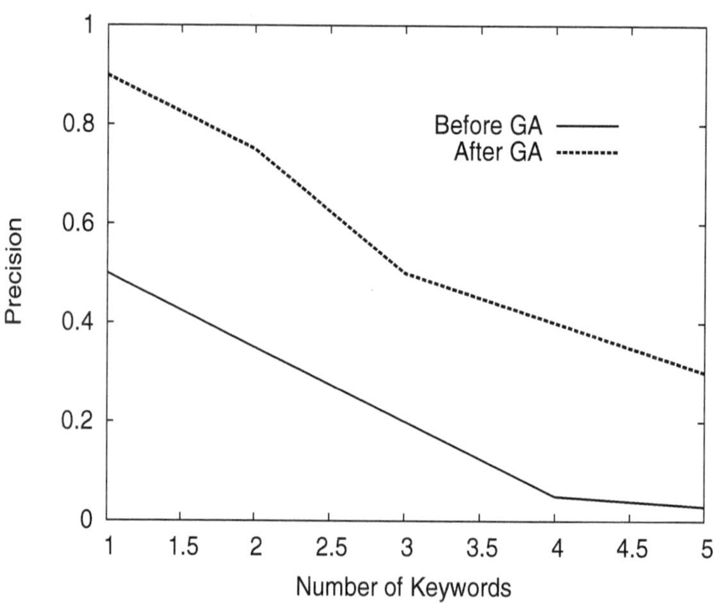

Fig. 13.14 Precision versus Number of Keywords

must be reduced. A measure of the accuracy of a search engine is its precision values. The precision is defined as:

$$Precision = \frac{Number\ \ of\ \ retrieved\ \ and\ \ relevant\ \ results}{Total\ \ number\ \ of\ \ retrieved\ \ results}$$

The precision metric is a measure of the user utility of the search results. If the search engine has high precision, the results are highly useful to the users and *vice versa.*

The precision values before and after applying of Genetic Algorithms to the UniProt database are shown in Figure 13.14. In the case of the Medical Abstract Database and the Cabos databases, it is observed that the precision of results obtained before and after the application of GA is near identical. This may be explained by the nature of these two databases, i.e., there exist more obvious semantic relationships than in the case of the protein database. For the protein database, on the other hand, we find that the terms in the data collection are diverse to a greater extent. We may then conclude that the gain in precision with the use of GA is best displayed in highly heterogenous databases.

13.10 Summary

This chapter proposes the application of semantic retrieval of information as an apposite technique for data mining over biological databases. It demonstrates the technique over a sample databases and shows that the application of GAs can significantly increase the accuracy of search results. The framework suggested can be further experimented with and applied to a wide range of biological data, and particularly to data collections of larger volumes, to test for the adaptability of the approach.

References

1. Singh, A.K.: Querying and Mining Biological Databases. Journal of Interactive Biology 7(1), 7–8 (2003)
2. Luk, R., et al.: A Survey of Search Engines for XML Documents. In: SIGIR Workshop on XML and IR (2000)
3. Cohen, S., Mamou, J., Kanza, Y., Sagiv, Y.: XSEarch: A Semantic Search Engine for XML. In: VLDB, pp. 45–56 (2003)
4. Letsche, T.A., Berry, M.W.: Large-Scale Information Retrieval with Latent Semantic Indexing. Information Sciences - Applications 100, 105–137 (1997)
5. Landauer, T.K., Dumais, S.T.: A Solution to Plato's problem: the Latent Semantic Analysis Theory of Acquisition, Induction and Representation of Knowledge. Psychological Review 104(2), 211–240 (1997)
6. Williams, H.E., Zobel, J.: Indexing and Retrieval for Genomic Databases. IEEE Transactions on Knowledge and Data Engineering 14(1) (January/February 2002)
7. Hammouda, K.M., Kamel, M.S.: Efficient Phrase-Based Document Indexing for Web Document Clustering. IEEE Transactions on Knowledge and Data Engineering 16(10), 1279–1296 (2004)

8. Bellettini, C., Marchetto, A., Trentini, A.: An Approach to Concerns and Aspects Mining for Web Applications. International Journal of Information Technology (IJIT) (2005)
9. Guo, L., et al.: XRANK: Ranked Keyword search over XML Documents. In: SIGMOD 2003 (2003)
10. Deerwester, S., Dumais, S.T., Landauer, T.K., Furnas, G.W., Harshman, R.A.: Indexing by Latent Semantic Analysis. Journal of the American Society of Information Science (1990)
11. Caid, W.R., Dumais, S.T., Gallant, S.I.: Learned Vector Space Models for Information Retrieval. Journal of Information Processing and Management (1995)
12. Berry, M., Dumais, S., O'Brien, G.: Using Linear Algebra for Intelligent Information Retrieval. SIAM Review 37(4), 573–595 (1995)
13. Cooper, R., et al.: Indexing Genomic Databases. In: Fourth IEEE Symposium on Bioinformatics and Bioengineering (2005)
14. Golub, G., Van Loan, C.: Matrix Computations, 2nd edn. Johns-Hopkins (1989)
15. Foltz, P.: Using Latent Semantic Indexing for Information Filtering. In: Proceedings of the ACM Conference on Office Information Systems (COIS), pp. 40-47 (1990)
16. Kikuchi, N., Kameyama, A., et al.: The Carbohydrate Sequence Markup Language (CabosML): an XML Description of Carbohydrate Structures, Bioinformatics 21(8), 1717–1718 (2005)

Chapter 14
Probabilistic Approach for DNA Compression

Abstract. Rapid advancements in research in the field of DNA sequence discovery has led to a vast range of compression algorithms. The number of bits required for storing four bases of any DNA sequence is two, but efficient algorithms have pushed this limit lower. With the constant decrease in prices of memory and communication channel bandwidth, one often doubts the need of such compression algorithms. The algorithm discussed in this chapter compresses the DNA sequence, and also allows one to generate finite length sequences, which can be used to find approximate pattern matches. DNA sequences are mainly of two types, Repetitive and Non-Repetitive. The compression technique used is meant for the non-repetitive parts of the sequence, where we make use of the fact that a DNA sequence consists of only 4 characters. The algorithm achieves bit/base ratio of 1.3-1.4(dependent on the database), but more importantly one of the stages of the algorithm can be used for efficient discovery of approximate patterns.

14.1 Introduction

The realization of the fact that the DNA is the prime genetic molecule, which carries the hereditary information in its chromosomes, focused attention on its structure. The late 1940s and 1950s saw a huge amount of research in this field when the structure of the DNA is analyzed. Researchers wanted to find out how the chromosomes replicated themselves to form two identical copies and how they carried genetic information. The discovery of the double helix is a path breaking revelation, when researchers expressed the structure as a three dimensional, helical form, where the difference between two genes is in the order and number of 4 nucleotide building blocks along complementary strands. Extensive research has shown that though the genetic structure remains to be the same, there is a lot of variation in the DNA of one organism to the other. For instance, the chromosomes of small viruses have single stranded DNA, instead of the usual double stranded molecules. The right-handedness in the twisting of the helix is also replaced by a left-handed twist in some organisms. The shape of the DNA molecules varies from linear to circular and from mono-coiled to super-coiled.

K.R. Venugopal, K.G. Srinivasa, L.M. Patnaik: Soft Comput. for Data Min. Appl., SCI 190, pp. 279–289.
springerlink.com

The evolution of such a huge plethora of DNA sequences have led to Bioinformatics scientists developing algorithms and methods to compress and store the sequences. The final goal though is not only in compressing the sequences but identifying the properties and patterns from the compressed sequences, saving us the overhead of decompressing the sequence. The four bases found in a genomic sequence are:

- Adenine — A
- Cytosine — C
- Guanine — G
- Thymine — T

A Genomic sequence in a programming-sense is a string of A,C,T,Gs of an approximate length of 3 billion. A genomic sequence can be identified using various properties it possesses. Repetition is a very important property, which is exploited in compression [1]. Sequences can be classified as highly repetitive (tandem repeats), moderately repetitive (interspersed repeats) and single copy (no repeats). Other properties include approximate matches, palindromes and reverse matches. A string that has a four-character alphabet requires 2 bits per base for storage. The bit/base ratio has been lowered to 1.76 - 1.5 [2], with the use of intelligent algorithms, which uses the redundancies in a genomic sequence to compress but also help in motif discovery. This concept brings down the bit/base ratio to around 1.4 and also generates subsequences, which can be used to find approximate matches. Scientists have often relied on statistical analysis to relate to identification of the various properties in a DNA sequence. The probability modeling approach has been used as an initial step to approximate the ratio of the four bases in the sequence. This is an important step as it indicates to the algorithm, what it should expect as it moves ahead in the various passes in the step-wise compression algorithm. The algorithm presents a compression model and also exposes its potential in Motif Discovery.

The very first algorithm for DNA sequence compression is that of Grumbach and Tahi [2] [3], called BioCompress, is mainly aimed at Nucleic Acid sequences. It used a basic Lempel Ziv [5] [6] style substitution algorithm that detected exact matches and complementary palindromes. A complementary palindrome is that in which, the reverse of a particular subsequence, along with the complementary base interchanges, is an exact match of the original subsequence. Such kind of redundancy is quite frequent in DNA sequences and is often subjected to high compression. The updated version of the first algorithm is called Bio-Compress-2. The only major difference is that, this algorithm could detect regions where redundancy is absent and applied Arithmetic Coding of order 2 in these areas.

An efficient encoding scheme, *Cfact* is brought forward by Rivals. It had a two-pass technique, where the first pass, is used to detect exact repeats, and the second pass is used to encode the repeats. Regions with no repeats are coded at two bits per base. Though the encoding scheme is useful, the algorithm, which used this scheme, did lossy compression [7], which is of little use for DNA sequences. GenCompress [1] [9] [10] developed by Chen, had a better performance than its previous algorithms, Cfact and BioCompress-2 [3]. The algorithm created a Suffix

Tree in the first pass, and the encoding is done in the second pass. The algorithm looked for optimal prefixes at every step, and gave a guaranteed gain in the bits/base ratio. If no gain is achievable, it used 2 bits/base. There are two variants of this algorithm, namely, GenCompress-1 that used hamming distances for the repeats, while GenCompress-2 used Edition Distance (Insertion, Deletion, Substitution) at each step. CTW+LZ [2] algorithm came about as a very efficient compression algorithm, but it has very slow execution time. Developed by Matsumoto, this algorithm uses CTW (Context Tree Weighting) method and local heuristics for resolving the Greedy Selection Problem. Though this algorithm is not practically used because of the high time consumption, Lempel-Ziv algorithms have formed the crux of most Gene Compression algorithms. DNACompress [11] developed by Chen, is a 2-phase algorithm, which used the Lempel-Ziv algorithm for compression. In the 1st phase, special software called PatternHunter [12] is used to find the exact repeats and the complimentary palindromes. The matches are sorted in descending order of size or some gain function. The second phase is finding non-repeating regions and approximately repeating regions. PatternHunter uses strings for non consecutive systems as a seed for search. The algorithm had a good execution time, and is popular.

A departure from the Lempel-Ziv method is the use of Normalized Maximum Likelihood. NMLComp developed by Tabus, encoded the NML Model for discrete regression. The algorithm is suitable for encoding approximate block matches, based on replacement operations. The algorithm has a low complexity level and is quite light in terms of computational requirements. The Sequitur used Digram Uniqueness and Rule Utility. Digram Uniqueness says no pair of adjacent symbols appear more than once in the grammar, while Rule Utility says that each rule is used at east twice (except for the start rule). The DNASequitur is an improvement on the previous Grammar-based compression algorithm, and it used Reverse Compliments. The usual approach used by most algorithms is to find exact repeats and approximate repeats. The largest subset of compatible repeats is generated and encoded using algorithm-specific methods. Our algorithm is a departure from the usual first step of searching. It uses three step encoding process, to compress the whole sequence. The advantage in not using repeats is that the sequence is compressed irrespective of whether the sequence has redundancy or not. The first step in our algorithm Gene-Compressor is a calculating the probability of the bases, followed by Huffman coding them. The second step encodes the sequence further using a transformation that aims at grouping the repetitions in the sequence. The DNA sequence is decompressed in the future for analysis and use; hence the transformation has been designed with an Inverse Transformation process in mind. The third step uses the localized repetition property that is present in the encoded sequence and uses a slightly modified version of Run Length Encoding [13] [14] [15] [16].

14.2 Probability Model

In this section, we present a model which is used to calculate the probability of A, C, T and G in the sequence. Calculating the probability is the first step in the algorithm, as the probability values of the 4 bases are used to encode the sequence

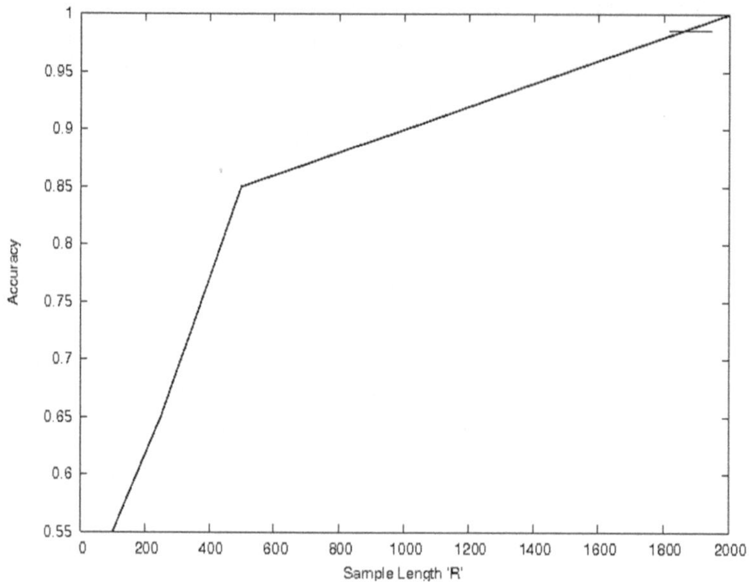

Fig. 14.1 Variation of accuracy of probability model with changing Sample Length 'R'

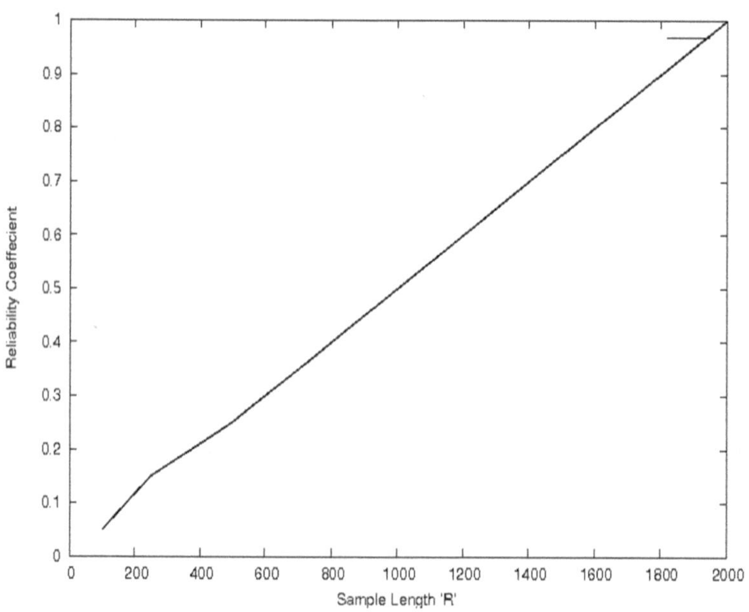

Fig. 14.2 Variation of accuracy with changing Sample length 'R'

in a Huffman Code manner. As the length of the sequences is of the order of billions, we cannot do a full sequential scan of the sequence. This brings to light the need of a probability model, which can be used to calculate the occurrences of the bases. The steps for finding the probability are given below.

1. Let the length of the sequence be L. Using statistical methods, generate N such that $1 < N < (L/10)$. N' denotes the partition length to be considered. The value of N' should be restricted to $L/10$ as we want to calculate using a minimum of 10 samples.
2. Generate another number, R' such that $1 < R < N$. R' denotes the sample size of each sample.
3. Partition the sequence into groups of N bases. Let the groups be called G_1, G_2, G_3 ... G_i ... G_N.
4. A sample of each partition of length R can be selected in $(N - (R - 1))$ ways. The sample is selected using a $Rand()$ function with seed value equal to system time (which assures uniqueness). The $Rand()$ function chooses any one of the Samples. The length of the sample R' affects the accuracy of the Probability Model. The graph (As shown in Figure 14.1) displays values corresponding to partition size of 2000.
5. The probability of the four bases is calculated in these samples of R. Let the probability be defined as
   $(PG_{1A}, PG_{1C}, PG_{1T}, PG_{1G})$ for the 1st group
   $(PG_{2A}, PG_{2C}, PG_{2T}, PG_{2G})$ for the 2nd group
   $(PG_{3A}, PG_{3C}, PG_{3T}, PG_{3G})$ for the 3rd group
   ...
   ...
   ...
   $(PG_{iA}, PG_{iC}, PG_{iT}, PG_{iG})$ for the ith group

6. The groups are classified into three types according to the conditions given below:
   - Condition 1: Let the difference between all pairs (PG_{iX}, PG_{iY}) where $X, Y = A, C, T, G$ and $X! = Y$ be equal to D. $'i'$ signifies the group number. Condition 1 state that all values of D should be less than 0.05. This condition signifies *No Base Dominance*, a condition where the probability of occurrence of all bases tends to 0.25 and they have equal share in the sequence. This group is assigned a weight of 5.
   - Condition 2: The probability of any two of the four bases PG_{iX}, where $X = A, C, T, G$ should be greater than 0.4. Condition 2 signifies *Two Base Dominance*, a situation in which where two complementary bases dominate the sequence. This group is assigned a weight of 3.
   - Condition 3: The probability of any 1 of the bases PG_{iX}, where $X = A, C, T, G$ should be greater than 0.8.
   - Condition 4: signifies *Single Base Dominance*, a situation where one base dominates the sequence. In the absence of a complementary pair, the base forms Hydrogen Bonds to attain stability. This group is assigned a weight 2.

7. Multiply the probability of each base in a group with the corresponding multiplier of that group, i.e., $(PG_{iX} * W_i)$ where $X = A, C, T, G$ and i' denotes the group number.
8. To find the probability of a particular base X, $P_x = (\Sigma(PG_{ix} * W_i))/\Sigma W_i$, where $X = A, C, T, G$

The group length is selected as 1000 and the subsequence in that group is of length 100. This gives a reliability co-efficient of 0.1. These values are not standardized and can be manipulated according to the reliability of statistical analysis required. The reliability coefficient with changing values of R' and constant $N' = 2000$(As shown in Figure 14.2).

14.3 Algorithm

Problem Definition: Given a genomic sequence, the objectives are: (i) Compress the sequence so as to obtain an effective bit/base ratio. (ii) Generate subsequences, which can be used for Pattern Matching.

Assumptions: The probability of each base calculated using the above probability model described in section 3, tends to the actual probability in the sequence. We calculate the probability of each base and then apply Huffman coding to the Sequence. Huffman Coding helps us convert the file into 0s and 1s, the basic requirement, if we want to bring down the bit/base ratio from 2 to 1.4. Algorithm for Huffman tree creation is given below.

1. Let the base probability values be leaf nodes or independent sub-trees.
2. Sort the sub-trees according to their values.
3. Combine the sub-trees with two lower-most values by a root node.
4. Eliminate the two sub-trees chosen from the list and insert the new sub-tree.
5. Repeat Steps 2 to 5 till we obtain a single Sub-tree.
6. Mark all the left edges as zero and all the right edges as one.
7. The path from the root to a particular base is the Huffman Code for the corresponding base.

The new file is approximately double the size of the original sequence, but it brings the alphabet size down to 2.

Algorithm PacketGen: Used to generate intermediate packets.

- The sequence is broken down into packets of length p'.
- All p' length rotations of each packet P is generated.
- The last base of each rotation is extracted and put in another packet P'.
- This packet is interposed into the sequence in place of Packet P.

Using the PacketGen algorithm, the sequence is then broken down into packets of length p'. The value of p' is an important tradeoff and deserves discussion. The various values of p' and its effect on the algorithm are discussed in section 5. For the time being, we assume that a p' value of eight is optimal and hence used. The packets P_i are stored in a buffer, along with all its p' length rotations. The rotations of a particular packet can be used in any motif discovery algorithm, which searches for approximate repeats.

Once all the rotations are generated, we take the last character of each rotation and form a new p' length packet, P_i'. This packet is put instead of the original packet. The new packet, P_i', has two very important properties which are exploited. P_i' has very high localized repetitions, because it has been constructed from a sorted order of strings. This property can be used to compress the packet, but the main question is how do we get back the original packet P_i'. The method used above to encode a packet using its rotations can be reversed to get back the original packet. This is done by a series of sorting operations.

Algorithm Gene-Compressor
Input: Genomic Sequence G
Output: Compressed Sequence GC

1. Calculate the probability of each base in the sequence using the probability.
2. Calculate encryption codes according to the probability to Huffman's coding algorithm.
3. Encrypt the input file G by replacing each of the bases by their corresponding Huffman Codes.
4. The new file G' consists only of $0's$ and $1's$ and $Length(G') = 2 * Length(G)$, approximately.
5. The sequence in G' is broken into small packets, P_i of length $'p'$.
6. Each packet P_i is put in a buffer B and all p rotations of the buffer are generated and sorted in ascending order.
7. The last character of each rotation is extracted and a string P_i' is made out of it.
8. The packet P_i is replaced by P_i' in the file G'.
9. The file G' now consists of packets which have a high percentage of repetitions of $0's$ and $1's$.
10. Using this to our benefit, let us do run length encoding of consecutive repetitions of four and five bases only.
11. Store the run-length encoded sequence in a file called G_c, which is the output file.

Once we generate the encoded sequence, we apply a modified version of run-length encoding, where only 4 and 5 length repetitions are encoded. This has two reasons. First, we cannot increase the number of alphabets used in the final compression. It has to remain at 4. Hence we just add two more characters, 4 and 5, along with the characters 0 and 1. Additionally, a close observation has showed that the frequency of localized rotations occurs rarely.

14.4 Optimization of P'

The value of p', or the length of the packets can vary from two, three, four to around sixteen. With increase in the value of p', the buffer length increases meaning we can generate motifs of a larger length. But on the other hand, using large value of p' increases the overhead of sorting and generating full-length rotations. Using very low values of p' means that the repetitions are of length two to three, which does not give any compression sense when we apply run length encoding. The algorithm is tested with p' values of four, six, eight and it is found that eight is the most optimal value for compression.

At $p = 8$, the buffer length to be sorted is not too high, and also it gives repetitions of length 4 to 5, which is the ideal for the modified run length encoding process that is used in this algorithm. At $p = 4$, the repeat sequences is of length 2, which did not provide any decrease in the bit/base ratio, when encoded using run-length encoding. At $p = 6$, the buffer length is too less for later use in motif discovery.

14.5 An Example

Let a sequence of length 250 bases be:
atattccgtttaattgcagatagtgtatcgccgggatcaacccctttgggagatatgcattcctagttgcagtactgatcgag
cggggcttcccttacgaatggtttagcgagaatcgggaccgagggctatattgcgacaaatcagcgccgagtctgcagc
tacctaaacgcatagatgtatactgataggcgtatgcagactcgcccccttgccaaagattgcaaacgtagcccaaagatg
agagctcg

Calculating the probabilities using statistical measures and the probability model given in section 3: $PA : 66/250 = 0.264$, $PC : 59/250 = 0.236$, $PT : 60/250 = 0.24$, $PG : 65/250 = 0.26$.

Taking the example of the first few bases: "atattccgttta...", The subsequence is Huffman coded as: "011101111110100011111101" and According to Figure 14.3, dividing into packets of Length $p' = 8$, Packet 1: 01110111, Packet 2: 11101000 and Packet 3: 11111101. Generating all p length rotations for Packet I, we get,

Random Order	Sorted Order
10111011	01110111
11011101	01110111
11101110	10111011
01110111	10111011
10111011	11011101
11011101	11011101
11101110	11101110
01110111	11101110

Extracting the last character from the Buffer, we get the transformed packets, Packet 1 : 11111100, Packet 2: 10100110 and Similarly Packet 3: 11111110. Applying modified Run-length encoding to the packets we obtain: Packet 1 : 51100,

Packet 2 : 10100110 and Packet 3 : 51110. Applying the algorithm to the sequence of 250 bases we get:

51100101001101110111011001010101001101100110010011015040101101101111
40100011101140101100101051110110011101115051010100011101015010001000
11110010110110101001101010510101501011011015000101501000110051110101
50101110001115010001000110110101115051000410010110010101010011051100
10101100101501041001140101001101041001011150101140110010101000110011
0101001011100014011010150l0

Finally we get, the Length of the above sequence as 367 and the Compression rate (bit/base ratio) as 1.472.

14.6 Performance Analysis

The Compression algorithm, Gene-Compressor, is used for compression of Repetitive and Non-repetitive Sequences. During the development of the algorithm, we used a DNA Sequence generator to generate sample input and test the code. After the development of the algorithm, it is tested using standard DNA databases. The results of the tests have been given in Table 14.1 and Table 14.2. Table 14.1 gives the comparative analysis of Gene-Compressor with other text-based compression algorithm. Table 2 gives the comparative analysis with other DNA-specific compression algorithms. All the values given are in bits/base. The algorithm performs better than most other algorithms in terms of compression. The algorithm does require a large amount of Buffer for storing the intermediate encoding stages. It also requires a high amount of Buffer for storing the full-length rotations of the packets.

Table 14.1 Comparison of Text Compression Algorithms

DNA Sequence name	Sequence Length	gzip -9	lz (1M)	arith +(1m)	PPMD+	adapted PPMD+	normal CTW	CTW -4	**Gene Compressor**
CHNTXX	121024	2.220	2.234	1.866	1.977	1.840	1.879	1.838	**1.473**
CHNTXX	155844	2.291	2.300	1.956	2.062	1.934	1.974	1.933	**1.403**
HEHCMVCG	229354	2.279	2.286	1.985	2.053	1.965	1.997	1.958	**1.523**
HUMDYSTROP	38770	2.377	2.427	1.948	2.237	1.921	1.960	1.920	**1.534**
HUMGHCSA	66495	1.551	1.580	1.438	2.077	1.694	1.376	1.363	**0.901**
HUMHBB	73308	2.228	2.255	1.911	2.116	1.921	1.917	1.892	**1.653**
HUMHDABCD	58864	2.209	2.241	1.950	2.130	1.948	1.909	1.897	**1.661**
HUMHPRTB	56737	2.232	2.269	1.942	2.130	1.932	1.922	1.913	**1.532**
MPOMTCG	186609	2.280	2.289	1.961	2.075	1.966	1.989	1.962	**1.715**
PANMTPACGA	100314	2.232	2.249	1.873	2.018	1.872	1.902	1.866	**1.722**
SCCHRIII	315339	2.265	2.268	1.935	2.023	1.950	1.976	1.945	**1.732**
VACCG	191737	2.190	2.194	1.862	2.002	1.910	1.897	1.857	**1.672**

Table 14.2 Comparison of DNA-Specific Compression Algorithms

DNA Sequence name	Sequence Length	Bio-Compress2	Gen-Compress	CTW-LZ	DNA-Compress	DNA-Pack	DNA-MEM	**Gene Compressor**
CHNTXX	121024	1.6848	1.6730	1.6690	1.6716	1.6602	1.660	**1.473**
CHNTXX	155844	1.6172	1.6146	1.6120	1.6127	1.6103	1.610	**1.403**
HEHCMVCG	229354	1.8480	1.8470	1.8414	1.8492	1.8346	1.834	**1.523**
HUMDYSTROP	38770	1.9262	1.9231	1.9175	1.9116	1.9088	1.908	**1.534**
HUMGHCSA	66495	1.3074	1.0969	1.0972	1.0272	1.0390	1.031	**0.901**
HUMHBB	73308	1.8800	1.8204	1.8082	1.7897	1.7771	1.776	**1.653**
HUMHDABCD	58864	1.8770	1.8192	1.8218	1.7951	1.7394	1.739	**1.661**
HUMHPRTB	56737	1.9066	1.8466	1.8433	1.8165	1.7886	1.788	**1.532**
MPOMTCG	186609	1.9378	1.9058	1.9000	1.8920	1.8932	1.892	**1.715**
PANMTPACGA	100314	1.8752	1.8624	1.8555	1.8556	1.8535	1.853	**1.722**
VACCG	191737	1.7614	1.7614	1.7616	1.7580	1.7583	1.758	**1.672**

14.7 Summary

The algorithm presented in this chapter is tested for around 12 sequences. It performs better than most of the text compression algorithm as it uses a upper limit of 2 bits/base. It performs better than most of the DNA compression algorithms also(as shown in Figure 14.3). It has a high execution time for Sequences of length greater than 2 billion. On the other hand, we have also seen that the Buffer used in the intermediate step can be used to generate full length and partial length rotations. These rotations can be used in Motif Discovery for finding approximate matches. We are trying to use this algorithm for compression of larger genes with a optimized execution time and also for Non-Repetitive genes like HUMDYSTROP, which failed to compress efficiently using most of the standard compressors.

References

1. Rivals, E., Delahaye, J.-P., Dauchet, M., Delgrange.: A Guaranteed Compression Scheme for Repetitive DNA Sequences. LIFL Lille I University Technical Report (1995)
2. Matsumuto, T., Sadakane, K., Imai, H.: Biological Sequences Compression Algorithms. Genome Information Ser. Workshop Genome Inform 11, 43–52 (2000)
3. Grumbach, S., Tahi, F.: Compression of DNA Sequences. In: Data Compression Conference, pp. 340–350 (1993)
4. Grumbach, S., Tahi, F.: A New Challenge for Compression Algorithms Genetic Sequences. Journal of Information Processing and Management 30, 866–875 (1994)
5. Ziv, J., Limpel, A.: Compression of Individual Sequences using Variable-Rate Encoding. IEE Transactions on Information Theory 24, 530–536 (1978)
6. Ziv, J., Limpel, A.: A Universal Algorithm for Sequential Data Compression. IEE Transactions on Information Theory 23(3), 337–343 (1977)
7. Sadel, I.: Universal Data Compression Algorithm based on Approximate String Matching. In: Probability in the Engineering and Informational Sciences, pp. 465–486 (1996)

8. Chen, X., Kwong, S., Li, M.: A Compression Algorithm for DNA Sequences and its Application in Genome Comparison. Genomic 12, 512–514 (2001)
9. Chen, X., Kwong, S., Li, M.: A Compression Algorithm for DNA Sequences. IEEE Engineering in Medicine and Biology Magazine 20(4), 61–66 (2001)
10. Li, M., Badger, J.H., Chen, J.H., Kwong, S., Kerney, P., Zhang, H.: An Information based Sequences Distance and its Application to whole Mitochondrial Genome. Bioinformatics 17(2), 149–154 (2001)
11. Chen, X., La, M., Ma, B., Tromp, J.: DnaCompress: Fast and Selective DNA Sequence Compression. Bioinformatics 18, 1696–1698 (2002)
12. Ma, B., Tromp, J., Li, M.: Patternhunter-faster and more sensitive homology search. Bioinformatics 18, 440–445 (2002)
13. Sata, H., Yoshioka, T., Konagaya, A., Toyoda, T.: DNA Compression in the Post Genomic Era. Genome Informatics 12, 512–514 (2001)
14. Willems, F.M.J., Shtralov, Y.M., Tjalkens, T.J.: The Context Tree Weighting Method: Basic Properties. IEE Transactions on Information Theory 41(3), 653–664 (1995)
15. Sadakane, K., Okazaki, T., Imai, H.: Implementing the Context Tree Weighting Method for Text Compression. In: DCC 2000: Proceedings of the Conference on Data Compression, USA (2000)
16. Rivals, E., Dauchet, M.: Fast Discerning Repeats in DNA Sequences with a Compression Algorithm. In: Proceedings of Genome Informatics Workshop, pp. 215–226. Universal Academy Press, Tokyo (1997)

Chapter 15
Non-repetitive DNA Compression Using Memoization

Abstract. With increasing number of DNA sequences being discovered the problem of storing and using genomic databases has become vital. Since DNA sequences consist of only four letters, two bits are sufficient to store each base. Many algorithms have been proposed in the recent past that push the bits/base limit further. The subtle patterns in DNA along with statistical inferences have been exploited to increase the compression ratio. From the compression perspective, the entire DNA sequences can be considered to be made of two types of sequences: repetitive and non-repetitive. The repetitive parts are compressed using dictionary-based schemes and non-repetitive sequences of DNA are usually compressed using general text compression schemes. In this chapter, we present a memoization based encoding scheme for non-repeat DNA sequences. This scheme is incorporated with a DNA-specific compression algorithm, *DNAPack*, which is used for compression of DNA sequences. The results show that our method noticeably performs better than other techniques of its kind.

15.1 Introduction

The bases found in DNA come in four varieties: adenine, cytosine, guanine, and thymine often abbreviated as A, C, G, and T, the letters of the genetic alphabet. Genome sequencing is finding the order of DNA nucleotides, or bases, in a genome the order of As, Cs, Gs, and Ts that make up an organism's DNA. Sequencing the genome is an important step towards understanding it. A genome sequence does contain some clues about where genes are, even though scientists are just learning to interpret these clues. The human genome is made up of over 3 billion of these genetic letters. The human genome is about 20-40 percent repetitive DNA, but bacterial and viral genomes contain almost no repetition. In repetitive DNA, the same short sequence is repeated over and over again. Somewhere in the genome the sequence GCA may be repeated 100 times in a row; elsewhere there may be 30 consecutive copies of the sequence ACTTCTG. For example, the following DNA sequence is just a small part of telomere located at the ends of each human chromosome:

K.R. Venugopal, K.G. Srinivasa, L.M. Patnaik: Soft Comput. for Data Min. Appl., SCI 190, pp. 291–301.
springerlink.com

```
...GGGTTAGGGTTAGGGTTAGGGTTAGGGTTAGGGT
TAGGGTTAGGGTTAGGGTTAGGGTTAGGGTTAGGGTT
AGGGTTAGGGTTAGGGTTAGGGTTAGGGTTAGGG...
```

An entire telomere, about 15 kb, is constituted by thousands of the repeated sequence "GGGTTA". Based on the repetition rate, DNA sequences are divided into three classes:

- *Highly repetitive:* About 10-15% of mammalian DNA reassociates very rapidly. This class includes tandem repeats.
- *Moderately repetitive:* Roughly 25-40% of mammalian DNA reassociates at an intermediate rate. This class includes interspersed repeats.
- *Single copy:* This class accounts for 50-60% of mammalian DNA.

Deriving meaningful knowledge from DNA sequence defines biological research through the coming decades and require the combined effort biologists, chemists, engineers, and computational scientists, among others. Some of the research challenges in genetics include Gene regulation, DNA sequence organization, Chromosomal structure and organization, Non-coding DNA types and functions, coordination of gene expression, protein synthesis, and post-translational events interaction of proteins in complex molecular machines, evolutionary conservation among organisms, Protein conservation, correlation of SNPs (single-base DNA variations) with health and disease, etc.

With increasing number of genome sequences being made available, the problem of storing and using databases has to be addressed. Conventional text compression schemes are not efficient when DNA sequences are involved. Since DNA sequences contain only 4 bases A, G, T, C, each base can be represented by 2 bits. However standard compression tools like *compress*, *gzip* and *bzip*2 have more than 2 bits per base when compressing DNA data. Consequently, DNA compression has become a challenge. Algorithms like *GenCompress* [1], *Biocompress* [2], *Biocompress-2* [3], that use the characteristics of DNA like point mutation or reverse complement achieve a compression rate of about 1.76 bits per base [4]. Many compression methods have been discovered to compress DNA sequences. Invariably, all the methods found so far take advantage of the fact that DNA sequences are made of only 4 alphabets, together with techniques to exploit the repetitive nature of DNA [5]. The algorithm given here is used to encode non-repetitive regions. The popular techniques that have been shown to be efficient in compressing non-repeat regions are Order-2 Arithmetic Coding and Context Tree Weighting Coding. Order-2 coding overcomes the constraint of Huffman coding that the symbol to be encoded has to be coded by round number of bits. The adaptive nature of coding has been an advantage. The adaptive probability of a symbol is computed from the context after which it appears. Order-2 algorithm usually have better compression ratios with high efficiency in general. The Context Tree Weighting Coding is proposed by Willems [6] and has a good compression ratio. The CTW encoder has two parts: a source modeler which is the actual CTW algorithm, which receives the uncompressed data and estimates the probability of thee next symbol and an encoder which uses the estimated probabilities to compress the data. The context tree is built dynamically during the encoding decoding process. The visited substring of shorter size than a fixed bound, exist as a path in the tree. Each node of the tree contains a probability. In order to encode a given bit, the following steps are performed: the path in the context tree which coincides with the current context is searched and if needed extended. For every node in this context path, an estimated probability of the next symbol is computed using weighting function on all the estimated probability values. The weighted probability is sent to the arithmetic encoder which encodes the symbol, and the encoder goes to the next symbol [7]. The repetitive regions are compressed using methods given in [8,9].

15.2 Related Work

The first DNA-specific algorithms are given by Grumbach and Tahi [2,3]. Two algorithm namely, *BioCompress* and $BioCompress-2$, based on Ziv and Limpel data compression method [10,11] are proposed. $BioCompress-2$ detects exact repeats and complementary palindromes located earlier in the target sequence, and then encodes them by repeat length and the position of a previous repeat occurrence. If no significant repetition is found then the arithmetic coding of order-2 is used. The use of arithmetic encoding is the only difference between *BioCompress* and $BioCompress-2$.

$Cfact$ algorithm is proposed by E. Rivals.et al, which searches the longest exact matching repeat using suffix tree data in an entire sequence. $Cfact$ is similar to $Biocompress-2$ except for being a two-pass algorithm, where the first pass involved building the suffix tree. In the second phase, repetitions are coded with guaranteed gain; else, two-bit per base encoding is used. The compression algorithm to detect the approximate tandem repeats in DNA sequences is later given in [8]. Approximate string matching is used to provide some lossy compression algorithms by [12]. However, lossy algorithms are of little use in DNA compression.

A better compression algorithm than *BioCompress* and $BioCompress-2$ is GenCompress [1, 13, 14]. The basic idea is to approximate repetitions. There are two variants of *GenCompress*-one that uses hamming distance for repeats and the other uses the edition distance(deletion, insertion and substitution) for the encoding of the repeats.

$CTW-LZ$ [4] algorithm is based on context tree weighting method. It basically works by combining LZ-77 type algorithm like *GenCompress* with CTW algorithm. The long repeats are coded by LZ77 while the short repeats are encoded by CTW. The execution time of $CTW+LZ$ is not impressive although it does achieve good compression ratios.

DNACompress [15] is a two-phase algorithm that employs the Ziv-Lampel compression scheme as $BioCompress-2$ and *GenCompress*. The first phase finds all approximate repeats including complementary palindromes, using a special software, PatternHunter [16]. The second phase involves coding of non-repeat regions and approximate repeat regions. The *DNACompress* achieves a better execution time in general than *GenCompress*.

DNAC [17] is another DNA compression algorithm that works in four phases. The first phase involves building of suffix tree to find exact repeats. The second phase involves extending exact matches to approximate matches using dynamic programming. In the third phase, it extracts the optimal non-overlapping repeats from overlapping ones. The repeats are encoded in the last phase. Some of the recent lossless compression for large DNA microimages compression is given in [18-20]. *DNAPack* that uses hamming distance for repeats and complementary palindromes, and either CTW or Arth-2 compression for non-repeat regions is proposed in [21]. The algorithm marginally performs better than the earlier mentioned algorithms due to selection of repeat regions using dynamic programming method rather than greedy approach. In this chapter, we propose another encoding algorithm based on memoization that is useful in coding non-repeats. Section 15.3 describes the idea behind the algorithm along with pseudocode. In Section 15.4, we describe the setup and evaluate the performance of the proposed encoding scheme by comparing the results obtained by incorporating our encoding scheme into *GenCompress* and *DNAPack* algorithms along with results obtained by other algorithms.

15.3 Algorithm

The encoding scheme works in two passes. Each pass is identical, except that the symbols being encoded are different. The following method is used to represent the bases:

- In the first pass alphabets A and G are represented by A; T and C are represented by T.
- In the second pass alphabets A and C are represented as A; G and T are represented by T.

For example, the sequence `AGTC` would be taken as `AATT` and `ATTA`. The decoding procedure requires both the sequences, with four possible combinations of A and T representing each base as shown in Table 15.1.

The above substitutions is made to the entire DNA sequence. Let the pass 1 substitution sequence be S_1 and pass 2 sequence be S_2. If the length of the sequences is l then the each linear sequence is transformed to a matrix of dimension $\alpha \times \beta$ where $\alpha * \beta = l$. The transformation is done row-wise and hence the entry in i^{th} row and j^{th} column (zero-indexed) would correspond to alphabet in $i * \beta + j$ position in the sequence.

Table 15.1 Mapping scheme for decoding

Sequence 1	Sequence 2	Base
A	A	A
A	T	G
T	T	T
T	A	C

The choice of α and β can be made in different ways. A simple approach would be to break the sequence into multiple sequences each of length of a perfect square, chosen greedily, and represent each sub-sequence as a square matrix. This is always possible since any number can be represented as sum of squares. An efficient way to determine the split is to consider all possible combinations of α and β and take the combination that leads to maximum compression ratio. This increases the complexity considerably but can be made non-prohibitive if the length of the longest sequence is restricted to a certain maximum value. The encoding and decoding of S_1 and S_2 are done independently. The compression technique works on the matrices obtained by the sequences. The encoding idea is based on the idea of recursively dividing the matrix into sub-matrices until each sub-matrix is composed of a single alphabet. The division of matrix can be done in one of the following ways:

- L : The matrix can be divided into left half and right half. If the number of columns is odd, the center column is included in the left half.
- U : The matrix can be divided into upper half and lower half. If the number of rows is odd, the center row is included in the upper half.
- C : The matrix can be split into even columns and odd columns. The leftmost column is considered even.
- R : The matrix can be split into even and odd rows. The first row is considered as even.

For example, the matrix

```
AAAATTTT
AAAATTTT
AAAATTTT
```

can be divided into two 3×4 sub-matrices :

```
AAAA    TTTT
AAAA    TTTT
AAAA    TTTT
```

the left sub-matrix can be encoded as A and the right sub-matrix can be encoded as T. Hence the entire division of matrix would be represented by "LAT". The letter A and T denote that each sub-matrix is composed only of alphabet A and T, respectively. If the matrix is a combination of both the alphabets, it is divided into one of the four ways mentioned above. The first letter gives the division type followed by encoding of left/upper/even submatrix followed by corresponding right/lower/odd submatrix. For example, the matrix :

```
ATATATAT
TATATATA
ATATATAT
TATATATA
```

would be encoded the best as `CRATRTA`. The first column split produces two submatrices:

```
AAAA    TTTT
TTTT    AAAA
AAAA    TTTT
TTTT    AAAA
```

The subsequent letters (RAT and RTA) further describe each of the subimages, until only 2×4 images of As and Ts are left.

```
AAAA    TTTT
AAAA    TTTT

TTTT    AAAA
TTTT    AAAA
```

Decompressing the encoded string requires the knowledge of the original size of the matrix. The dimension is not encoded in the string. This is stored as a separate array of integers and is used at the time of decompression. The compression ratio is given taking into account the space needed to store the sizes of each matrix.

The simplest implementation(Algorithm 1) is a recursive function that tries all possible ways of dividing the matrix and keeps the solution having the minimum encoded length, with memoization to keep it efficient. The main function is the *compress* which takes the matrix as the argument and returns the best encoding string possible. The matrix is considered as array of strings. The termination condition occurs when all the entries in the matrix have the same alphabet. Otherwise compress makes use of four functions *splitrow*, *splitcolumn*, *splitupper* and *splitleft* in order to try all possible combinations of division of matrix recursively. *mem* is the map data structure that holds the values for each matrix for which the best encoding string has been found.

For simplicity, the algorithm is shown without incorporating any of the optimizations that reduce the running time significantly. The functions in algorithm takes the matrices themselves as arguments thus maintaining copies of matrices of their own. This can be avoided by representing the sub-matrices by 6 parameters and keeping a single copy of the matrix.

- *row*: the topmost row in the sub-matrix
- *col*: the leftmost column in the sub-matrix
- *rowC*: the number of rows in the sub-matrix
- *colC*: the number of columns in the sub-matrix
- *rowS*: the distance between adjacent rows in the sub-matrix, relative to the original matrix
- *colS*: the distance between adjacent columns in the sub-matrix, relative to the original matrix

Initially, *row* and *col* are the top left corner of the original matrix, and both step sizes are 1.

Algorithm 1. Encoding algorithm

Input : Matrix consisting of As and Ts
Output: Encoded string corresponding to input matrix
mem is the data structure used to store matrices and their corresponding encoded strings if already found
t_s is local string ; s_1 and s_2 are matrices local to the functions

```
function compress ( matrix mat )
if mem[mat] not empty then
   return mem[mat]
end if
if mat contains only single alphabet(τ) then
   mem[mat] ← τ
   return mem[mat]
else
   if mat has more than one column then
      t_s ← "C" + splitcolumn(mat)
      if mem[mat] is empty OR len(t_s) < len(mem[mat]) then
         mem[mat] ← t_s
      end if
      t_s ← "L" + splitlower(mat)
      if mem[mat] is empty OR len(t_s) < len(mem[mat]) then
         mem[mat] ← t_s
      end if
   end if
   if mat has more than one row then
      t_s ← "R" + splitrow(mat)
      if mem[mat] is empty OR len(t_s) < len(mem[mat]) then
         mem[mat] ← t_s
      end if
      t_s ← "U" + splitupper(mat)
      if mem[mat] is empty OR len(t_s) < len(mem[mat]) then
         mem[mat] ← t_s
      end if
   end if
   return mem[mat]
end if

function splitcolumn ( matrix mat )
return compress(even columns of mat)+compress(odd columns of mat)

function splitrow ( matrix mat )
return compress(even rows of mat)+compress(odd rows of mat)

function splitleft ( matrix mat )
return compress(first half columns of mat)+compress(second half columns of mat)

function splitupper ( matrix mat )
return compress(first half rows of mat)+compress(second half rows of mat)
```

- *Left-Right*: Sets colC to half for the left sub-matrix and *colC*-half for the right sub-matrix, where half is (*colC*+1)/2. Also sets col to col+half*colS for the right sub-matrix. Similar method is used for *rowC* for Upper-lower split.
- *Even-Odd Columns*: Sets *colC* to half for the even sub-matrix and *colC*-half for the odd sub-matrix, where half is (colC+1)/2. Also sets col to *col+colS* for the odd sub-matrix, and *colS* to 2**colS* for both sub-matrices. Similar method is used for *rowC* and *rowS* in even-odd row split.

15.4 Experimental Results

Since the compression algorithm proposed is specifically made to compress non-repeat DNA sequences, a fair method of evaluation of performance can be made only by combining the compression scheme with another DNA-specific algorithm that exploits repeating sequences. For this purpose we use *DNAPack*, which is found to outperform most other methods available. We briefly describe the working of *DNAPack* and the method of incorporating memoization algorithm into it to achieve better results. DNAPack is based on dynamic programming for selection of segments as opposed to greedy methods of selection [22]. Let s be the input sequence. Let $BestComp[i]$ be the smallest compressed size of prefix $s[1\ldots i]$. The recurrence given in Fig 15.1 is the general scheme of dynamic programming.

Initialization: $BestComp[0] = 0$
Recurrence:

$$BestComp[i] = min \left\{ \begin{array}{l} BestComp[j] + CopyCost(j,i,k)\ \forall k \forall 0 < j < i \\ BestComp[j] + PalinCost(j,i,k)\ \forall k \forall 0 < j < i \\ BestCopy[j] + MinCost(j+1,i)\ \forall 0 < j < i \end{array} \right\}$$

Fig. 15.1 Dynamic programming scheme for finding best compression

Table 15.2 Comparison with text-compression algorithms

DNA Sequence name	sequence length	gzip-9	lz(1M)	arith(1M)	PPMD+	adapted PPMD+	normal CTW	CTW-4	DNAMem
CHNTXX	121024	2.220	2.234	1.866	1.977	1.840	1.879	1.838	1.6601
CHNTXX	155844	2.291	2.300	1.956	2.062	1.934	1.974	1.933	1.6101
HEHCMVCG	229354	2.279	2.286	1.985	2.053	1.965	1.997	1.958	1.8349
HUMDYSTROP	38770	2.377	2.427	1.948	2.237	1.921	1.960	1.920	1.9084
HUMGHCSA	66495	1.551	1.580	1.438	2.077	1.694	1.376	1.363	1.0311
HUMHBB	73308	2.228	2.255	1.911	2.116	1.921	1.917	1.892	1.7765
HUMHDABCD	58864	2.209	2.241	1.950	2.130	1.948	1.909	1.897	1.7395
HUMHPRTB	56737	2.232	2.269	1.942	2.130	1.932	1.922	1.913	1.7884
MPOMTCG	186609	2.280	2.289	1.961	2.075	1.966	1.989	1.962	1.8925
PANMTPACGA	100314	2.232	2.249	1.873	2.018	1.872	1.902	1.866	1.8533
SCCHRIII	315339	2.265	2.268	1.935	2.023	1.950	1.976	1.945	1.8331
VACCG	191737	2.190	2.194	1.862	2.002	1.910	1.897	1.857	1.7582

$CopyCost(j,i,k)$ is the number of bits needed to encode the substring of size k starting at position i if it is an approximate repeat of the substring of size k starting at j. The *PalinCost* is similarly defined for reverse complementary substrings. The function $MinCost(j+1,i)$ is the number of bits needed for compression of the segment $s[j+1,i]$. It depends on the size of the substring as well as the compression ratio obtained for the algorithm by arithmetic coding or CTW. *MinCost* allows the creation of repeat segment if it yields a benefit in the compression ratio [22]. We replace the CTW or arithmetic coding with our compression algorithm. The modified algorithm is referred as *DNAMem*. The performance of the *DNAMem* is evaluated along with other popular algorithms of its kind.

Comparison: The comparison of results are made between the conventional text compression algorithms with *DNAMem*. Table 15.2 gives the compression ratios expressed as *bitsperbase*. The first column gives the DNA sequence name. The second column gives the sequence length. Columns from 3-10 gives the *bitsperbase* value of all the algorithms. The *DNAMem* algorithm performs better than all the algorithms in all cases. Table 15.3 shows the comparison of DNA-specific algorithms. The *DNAMem* performs slightly better than all others in 7 out of 11 sequences considered. BC2, GC and DNAC refer to BioCompress2, GenCompress, DNACompress respectively.

Execution Time Analysis: The experiment is conducted on Pentium V running Red Hat Linux 9. On an average the execution time taken by *DNAMem* is 26mins, while that of *DNAPack* is 1min. The high execution time is the result of high asymptotic complexity of the memoization algorithm. Despite the improvements made the complexity remains $O(n^5)$, where n is the size of longest non-repeat sequence.

Table 15.3 Comparison with DNA-specific compression algorithms

DNA Sequence name	sequence length	*BC2*	GC	CTW-LZ	DNAC	DNAPack	DNAMem
CHMPXX	121024	1.6848	1.6730	1.6690	1.6716	1.6602	**1.6601**
CHNTXX	155844	1.6172	1.6146	1.6120	1.6127	1.6103	**1.6101**
HEHCMVCG	229354	1.8480	1.8470	1.8414	1.8492	**1.8346**	1.8349
HUMDYSTROP	33770	1.9262	1.9231	1.9175	1.9116	1.9088	**1.9084**
HUMGHCSA	66495	1.3074	1.0969	1.0972	**1.0272**	1.0390	1.0311
HUMHBB	73308	1.8800	1.8204	1.8082	1.7897	1.7771	**1.7765**
HUMHDABCD	58864	1.8770	1.8192	1.8218	1.7951	**1.7394**	1.7395
HUMHPRTB	56737	1.9066	1.8466	1.8433	1.8165	1.7886	**1.7884**
MPOMTCG	186609	1.9378	1.9058	1.9000	**1.8920**	1.8932	1.8925
PANMTPACGA	100314	1.8752	1.8624	1.8555	1.8556	1.8535	**1.8533**
VACCG	191737	1.7614	1.7614	1.7616	1.7580	1.7583	**1.7582**

15.5 Summary

We have presented a new algorithm for encoding non-repeat parts of DNA sequences. Twelve sequences are used for experimentation. The algorithm proposed is combined with *DNAPack* compression algorithm to evaluate the performance of compression with other algorithms of its kind. The results obtained show that *DNAMem* clearly outperforms conventional text compression algorithms and marginally does better that DNA-specific algorithms. The compression ratio of *DNAMem* is the best for 7 out of 11 sequences considered. However, the time taken by *DNAMem* is around 25 times slower than other similar algorithms.

References

1. Chen, X., Kwong, S., Li, M.: A Compression Algorithm for DNA Sequences and its Application in Genome Comparison. Genomic 12, 512–514 (2001)
2. Grumbach, S., Tahi, F.: Compression of DNA Sequences. In: Data Compression Conference, pp. 340–350 (1993)
3. Grumbach, S., Tahi, F.: A New Challenge for Compression Algorithms Genetic Sequences. Journal of Information processing and Management 30, 866–875 (1994)
4. Matsumuto, T., Sadakane, K., Imai, H.: Biological Sequences Compression Algorithms. In: Genome Information Ser. Workshop Genome Inform, vol. 11, pp. 43–52 (2000)
5. Rivals, E., Delahaye, J.-P., Dauchet, M., Delgrange.: A Guaranteed Compression Scheme for Repetitive DNA Sequences. LIFL Lille I Univerisity Technical Report (1995)
6. Willems, F.M.J., Shtralov, Y.M., Tjalkens, T.J.: The Context Tree Weighting Method: Basic Properties. IEE Transaction on Information Theory 41(3), 653–664 (1995)
7. Sadakane, K., Okazaki, T., Imai, H.: Implementing the Context Tree Weighting Method for Text Compression. In: DCC 2000: Proceedings of the Conference on Data Compression, USA (2000)
8. Rivals, E., Dauchet, M.: Fast Discerning Repeats in DNA Sequences with a Compression Algorithm. In: Proc. Genome Informatics Workshop, pp. 215–226. Universal Academy Press, Tokyo (1997)
9. Sata, H., Yoshioka, T., Konagaya, A., Toyoda, T.: DNA Compression in the Post Genomic Era. Genome Informatics 12, 512–514 (2001)
10. Ziv, J., Limpel, A.: Compression of Individual Sequences using Variable-Rate Encoding. IEE Transactions on Information Theory 24, 530–536 (1978)
11. Ziv, J., Limpel, A.: A Universal Algorithm for Sequential Data Compression. IEE Transactions on Information Theory 23(3), 337–343 (1977)
12. Sadel, I.: Universal Data Compression Algorithm based on Approximate String Matching. Journal of Probability in the Engineering and Informational Sciences, 465–486 (1996)
13. Chen, X., Kwong, S., Li, M.: A Compression Algorithm for DNA Sequences. IEEE Engineering in Medicine and biology Magazine 20(4), 61–66 (2001)
14. Li, M., Badger, J.H., Chen, J.H., Kwong, S., Kerney, P., Zhang, H.: An Information based Sequences Distance and its Application to Whole Mitochondrial Genome. Bioinformatics 17(2), 149–154 (2001)
15. Chen, X., La, M., Ma, B., Tromp, J.: DnaCompress: Fast and Effective DNA Sequence Compression. Bioinformatics 18, 1696–1698 (2002)

16. Ma, B., Tromp, J., Li, M.: PatternHunter-Faster and more Sensitive Homology Search. Bioinformatics 18, 440–445 (2002)
17. Chang, C.: Dnac: A Compression Algorithm of DNA Sequences by Non-Overlapping Approximate Repeats. Master Thesis (2004)
18. Modegi, T.: Development of Lossless Compression Techniques for Biology Information and its Application for Bioinformatics Database Retrieval. Genome Informatics (14), 695–696 (2003)
19. Zhang, Y., Parthe, R., Adjeroh, D.: Lossless Compression of DNA Microarray Images. In: CSBW, pp. 128–132 (2005)
20. Tan, Z., Cao, X., Ooi, B.C., Tung, A.K.H.: The Ed-Tree: An Index for Large DNA Sequence Databases. In: SSDBM (2003)
21. Behzadi, B., Le Fessant, F.: DNA Compression Challenge Revisited: A Dynamic Programming Approach. In: Apostolico, A., Crochemore, M., Park, K. (eds.) CPM 2005. LNCS, vol. 3537, pp. 190–200. Springer, Heidelberg (2005)
22. Apostolico, A., Lonardi, S.: Compression of Biological Sequences by Greedy Off-Line Textual Substitution. In: DCC (2000)

Chapter 16
Exploring Structurally Similar Protein Sequence Motifs

Abstract. Protein sequence motifs are short conserved subsequences common to related protein sequences. Information about motifs is extremely important to the study of biologically significant conserved regions in protein families. These conserved regions can determine the functions and conformation of proteins. Conventionally, recurring patterns of proteins are explored using short protein segments and classification based on similarity measures between the segments. Two protein sequences are classified into the same class if they have high homology in terms of feature patterns extracted through sequence alignment algorithms. Such methodology focuses on finding position specific motifs only. In this chapter, we propose a new algorithm to explore protein sequences by studying subsequences with relative-positioning of amino acids followed by K-Means clustering of fixed-sized segments. The dataset used for our work is most updated among studies for sequence motifs. The various biochemical tests that are found in literature are used to test the significance of motifs and these tests show that motifs generated are of both structural and functional interest. The results suggest that this method may also be applied to closely-related area of finding DNA motifs.

16.1 Introduction

Motif Identification is one of the most important problems covering many applications in protein sequence analysis. It is related to the discovery of portions of protein strands of major biological interest with important structural and functional features. These functional and structural features may include enzyme-binding sites, prosthetic group attachment sites, or regions involved in binding other small molecules. There exists a strong relationship between the amino-acid sequence and corresponding structure. Many biochemical tests suggest that sequence determines conformation completely [1]. For example, conserved blocks within groups of related sequences can often highlight features which are responsible for structural similarity between protein that can be used to predict the dimensional structure of protein. Motifs also carry useful information for generation rules for determining whether a

K.R. Venugopal, K.G. Srinivasa, L.M. Patnaik: Soft Comput. for Data Min. Appl., SCI 190, pp. 303–318.
springerlink.com

Table 16.1 An example of Position Weighted Matrix. The columns denote the positions in the pattern.

	1	2	3	4
A	0.01	0.02	0.01	0.05
C	0.03	0.04	0.01	0.03
D	0.07	0.04	0.03	0.05
E	0.01	0.07	0.01	0.07
...	...	...	...	...
Y	0.00	0.01	0.00	0.02

given sequence belongs to a characteristic protein family. The PROSITE database [2] consists of large number of collection of such patterns used to identify protein families. Core PROSITE sequence patterns are created from biologically significant protein families. Analysis of three-dimensional structure of PROSITE patterns suggests that recurrent sequence motifs imply common structure and function [3].

However, no currently existing definition of "motif" is fully satisfying for the purposes of accurately identifying features that motifs are supposed to represent. There are two popular ways of defining a motif that are widely used today. The first one works with defining representation of motifs given in the form of PSSM(Position Specific Scoring Matrix), also called Position Weight Matrix(PWM) [4,5]. A PWM is a matrix of score values that gives a weighted match to any given substring of fixed length as shown in Table 16.1. It has one row for each symbol of the alphabet, and one column for each position in the pattern. PWM score is defined as $\sum_{j=1}^{N} m_{i(j),j}$, where j represents position in the substring, $i(j)$ is the symbol at position j in the substring, and $m_{i,j}$ is the score in row i, column j of the matrix. In other words, a PWM score is the sum of position-specific scores for each symbol in the substring. Interesting PWMs are those that have high information content, which says how different a PWM is from a uniform distribution.

The second technique defines a motif as a consensus [6,7], a pattern that appears repeatedly with certain number of differences. The methodology of considering motifs are patterns helps explore motifs identification problem in a more exhaustive way, although the relative algorithmic complexity remains high. A motif may be composed of various submotifs separated with variable-length distances that may be found in reasonably efficient way [8]. It is observed that when motifs are extracted as patterns, there is scope for finding further structure from the set of motifs found even when the set is huge. This is generally explored by means of clustering or by combining other models of motifs like PSSMs. It is generally accepted that PSSMs are more appropriate for modeling well-characterized biological feature for the purpose of identifying other occurrences of the feature, however identifying PSSMs themselves is a difficult problem when large data sets are involved.

In this chapter, we explore motifs by defining a new measure of similarity. The segments are clustered by similarity with respect to distances between each pair of amino-acids rather than consensus between the amino-acid positions. The protein sequences are broken down into fixed length segments (non-overlapping) and are not of sliding window type as used in [9-11]. The segments are then grouped using K-Means clustering with respect to similarity of secondary structure. Clusters with similarity higher than a pre-determined threshold are taken to obtain sequence motifs.

16.2 Related Work

Many computational approaches have been introduced for the problem of motif identification in a set of biological sequences which are classified based on type of motifs discovered. Surveys by [6,12,13] in literature cover several motif discovery techniques. Finding PWM-based motifs of complicated nature, that allow gaps, insertions/deletions such as profiles and Hidden Markov Models are studied in [14]. Methods for finding multiple shared motifs within a set of unaligned sequences are given by Gibbs sampling, MEME, probabilistic suffix trees, etc. Finding recurring sequence motifs through K-Means clustering is proposed by Han and Baker [9,10]. In order to overcome the problem of sensitivity of initial points for cluster centers, a greedy method proposed by Zhong et al., [11] is used for initialization method for clustering.

16.3 Motifs in Protein Sequences

Protein motifs may be defined by their primary sequence or by the arrangement of secondary structure elements. The term motif is used in two different ways in structural biology. The first refers to a particular amino-acid sequence that is characteristic of a specific biochemical function. An example is the zinc finger motif, CXX(XX)CXXXXXXXXXXXXHXXXH, which is found in a widely varying family of DNA-binding proteins. The conserved cysteine and histidine residues in this sequence motif form ligands to a zinc ion whose coordination is essential to stabilize the tertiary structure. The second, equally common, use of the term motif refers to a set of contiguous secondary structure elements that either have a particular functional significance or define a portion of an independently folded domain. Along with the functional sequence motifs, the former are known generally as functional motifs. An example is the helix-turn-helix motif found in many DNA-binding proteins [15].

Examples of well-known Motifs: Described below are some commonly found motifs along with their characteristics. Most of the motifs obtained in our experimental results are indicative of conserved helix turns.

Helix-turn-helix Motifs: The helix-turn-helix motif was the first protein motif to be discovered for site specific DNA recognition. This motif has been widely investigated and there exists substantial knowledge of the chemical interactions of specific residues. It characterizes a family of transcription factors.

The motif is characterized by two α-helices connected by a loop. Transcription factors of this type are typically dimeric, each with one helix containing basic amino acid residues that facilitate DNA binding. One helix is typically smaller and due to the flexibility of the loop, allows dimerization by folding and packing against another helix. The larger helix typically contains the DNA binding regions. Basic Helix-turn-Helix proteins typically bind to a consensus sequence called an E-box, CANNTG. The canonical E-box is CACGTG, however some transcription factors bind to different sequences, which are often similar to the E-box Features of the helix-turn-helix motif have been reviewed extensively by Pabo and Sauer [16] and Nelson [17].

One of the two helices is responsible for binding to DNA in a sequence-specific manner, and is referred to as the recognition helix. In most proteins, the second of the two helices is the recognition helix. Residues of the recognition helix interact directly with bases in the major groove of thc DNA. A combination of residues in both the helices are believed to be responsible for maintaining the appropriate angle between the two helices. Proteins with helix- turn-helix motifs share only limited sequence homology in the motif region; the dissimilarity is attributed to the sequence-specific interactions with the bases in the DNA. Most proteins have at most one helix-turn-helix motif.

Beta ribbon is an extremely common motif. Structurally two antiparallel beta strands are connected by a tight turn of a few amino acids between them.

Zinc finger: Two beta strands with an alpha helix end folded over to bind a zinc ion. The structure of each individual finger is highly conserved and consists of about 30 amino acid residues, constructed as a $\beta\beta\alpha$-fold and held together by the zinc ion. The α-helix occurs at the C-terminal part of the finger, while the β-sheet occurs at the N-terminal part. Many transcription factors, regulatory proteins, and other proteins that interact with DNA contain zinc fingers. These proteins typically interact with the major groove along the double helix of DNA in which case the zinc fingers are arranged around the DNA strand in such a way that the α-helix of each finger contacts the DNA, forming an almost continuous stretch of α-helices around the DNA molecule.

Greek key is a type of structural motif with 4 beta strands folded over into a sandwich shape.

Homeodomain Motifs Proteins containing the homeodomain motif play an important role in plant and animal development. The homeodomain [18] motif is made up of three α-helices and an extended N-terminal arm. The first and second α-helices pack against each other in an anti-parallel arrangement, while the third α-helix lies perpendicular to them. The third helix is the recognition helix; like its counterpart in the helix-turn-helix motif, it interacts with DNA in the major groove and provides the DNA-binding specificity. However, unlike the helix-turn-helix unit, the 60-residue homeodomain forms an independent folded structure and can independently bind to DNA. It is interesting to note that the homeodomain motif contains a canonical helix-turn-helix structure. Mutational and evolutionary analysis and crystal structures of these domains are also available.

Motif detection: There are many definitions for the protein sequence motif. One of the definition for the protein sequence motif is given as follows [19]:

Definition 1. Given an integer h, a set of m sequences $S_1, S_2, S_3, \ldots, S_m$ on the alphabet $A, C, D, E, F, G, H, I, K, L, M, N, P, Q, R, S, T, V$, and threshold $k \leq m$, a motif is a pattern occurring in at least k sequences of $S_1, S_2, S_3, \ldots, S_m$ with at the most h errors.

Definition 2. Given a sequence S of length m, i.e., $S = s_{1\ldots m}, S_{i\ldots j}, 1 \leq i \leq j \leq m$ denotes the subsequence $s_i \ldots s_j$ starting from position i and ending at j.

The motifs to be extracted are usually a collection of subsequences that are constrained and represented by certain expressions. There are basically two types of sequence motifs based on their representation –deterministic motifs and probabilistic motifs. Deterministic motifs either match or do not match a sequence while probabilistic motifs is given a score of probability of matching against a sequence. The definition given is suited for a deterministic motif. For the probabilistic motif, motifs can be considered as the patterns ranking highest when matched against the given set of proteins among all possible patterns. To extract deterministic motifs, we consider a sequence motif as the common subsequence of a set of protein sequences. The most simple way to represent a motif is to use a consensus string that is a single string composed of most likely residues on each position with substitutions and wildcards. A regular expression is a natural choice to represent wider range of motifs and is a powerful notational algebra available to describe strings and sequences [20].

The PROSITE [2] database is a well-established source of protein motifs. It records a large number of classified motifs into different families. For each family, some motifs are given to characterize the family. PROSITE uses regular expressions to represent the motifs. Database is a well-established source of protein motifs. It records a large number of classified motifs into different families. For each family, some motifs are given to characterize the family. PROSITE uses regular expressions to represent the motifs. For example, a PROSITE pattern may look like $< [AH] - x - A - x(4) - EP >$. In the regular expression used by Prosite, [ ] means a match of included amino-acid in that position, x is for any amino acid, () is used to specify that range, is to show the excluded amino acids, ‘,’ denotes the end of the pattern, ‘<’ is for N-terminal and ‘>’ is for C-terminal.

16.4 Algorithm

The presence of functionally conserved regions in proteins by itself confirms the fact that amino-acids do not appear independently. The structure of motifs is decided by the arrangement of amino-acids, which give them their characteristic function. The conserved regions contain amino-acids that seem to occur in groups, with some gaps between them. With this assumption, the algorithm tries to detect conserved regions using similarity between relative positions of amino-acids across protein families.

We use a Pairwise Relative Distance Matrix(PRDM) for distance measures between protein segments. The PRDM is calculated for each protein segment. The PRDM is a three dimension matrix with two dimensions used to represent each amino-acid pair and the other is used for distance between corresponding pairs. For example, $p[A][S][3] = 2$ denotes that there are two instances of regular expression $A - x(2) - S$ in the corresponding segment.

Algorithm 1. Motif Extraction Algorithm

Input Protein sequence and corresponding secondary structure
Output Structurally similar motifs
1. Divide the protein sequence and corresponding secondary structure sequence into segments of size s
2. Initialize pairwise-relative distance matrices(PRDMs) $p[i][j][k] \leftarrow 0$, $1 \leq i, j \leq 20$ and $0 < k < s$
3. Foreach segment increment $p[X][Y][d]$ if position of amino-acid Y – position of amino-acid X = d
4. Compute the distances between segments from their corresponding PRDM
5. Apply Improved K-Means clustering and partition the initial segments
6. Calculate the secondary structure similarity among segments within one cluster
7. Extract consensus sequences from clusters with structural similarity $> threshold$

The Improved K-Means algorithm proposed by Zhong et al., is used for clustering segments. The greedy initialization overcomes the potential problems of random initialization and tries to choose suitable initial points so that final partitions can represent the underlying distribution of the data samples more consistently and accurately. Each point is represented by one PRDM corresponding to one segment. In greedy initialization method, the clustering algorithm is performed for several iterations during each run. After each run, initial points, which can be used to form cluster with good structural similarity are chosen and their distance is checked against that of all points already selected in the initialization array. If the minimum distance of new points is greater than specified distance, these points is added to the initialization array. Satisfaction of the minimum distance can guarantee that each newly selected point is well separated from all the existing points in the initialization array and will potentially belong to different natural clusters. This process is repeated several times until the specified number of points is chosen. After this procedure, these selected points are used as initial centers for K-Means clustering algorithm [11].

16.5 Experimental Setup

This section introduces the experimental parameters, the dataset, distance measures between segments and similarity measure of clusters. A measure for comparing the results obtained by using city-block distance metric for calculating distance between segments and the relative-distance metric is given.

Algorithm 2. Improved K-Means algorithm

while the number of initial points discovered is less than the total number of clusters **do**
  Run the traditional K-means algorithm for a fixed number of iterations on the whole sample space
  Assess the structural similarity of clusters created by each initial point
  if the structural similarity for one cluster is bigger than or equal to a given threshold **then**
    Check the minimum distance of the point creating this cluster with existing points in the initialization array
    if the minimum distance is bigger than threshold **then**
      This new point is included into the initialization array
    end if
  end if
end while

Dataset: The dataset used in this work is obtained from the Protein Sequence Culling Server(PISCES) [21]. Two thousand protein sequences are taken with identity cutoff of 30%, resolution cutoff of 2.1 and the R-factor cutoff of 1.0. The PISCES produces a satisfactory non-redundant database.

Generation of Segments: The protein segments are obtained from the protein sequences by taking eight successive residues at a time. Each segment therefore represents eight continuous positions from the protein sequence. Sixty thousand sequence segments from 2000 protein sequences are obtained. These segments of eight positions are classified into different groups using improved K-Means algorithm. Each segment is defined by a Position Relative Distance Matrix(PRDM). For each pair of amino acids (20×20) along with the possible distance between them (< 8) the matrix has a value to denote the number of such instances in the corresponding segment. Hence each segment is represented by a $20 \times 20 \times 8$ matrix.

Distance Measures: Each centroid of a sequence cluster is represented by a PRDM. The values of centroid is first calculated on an average measure of frequency and then the PRDM of the segment chosen as a closest point to the centroid. This is similar to what is done in K-Menoids method. The approach is reasonably resistant to outliers.

Definition 3. Let $\mathscr{S}_1 = A_1, A_2, \ldots, A_8$ and $\mathscr{S}_2 = B_1, B_2, \ldots, B_8$ be two segments. Let $A_i, A_j (i < j)$ and $B_k, B_l (k < l)$ be any two amino acid pairs taken from $\mathscr{S}_1$ and $\mathscr{S}_2$ respectively. If $A_i = B_k$, $A_j = B_l$ and $j - i = l - k$ then $\mathscr{S}_1$ and $\mathscr{S}_2$ are said to have a *pairmatch* for $< A_i, A_j, j - i >$.

Definition 4. The $cumulative pairmatch(c_p)$ between two segments is the total number of *pairmatches* considering combinations of all amino-acids pairs and distances.

For example, c_p between *AACC* and *ACAA* is 2 (for *AC* and *AA* both of distance 1). The maximum c_p for segment of size n is $n(n+1)/2$. Since the size of segment in this case is 8 the maximum c_p is 28. Therefore, the value of the c_p increases with more number of similar segments. Since the convention is to use lower values to denote closer distances between points, we calculate the distance between two segments as follows:

$$distance = 2^{28} - 2^{c_p}$$

The exponential scale is used instead of linear since the probability of pairmatches between segments decrease exponentially.

Secondary Structure Assignment: Definition for Secondary Structure of Proteins(DSSP) [22] is used for determining the secondary structure of proteins. We follow the convention of using H to represent helices, E for sheets and C for random coils.

Intra-cluster Similarity Measures: The following formula calculates the level of structural similarity. Structural similarity for a given cluster (%) is [9]

$$\frac{\sum_{i=1}^{l} max(P_{i,H}, P_{i,E}, P_{i,C})}{l}$$

l is the window size. $P_{i,H}$ is the frequency of occurrence of helices among the sequence segments for the cluster in position. $P_{i,C}$ and $P_{i,E}$ are similarly defined. The secondary structure with the maximum frequency is used for representing the common structure in that position. The average results of the maximum frequency from all positions of a given window show the structural similarity level for a given cluster. If the structural similarity for secondary structure within the cluster exceeds 70%, the cluster can be considered structurally similar [23]. If the structural similarity for secondary structure within the cluster is between 60% and 70%, the cluster can be considered weakly structurally similar.

Evaluation of Performance: The percentage of sequence segments belonging to clusters with high structural similarity and the number of clusters with high structural similarity are two measures to evaluate the performance of clustering algorithm [11]. The number of segments belonging to clusters with high structural similarity and number of clusters with high structural similarity are averaged from results of three runs. The percentage of sequence segments belonging to clusters with the structural similarity greater than 60% is calculated by dividing the sum of all sequence segments belonging to clusters with structural similarity greater that 60% by total number of sequence segments in the database.

16.6 Experimental Results

In this section, we compare the experimental results of the city-block distance based algorithm and relative-distance measure algorithm. The motifs generated by relative-distance measure algorithm is shown along with brief explanations of their structural and biological interest based on biochemical experiments published in literature like [24-28], etc..

Comparison of PWM-based algorithm with PRDM algorithm: In Table 16.2, the average percentage of sequence segments belonging to clusters with high-structural similarity for the City-block based algorithm and PRDM algorithm is given. The first column of Table 16.2 shows the algorithm with different distance measures. Four distance measures given for each approximately correspond to each other. The first column in PRDM refers to the *cumulativepairmatch*(c_p) value. Lower the c_p larger is the distance between the segments, hence the decreasing trend in numbers. For PWM, the first column is the value of the minimum distance between initial cluster centroids for K-Means algorithm, measured with city-block metric.

The second column in Table 16.2 gives the average percentage of sequence segments belonging to clusters with structural similarity greater than 60% from three runs. The third column of Table 16.2 gives the standard deviation of the percentage of segments belonging to structural similarity greater than 60%. The fourth column to Table 16.2 gives the average percentage of sequence segments belonging to clusters with structural similarity greater than 70% from three runs. The fifth column of Table 16.2 gives the standard deviation of the percentage of segments belonging to structural similarity greater than 70%. Analysis of the clustering process of the both the algorithms reveals that the initial points chosen are very close to each other. However, the results show that the number of high similarity clusters steadily improve with increase in distance for both the algorithms. Compared to PWM, the PRDM shows slightly better performance with respect to the percentage of segments included, however the standard deviation measure is better for PWM. For clusters

Table 16.2 Comparison of percentage of sequence segments belonging to clusters with high structural similarity

		CITY-BLOCK		
Distance Measures	$> 60\%$	$> 60\%$	$> 70\%$	$> 70\%$
1100	26.43%	1.31	10.61%	0.68
1200	30.74%	0.51	11.26%	0.49
1300	31.91%	0.61	13.96%	0.38
1500	35.18%	0.53	15.16%	0.39
		PRDM		
26	24.05%	1.21%	8.28%	1.71
24	29.69%	1.12%	11.18%	1.01
22	33.33%	1.42%	13.18%	1.11
20	36.14%	1.22%	15.49%	1.10

Table 16.3 Comparison of Number of Clusters with High Structural Similarity

Distance Measures	> 60%	> 60%	> 70%	> 70%
		CITY-BLOCK		
1100	191	4.34	61	3.18
1200	181	4.47	66	3.69
1300	196	3.44	78	3.08
1500	221	3.38	81	3.10
		PRDM		
26	151	4.01	44	2.27
24	190	4.92	47	2.51
22	213	4.22	48	2.92
20	234	3.82	51	2.55

with structural similarity greater than 70% the trend is nearly the same as that of structural similarity greater that 60%. Table 16.3 shows the number of clusters exceeding given structural similarity thresholds for the city-based and PRDM-based algorithms. The first column shown is the same as in Table 16.2. The second column shows the average number of clusters with structural similarity greater than 60% from three runs rounded off to nearest whole number. The third column indicates the standard deviation for the number of clusters with structural similarity greater than 60% for three runs. The fourth column shows the average number of clusters with structural similarity greater than 70% from three runs rounded off to nearest whole number. The fifth column indicates the standard deviation for the number of clusters with structural similarity greater than 70% for three runs. The improvement in the number of clusters for PRDM is much higher than of PWM. For PRDM the increase in clusters from distance $c_p = 26$ to $c_p = 20$ is 55%, while for city-block the increase in clusters from distance 1100 to 1500 is only 15%. However the standard deviation of the number of clusters for PRDM is lower than that of city-block. The comparison between results obtained shows that the PRDM-based algorithm is not significantly better than city-block metric in all respects. However, the purpose of the algorithm is not just to increase the number of high similarity clusters or the percentage of segments belonging to high similarity clusters but to intuitively explore local sequence motifs that are expressed differently.

Sequence Motifs: About 70 local sequence motifs indicating common structure are discovered in this work. Various biochemical tests that are found in the literature are applied to the best among these for uncovering their biological properties. Some of the motifs that appeared nearly the same as the ones already found are not reported here. Most of the motifs have helices as predominant secondary structure as expected. Analysis of some of the motifs obtained suggest that PRDM uncovers subtle motifs which are sensitive to relative positioning of amino-acids that are not obtained using the PWM city-block based approach.

Among the motifs obtained 15 of those that are biologically meaningful are given in the tables that follow. The following format is used to represent each sequence in motif table.

- The first row gives the Motif pattern in PROSITE format. Each of these patterns are of size eight. Only significant amino-acid residues are shown explicitly($>$ 15%).
- The second row gives the representative secondary structure, i.e., the secondary structure sequence that best describes the motif
- The third row shows the Hydrophobicity index. The Hydrophobicity is the sum of the frequencies of occurrence of alanine, valine, isoleucine, leucine, methonine, proline, phenylalanine and tryptophan. There are denoted by high(H),medium(M) or low(L) depending on the Hydrophobicity index. H if greater than 0.7, L if less that 0.3 and M otherwise.
- The fourth row gives the number of segments present in the cluster from which the motif is obtained.
- The fifth row gives the structural similarity measure for the corresponding cluster
- The sixth row indicates the statistical significance of motif expressed as Z-score (percentile based).

Pattern 1 (Table 16.4) has conserved Serine residues for the first half of the motif. They are hydrophilic in nature and outer regions of proteins are found to be rich with them. Pattern 2(Table 16.5) has conserved regions of leucine and is a amphipathic helix motif and is one of the common structural motifs present in proteins [29]. Possible functions of amphipathic helices have been tested in [30]. Pattern 3

Table 16.4 Pattern 1: Conserved Serine Residues for the First Half of the Motif

Motif	[ES]-S-S-[DS]-S-x(3)
Sec.Str	C-C-C-C-H-H-H-H
H	L-L-L-L-L-L-L-M
N.Seg	101
Struct.Sim	63.31%
Z-Score	97.45

Table 16.5 Pattern 2: Conserved Regions of Leucine

Motif	[LV]-[LV]-[LE]-L-x-L-x(2)
Sec.Str	H-H-H-H-H-H-H-H
H	H-H-H-H-H-H-M-M
N.Seg	67
Struct.Sim	71.3%
Z-Score	96.12

Table 16.6 Pattern 3: Conserved Glutamic Acid and Lysine Bases

Motif	E-E-x(4)-[EK]-[EK]
Sec.Str	H-H-H-H-H-H-H-H
H	L-L-H-H-H-H-L-M
N.Seg	162
Struct.Sim	66.6%
Z-Score	98.12

Table 16.7 Pattern 4: Conserved Glutamic Acid and Lysine

Motif	[AK]-[AK]-[K]-x-[EK]-x-K-K
Sec.Str	H-H-H-H-H-H-H-H
H	L-L-H-H-H-L-M-L
N.Seg	151
Struct.Sim	65.1%
Z-Score	97.61

Table 16.8 Pattern 5: Motif with Complete Helix and Low Hydrophobicity

Motif	E-E-K-L-E-x-K-K
Sec.Str	H-H-H-H-H-H-H-H
H	L-L-L-M-L-L-L-L
N.Seg	234
Struct.Sim	64.1%
Z-Score	95.12

Table 16.9 Pattern 6: Coil to Helix Transitional Motif

Motif	L-G-E-V-[EK]-L-x(2)
Sec.Str	C-C-C-C-H-H-H-H
H	L-M-H-L-M-L-L-M
N.Seg	100
Struct.Sim	60.56%
Z-Score	97.16

(Table 16.6) shows motif that has conserved Glutamic acid and Lysine bases. Glutamic acids are acidic while lysine is basic in nature. The ability to form ionic bonds with charged molecules suggest its function in salt bridges [24]. Pattern 4 (Table 16.7) shows motifs with conserved glutamic acid and lysine. The secondary structure is completely helical just like in Pattern 3. This pattern can be classified as a polar helix. Major functional locations on protein surfaces are the result of charged amino acids [25].

Table 16.10 Pattern 7: Coil to Sheet Transitional Motif with varying Hydrophobicity

Motif	L-G-L-L-V-V-V-x
Sec.Str	C-C-C-C-E-E-E-E
H	L-M-L-M-H-H-M-L
N.Seg	110
Struct.Sim	63.56%
Z-Score	95.34

Table 16.11 Pattern 8: Conserved Region of Glycine

Motif	L-[SP]-[EP]-D-P-D-[DP]-V
Sec.Str	C-C-C-C-H-H-H-H
H	M-H-M-M-M-L-L-M
N.Seg	211
Struct.Sim	70.45%
Z-Score	96.15

Table 16.12 Pattern 9: Completely Conserved Regions of Glycine

Motif	G-G-G-G-[GS]-[GS]-G-[GS]
Sec.Str	C-C-C-C-C-C-C-C
H	H-H-H-M-M-M-M-H
N.Seg	48
Struct.Sim	77.1%
Z-Score	98.89

Table 16.13 Pattern 10: Sheet-Coil-Helix Motif

Motif	[LV]-x-[LV]-x-V-V-D-x
Sec.Str	E-E-E-E-C-C-C-H
H	M-M-M-H-M-M-H-M
N.Seg	51
Struct.Sim	71.1%
Z-Score	97.67

Pattern 5(Table 16.8) has a motif with complete helix and low Hydrophobicity. This is due to the presence of glutamic acids and lysine. Pattern 6 (Table 16.9) shows a coil to helix transitional motif. The conserved glycine is found to be predominant in the second position with frequency of occurrence exceeding 30%. Pattern 7(Table 16.10) is a coil to sheet transitional motif with varying Hydrophobicity. The conserved region of glycine is responsible of disruption of sheets and helices– Pattern 8(Table 16.11).

Table 16.14 Pattern 11: Proline which Contains an Unusual Ring

Motif	[P]-[ST]-T-[ST]-[S]-T-[PST]-x
Sec.Str	C-C-C-C-C-C-C-C
H	M-L-L-L-L-L-L-M
N.Seg	34
Struct.Sim	71.6%
Z-Score	98.10

Table 16.15 Pattern 12: Sheet-Coil-Helix Motifs

Motif	V-V-[IV]-G-R-[AR]-R-A
Sec.Str	E-E-E-C-C-H-H-H
H	H-H-H-M-L-L-L-L
N.Seg	80
Struct.Sim	73.65%
Z-Score	95.6

Table 16.16 Pattern 13: Sheet-Coil-Helix Motifs Hydrophobicity Transition

Motif	G-x-x-A-[AR]-[AV]-V-V
Sec.Str	E-E-E-C-C-H-H-H
H	L-L-M-M-M-M-H-H
N.Seg	21
Struct.Sim	73.65%
Z-Score	94.6

Table 16.17 Pattern 14: Alternating Patterns of Sheets, Helices and Coils

Motif	[ILV]-x(5)-D-x
Sec.Str	E-H-E-H-E-E-C-C
H	H-M-H-M-M-L-L-L
N.Seg	220
Struct.Sim	67.66%
Z-Score	98.81

Pattern 9(Table 16.12) has completely conserved regions of glycine and is essentially made up only of coils as expected. Since glycine is the smallest amino acid, rotates easily, adds flexibility to the protein chain. However too much fragility makes the motif structurally less rigid. A related pattern is Pattern 11 (Table 16.14) here the effect is due to proline which contains an unusual ring to the N-end amine group, which forces the CO-NH amide sequence into a fixed conformation. It disrupt protein folding structures like α-helix or β-sheet. Common in collagen, where it undergoes a post-translational modification to hydroxyproline.

Table 16.18 Pattern 15: Helix-sheet-helix with conserved alanine

Motif	[AL]-x(6)-A
Sec.Str	H-E-E-E-E-E-E-H
H	L-L-M-M-M-M-L-L
N.Seg	21
Struct.Sim	70.67%
Z-Score	94.15

Pattern 10 is again a sheet-coil-helix motif with high Hydrophobicity. Both Pattern 12(Table 16.15) and Pattern 13(Table 16.16) are sheet-coil-helix motifs with smooth Hydrophobicity transition. Pattern 14 (Table 16.18) is an interesting structural motif with alternating patterns of sheets, helices and coils. It has a hydrophobic aminoacids on one side and an acidic residue at the other. Pattern 15(Table 16.18) is a Helix-sheet-helix with conserved alanine on both ends.

16.7 Summary

In this chapter, a new algorithm is proposed to explore protein sequences by studying subsequences with relative-positioning of amino acids followed by K-Means clustering of fixed-sized segments. The purpose of this approach is to find different motifs that have hidden relative-distance significance between its residues along with consensus of secondary structure. The most updated dataset from PISCES is used to generate protein segments. The size of motifs discovered are limited to eight which is the size of segment chosen. A decrease in segment size would reduce the structural significance while an increase would suffer statistical support. A comparison to PWM based algorithm is also made. The experimental results show that motifs generated are of structural and functional interest and hence are biologically meaningful.

References

1. Karp, G.: Cell and Molecular Biology(Concepts and Experiments), 3rd edn. Wiley, New York (2002)
2. Hulo, N., et al.: Recent Improvements to the Prosite Database. Nucl. Acids Res. (1994)
3. Kasuya, A., Thornton, J.M.: Three-Dimensional Structure Analysis of Prosite Patterns. Journal of Molecular Biology 286(5), 1673–1691 (1999)
4. Gribskov, M., McLachlan, A., Eisenberg, D.: Prole Analysis: Detection of Distantly Related Proteins. Proceedings of National Academy of Sciences 84(13), 4355–4358 (1987)
5. Hertz, G.Z., Stormo, G.D.: Escherichia Colipromoter Sequences: Analysis and Prediction. Methods in Enzymology 273, 30–42 (1996)
6. Brazma, A., Jonassen, I., Edhammer, I., Gilbert, D.: Approaches to the Automatic Discovery of Patterns in Biosequenes. Journal of Computational Biology 5, 279–305 (1998)
7. Vanet, A., Marson, L., Sagot, M.F.: Promotor Sequences and Algorithmical Methods for Identifying Them. Research in Microbioloby 150, 779–799 (1999)
8. Marson, L., Sagot, M.F.: Algorithms for Extracting Structured Motifs using Suffix Tree with an Application to Promoter and Regulatory Site Consensus Identification. Journal of Computational Biology 7, 345–362 (2000)

9. Han, K.F., Baker, D.: Recurring Local Sequence Motifs in Proteins. Journal of Molecular Biology 251(1), 176–187 (1995)
10. Han, K.F., Baker, D.: Biophysics - Global Properties of the Mapping Between Local Amino Acid Sequence and Local Structure in Proteins. Proceedings of National Academy Sciences, USA 93, 5814–5818 (1996)
11. Zhong, W., Altun, G., Harrison, R., Tai, P.C., Pan, Y.: Improved K-Means Clustering Algorithm for Exploring Local Protein Sequence Motifs Representing Common Structural Property. IEEE Transactions on Nanobioscience 4(3) (2005)
12. Brejova, B., DiMarco, C., Vinar, T., Hidalgo, S.R., Holguin, C., Patten, C.: Finding Patterns in Biological Sequences - Project Report for CS79g. University of Waterloo (2000)
13. Rigoutsos, L., Floratos, A., Parida, L., Gao, Y., Platt, D.: The Emergency of Pattern Discovery Techniques in Computational Biology. Metabolic Engineering 2, 159–177 (2000)
14. Durbin, R., Eddy, S., Krough, A., Mitchison, G.: Biological Sequence Analysis: Probabilistic Models of Protein and Nucleic Acid. Cambridge University Press, Cambridge (1998)
15. Petsko, G.A., Ringe, D.: Proteins Structure and Function. New Science Press (2003)
16. Pabo, C.O., Sauer, R.T.: Transcriptional Factors: Structural Familes and Principle of DNA Recognition. Annals of Revolutionary Biochemistry 61, 1053–1095 (1992)
17. Nelson, H.C.M.: Structure and Function of DNA-Binding Proteins. Current Opinion in Genetics and Development 5, 180–189 (1995)
18. Scott, M.P., Tamkun, J.W., Hartzell, G.W.: The Structure and Function of the Homeodomain. Biochemistry Biophysics Acta 989(1), 25–48 (1989)
19. Crochemore, M., Sagot, M.: Motifs in Sequences: Localization and Extraction. In: Hand book of Computational Chemistry. Marcel Dekker Inc., New York (2001)
20. Heger, A., Lappe, M., Holm, L.: Accurate Detection of Very Sparse Sequence Motifs. In: Proceedings of RECOMB, pp. 139-147 (2003)
21. Wang, G., Dunbrack Jr., R.L.: Pisces: Recent Improvements to a PDB Sequence Culling Server. Nucleic Acids Research 33 (2005)
22. Kabsh, W., Sander, C.: Dictionary of Protein Secondary Structure: Pattern Recognition of Hydrogen-Bonded and Geometrical Features. Biopolymers 22, 2577–2637 (1983)
23. Sander, C., Schneider, R.: Database of Homology Derived Protein Structures and the Structral Meaning of Sequence Alignment. Proteins Structural Functional Genetics 7(2), 121–135 (1967)
24. Berg, J.M., Tymoczko, J.L., Stryer, L.: Biochemistry, 5th edn. W H Freeman, New York (2002)
25. Robertson, A.D.: Intramolecular Interactions at Protein Surfaces and Their Impact on Protein Function. Trends Biochemistry Sciences 27, 521–526 (2002)
26. Kyte, J., Doolitle, R.F.: A Simple Method for Displaying the Hydropathic Character of Protein. Journal of Molecular Biology (157), 105–132 (1982)
27. Zimmerman, J.M., Eliezer, N., Simha, R.: The Characterization of Amino Acid Sequences in Proteins by Statistical Methods. Journal of Theoretical Biology (2001)
28. Finer-Moore, J., Stroud, R.M.: Amphipathic Analysis and Possible Formation of the Ion Channel in an Acetocholine Receptor. Proceedings of National Academy of Sciences, USA 81(1), 155–159 (1984)
29. Segrest, J.P., De Loof, H., Dohlman, L.G., Brouilette, C.G., Anantharamaiah, G.M.: Amphipathic Helix Motif: Classes and Properties. Protein Structural Functional Genetics 8(2), 103–117 (1990)
30. Kaiser, E.T., Kezdy, F.J.: Amphiphilic Secondary Structure: Design of Peptide Hormones. Science 223, 249–255 (1984)

Chapter 17
Matching Techniques in Genomic Sequences for Motif Searching

Abstract. Sequence retrieval serves as a "preprocess" for a number of other processes including motif discovery, in which obtained sequences are scored against a consensus before being recognized as a motif. This depends on the way sequences are stored prior to retrieval. The usage of two bits for representing genomic characters is optimal storage wise, however does not provide any details regarding length of repetitive characters or other details of positional significance. The intent of the chapter is to showcase an alternative storage technique for the sequence and its corresponding retrieval technique. We represent our technique with the use of integers for clarity of understanding. With the bit equivalent of the integers used in actual representation we could minimize storage complexity significantly. We give a clear picture of the requirements of a storage technique from a motif discovery perspective before showcasing our proposal.

17.1 Overview

Bioinformatics is a vast field that includes integrating the fundamentals of genomic and proteomic sciences with advanced mathematics and computational sciences [1]. The field of bioinformatics which we address here is Motif Discovery. A simple explanation is given by Leung-Chin [2] where they state, in order to start the decoding process (*gene expression*), a molecule called the *transcription factor* binds to a short region (*binding site*) preceding the gene. One kind of transcription factor can bind to the binding sites of several genes to cause these genes to co-express. These binding sites have similar patterns called *motifs*.

Motivation: Among existing methodologies, the storage technique involves the storage of characters in plaintext. Each of the subsequences which are retrieved from the sequence are then evaluated or scored with respect to a motif and returned. For the purpose of subsequence retrieval there exist a number of algorithms which utilize integer evaluation to find the matches. One such method is the dated Needleman-Wunsch Algorithm [10] which makes use of simple arithmetic to evaluate sequences

K.R. Venugopal, K.G. Srinivasa, L.M. Patnaik: Soft Comput. for Data Min. Appl., SCI 190, pp. 319–330.
springerlink.com

for the purpose of retrieving subsequences. With the intention of further promoting this method of arithmetic evaluation in order to retrieve subsequences we propose the work which is reflected in this chapter.

While high performance methods do exist in retrieving substrings, our approach looks at the problem from a storage perspective. The need is to have a storage technique for existing sequence databases which could result in a more informative method of storing sequences. A storage method with the ability to provide us with information of the genomic sequences, to favor compression and also compress to an extent at runtime, to be able to retrieve subsequences from it and so on. The proposed method provides us with these fundamental benefits and also promote the use of simple arithmetic evaluation of the subsequence and the main sequences.This chapter proposes such a method in which integers or bit sequences are used to represent the genomic characters and which also signify various positional details. For example, the number of consecutive characters is readily available as an entry in the table. Also, in cases where consecutive characters are present, the entire length gets reduced down to one single integer value thereby saving memory. The method provides a table storage for the main sequence. The entries in the table, as mentioned earlier, can be integers as well as bit sequences. Each of the entries made in the main table signify two main things:

- If consecutive genomic characters are present, the length of such a pattern.
- Integer redundancy if there is no change in the current entry in the table to the next with respect to the substring, thereby favoring compression since redundancy can be accounted for in popular compression algorithms.

The method also provides the critical task of retrieving subsequences with the use of simple arithmetic comparison involving the interdifferences between integers of the main sequences (entries in the table can be looked as an array of integers) and the integers of the subsequence. Method for mutations can also be incorporated by having a tolerance between the differences between the integer entries.

17.2 Related Work

Motif discovery has attracted a lot of attention from scientists from various computational backgrounds. Numerous approaches have been taken to motif discovery considering different constraints and have different trade-offs based on the constraints. Motif discovery algorithms can be broadly classified into two types: deterministic and non-deterministic. The deterministic algorithms are usually regular expression based and are exhaustive, that is the flexibility of tolerance is minimal in these algorithms. On the other hand non-deterministic algorithms are non-exhaustive and are stochastic in nature. That is, the flexibility of tolerance is greater in non-deterministic algorithms and thus can result in different sequences matches in different runs. Pratt [3] and TEIRESIAS [4] are examples of a deterministic algorithm. These algorithms, like many other, use regular expressions to represent sequences. Other enumerated approaches which are based more on mathematically intensive methods also exist, such as MDScan [5]. MDScan use higher order Markov model and a scoring technique to identify motifs.

Non-deterministic algorithms as earlier mentioned use algorithms which are non-exhaustive. Most of the methods which come in this category use positional weight matrices or matrices whose score depend on positions. These are probabilistic models. It has been a trend to prefer stochastic algorithms over exhaustive, deterministic algorithms simply because they are more capable of containing more information than regular expressions. Some of the popular motif discovery tools are Gibbs Sampling [6], CONSENSUS [7], MEME [8] and BioProspector [9] among others.

Genetic algorithms is another method which is employed for the purpose of motif discovery. They are stochastic in nature and utilize different forms of fitness functions usually assuming that only one motif per sequence is present. An evolutionary concept is utilized in genetic algorithms which consider populations, which in turn further evolve incorporating various mutations. Some populations continue to evolve resulting closer to the desired solution while other weaker populations become extinct.

17.3 Introduction

Sequence analysis is used extensively in motif discovery. A motif can be thought of as a sequence of characters that is widespread and is, in some way, biologically significant. Given a main string *M*, of a certain length, our task is to find substrings *Si* , that are similar if not identical to a subset of *M*. The alphabet derives its name from the four bases: adenine (A), cytosine (C), guanine (G) and thymine (T). Thus we refer to it as the *genetic alphabet*. These characters in a sequence define an organism's DNA. It is estimated that over 3 billion of these genetic letters constitute the human genome. A point of consideration is the repetitiveness of sequences. This observations help in various ways: compression and motif discovery for example. The human genome is about 20-40 percent repetitive. The extent of repetitiveness varies from organism to organism.

Before, exploring our algorithm we briefly discuss the relevance of the *(l, d)* problem. We primarily use the notations used in the *(l, d)* problem analysis. Given a length of string *l*, and each substring (that is of length *l* and is some S_i) can vary from the similar pattern found in the main string by *d* characters. That is given the main string *M*=GTACAGTTACAT and *S*=GTAC. Here *l*=4 and say that *d*=1 then, potential similar strings could be all variations of G,T,A,C which are of length 4 and differ from *S* by one character (GTAC, GTAA, GAAC and so on). In our algorithm the substring matches are done based on the value of *d*. If *d*=1 then we produce substrings from the main string which differ by 1 character, as output. Thus, resulting in similar sequences with a flexibility tolerance of *d* (in this case *d*=1).

Notations: We now define some important terms and sets:
Substring S: This is the input string for which we search. The actual variations of this substring result in S_i where i = 1,2,3...
Main string M: This is the string in which we search for S_i.

(l,d): l is the length of the substring and each of its variations. The number of variations depend on the value of d. l has to be lower than the length of M and should accommodate for d, i.e., $l \pm d$.
The character set: considered here is from the set {G,A,T,C}.

Degenerative Motifs: Since the substrings returned by the algorithm depend on the value of d, we can use a state like diagram to generate the substrings. Consider that the value of S={GTAGCGTA} and d=1, we thus can trace and obtain the following possible matches. We explain taking, for example,
M={ATCGGTTCGGTACGGTAC}.

At the first scan *S[0]* does not match with *M[0]*, so a count variable increases by 1. Beginning the scan from *M[1]*, this results in yet another mismatch and the count variable has to be incremented by 1 again. However, since the user specified d=1 this increment is not allowed. It is to be noted that at each iteration the count variable must be reset to 0. On the tenth iteration where *M[9]*=G, we proceed with the state like diagram as shown in Figure 17.1.

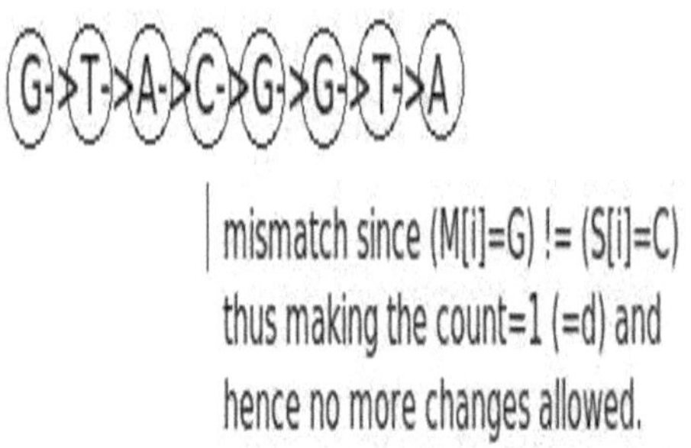

Fig. 17.1 Example of d-alternate (d=1) substring returned as result

Brute Force Analysis of Subsequence Retrieval: After discussing degenerative motifs we now look at the algorithm for obtaining the patterns which differ from the main substring S by at most d mutations. This is a straightforward approach to the *(l,d)* problem and incurs huge costs if the length of M is large.

We take an extra input from the user denoting the start and end points between which the user wishes the degenerative motifs to exist. If a user does not enter any *(start,end)* then the entire string is scanned. Algorithm 7 is a brute force approach to obtaining the subsequences. Some runs of the algorithm, with an example, are shown below:

M={GTACGCGACTGCACGTGCACGTGCACGGTGCA
CGTGCACGTGGCCCAACACAGATCGATGCATAGC
TAGATCGATAGCTAGATCGATCAGATCGATAGAT
CGATAGCTAGCATAGAGAGACTCGATAGATCGAT
AGCTAGACTAGCATCGATCAGCATCGACTACGAT
CAGCTACGACTAGCACTCGACAG}
S={AGCATAA}
d=1

Results in one subsequence returned : {AGCATAA}
With d=2 the subsequences returned are :
TGCATAG
AGCATAG
AGCATCG
AGCATCG

From the two *for* loops it is clear that the complexity of the algorithm is $O(n \ x \ n)$. It is observed that as d increases, the distance between the degenerative motifs and the desired sequence increases. The following graph in Figure 17.2 indicates the variance associated between the parameters: length (of S), d and the time taken to search M for S.

Algorithm 1. *Brute Force Approach (n)*

```
len1 ← length of M
len2 ← length of S
if start, end not defined then
    start ← 0
    end ← len1
end if
for i ← start; i ← end; i++ do
    for k ← 0; k ← len2; k++ do
        if M[i+k] == S[k] then
            skip
        else
            if M[i+k]! = S[k] and count < d then
                count++;
            end if
        else
            break;
        end if
    end for
end for
if k == len2 then
    return detected string
end if
count ← 0
```

17.4 Alternative Storage and Retrieval Technique

Table Construction: Here we discuss an alternative method that could be adopted for storing the content of M. Since the character set is specific to 4 bases, G,T,A,C, we use each of these four characters as four main indexes. We maintain a table where each row is headed by one of the four characters. The rest of the table contains integers which denote their occurrence (even if consecutive occurrences are present) and also hold information of preceding and succeeding characters which helps us

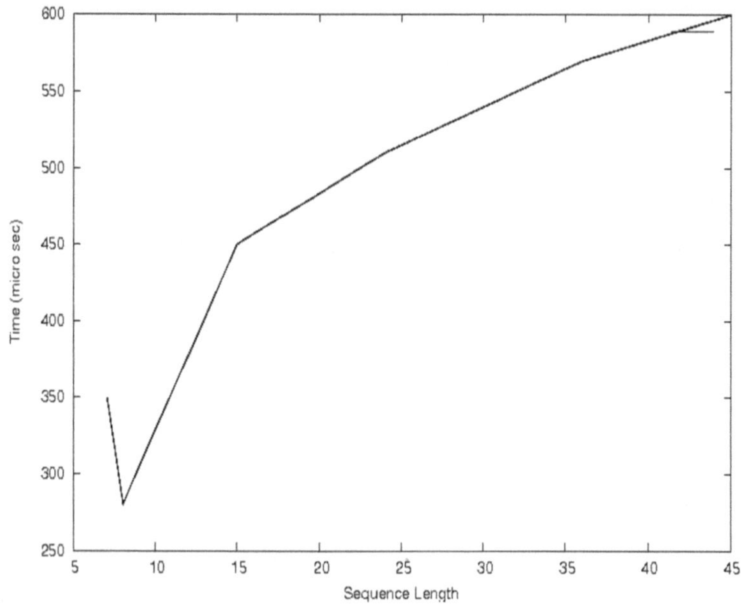

Fig. 17.2 Analysis of brute force approach

when we scan *M* for subsequences. It is observed that the method is more suitable for finding existing subsequences than degenerative ones. Also, with integers being readily related to the character set, compression techniques can be readily applied to the storage. We explain the construction of the digest table before going into the algorithm.

Consider *M*=GTACGGTCTGCATC. Now, we scan through *M* once and make entries into our table at each scan as we proceed. At *M[0]*=G which is the first character, and occurs once, so we enter 1 in the table. This trend follows on until we reach *M[4]*=G. Here since there are two consecutive G_s we enter the last value (4) plus 2, i.e., we enter 6. As we progress, *M[7]*=A and the present value is incremented and stored (in this case 7). Thus, we ensure that for a change in the sequence we increment the previous value by one. Also, if the next character does not match the current index in the table, we re-enter the previous value as in the case of *M[8]*=T. The algorithm for constructing the digest table is given below:

Table 17.1 Table construction for considered M

G	T	A	C
1	2	3	4
6	7	7	8
8	9	9	9
10	10	10	11
11	11	12	12
12	13	13	14

For our example, tracing M=GTACGGTCTGCATC results in the Table 1. It is observed that the table presented here is for the purpose of understanding the concept of our technique. Usage of integers naturally incurs high cost in terms of memory. In turn one can use bit patterns which convey the same message in the form of bits. We used *LEU1* and *LEU2* as sample to test our algorithm. The partial tables (since the tables would occupy too much space) are given below for verification purposes.

Subsequence Retrieval: Having constructed the table using the above algorithm, we now discuss how one can check whether the subsequence entered by the user exists in the table. Again, we illustrate the process with an example taking M to be the same as before in order to maintain continuity.

Given M=GTACGGTCTGCATC, let us say that we want to check if the subsequences CGGTC and CTGC exist. Looking at our table, we start with the index being G. We continue scanning till we reach C for the first time. Having reached C

Table 17.2 Table Construction for LEU1 - Initial Part

G	T	A	C
0	0	2	2
2	3	4	4
4	5	5	5
6	6	8	10
10	10	11	12
12	12	13	13
14	15	16	17
17	17	18	18
18	19	19	20
20	20	21	21
21	22	23	23
23	25	27	27
28	28	29	30
31	32	33	33
34	35	36	36
36	38	38	39
40	40	41	41
41	42	42	42
43	43	44	44
44	46	46	46
47	47	49	49
51	51	51	53
54	54	54	56
56	57	57	59
60	60	60	61
62	62	65	65

Table 17.3 Table Construction for LEU1 - Final Part

G	T	A	C
469	470	471	471
471	472	472	473
474	474	475	475
475	480	481	481
481	483	484	484
484	487	487	487
488	488	488	489
489	490	490	490
491	495	498	498
498	499	499	500
500	500	501	501
501	502	502	502
504	507	508	509
509	509	510	511
511	512	512	514
514	514	515	515
515	516	516	518
518	518	520	520
523	524	524	526
526	526	528	528
529	529	531	532
532	533	533	534
534	537	538	539
540	540	541	541
541	542	544	544

Algorithm 2. *Construction of Storage Table*

```
index ← the row belonging to one of G,A,T,C.
cur ← current value being entered into table
prev ← previous value entered into table.
prev ← 0, cur ← 0 index ← G.
Input: M
Output: Table consisting of integer values of positional significance
for each character in M do
    if M[i] == table[index] then
        check for consecutive occurrences of M[i]
        Let this value be count
        prev ← cur
        cur ← prev + count
        enter value into table.
        count ← 0
    else
        enter prev as value into table
    end if
end for
```

Table 17.4 Table Construction for LEU2 - Initial Part

G	T	A	C
0	1	2	2
4	4	5	5
5	6	8	8
8	10	11	11
11	12	13	14
14	15	15	16
16	17	18	18
18	21	21	22
22	23	23	24
24	24	26	27
27	27	29	29
30	31	33	33
33	35	35	35
37	39	39	39
40	43	43	43
45	45	45	47
48	48	49	49
50	50	50	51
53	54	54	55
55	56	58	58
60	60	60	61
62	62	62	64
64	65	65	65

Table 17.5 Table Construction for LEU2 - Final Part

G	T	A	C
487	487	488	488
488	489	490	490
490	491	492	494
494	494	495	495
495	497	497	498
498	499	501	501
501	502	502	502
503	504	504	505
505	506	506	506
507	507	507	511
511	512	514	514
515	515	517	517
518	518	519	519
519	520	520	521
522	523	523	524
525	529	529	529
530	530	530	532
532	532	533	533
535	536	536	536
537	537	538	540
540	540	541	542
543	545	545	545
547	548	548	549

where the value is 4, we need the next character in the subsequence (G) to be present next with an increment. This is true since the next index is G and the value is more than 4. Since two Gs are present in the subsequence as well as *M*, this is automatically validated by the increase of 2 in the value. Thus, at each scan, we need to check if the bases being compared are the same and the value is incremented correctly.

Algorithm 3. *Subsequence Retrieval Algorithm*

```
index ← the row belonging to one of {G,A,T,C}.
cur ← current value in table
prev ← previous value in table.
prev ← 0, cur ← 0 index ← G.
Input: M
Output: In this case, whether the subsequence is present or not
prev ← initial table value
cur ← initial table value
index ← G
construct table for S
for each character in M do
    if M[i]==table[index] then
        check if cur > prev same as S[cur] > S[prev]
        i.e., the consecutive numbers increase in the same order
    else
        move index to next location of M[i] in table
    end if
end for
if all_characters_matched then
    return true
else
    return false
end if
```

17.5 Experimental Setup and Results

All algorithms are implemented and executed on Pentium 5 running Red Hat Linux 9. Simple C/C++ code is used in order to better gauge the speed of algorithms. The compiler used is the Gnu C Compiler which came bundled with Red Hat Linux 9. Figure 17.2 is the graph plotted with the results obtained from an equivalent C code implementation of the brute force method. The figure shows the time taken for subsequence retrieval from relatively smaller length of strings unlike the existing ones. Figure 17.3 is the graph that is plotted for the proposed algorithm. The graph also included the *LEU1* and *LEU2* table construction time. The possible reason for large time values is the maintenance of indexes and the process of evaluating the entry of each cell with the previous cell. Implementation on large scale systems result in a significantly lesser running time.

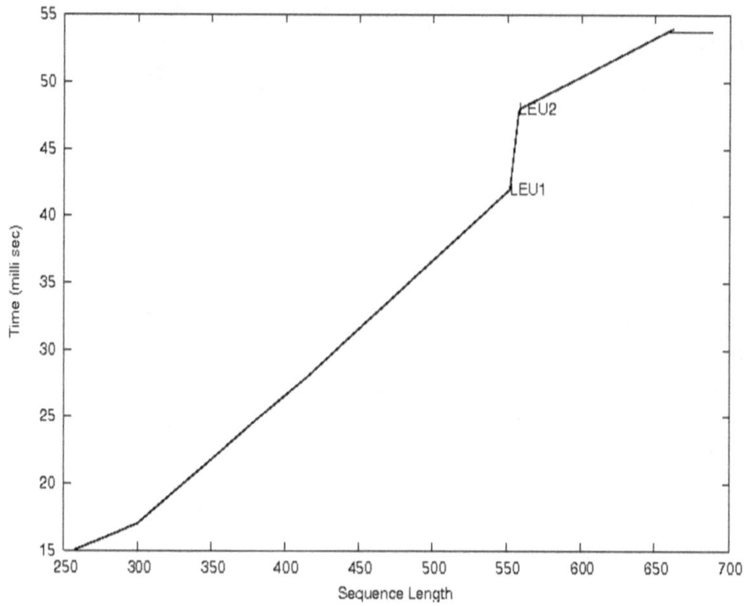

Fig. 17.3 Results of proposed approach of table construction

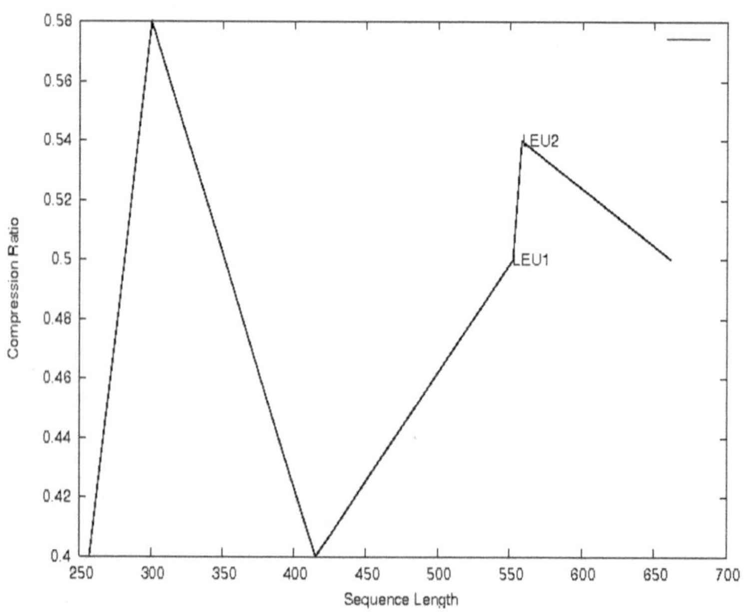

Fig. 17.4 Compression ratio for standard character storage

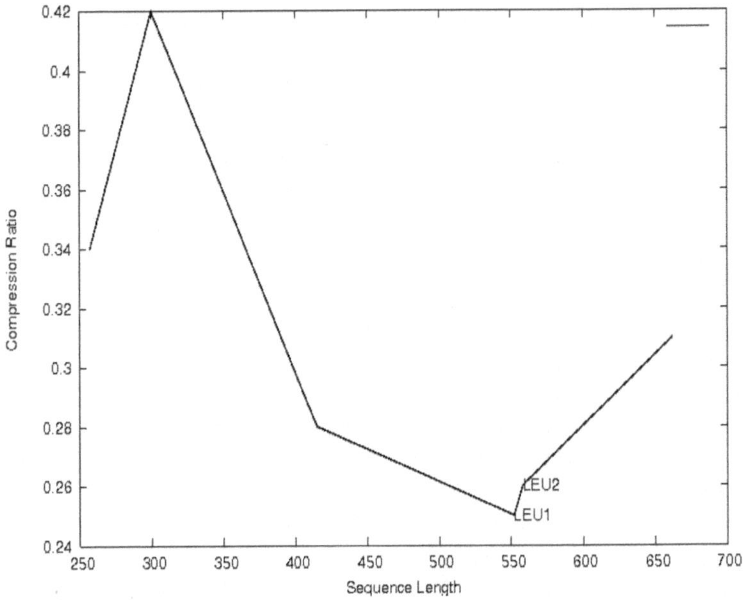

Fig. 17.5 Compression ratio for proposed technique

Figure 17.4 and Figure 17.5 indicate the compression ratio when the sequenced where compressed using bzip2. The compression ratio is calculated using compressed file size divided by regular file size and factors such are repetitiveness as well as type of the data in the file, integer or character are considered. Our file sizes where considerably larger than the character files since we utilized integers in our calculation. The usage of bits to represent bases would greatly reduce the file sizes associated with our technique.

17.6 Summary

We investigated the problems associated with retrieving patterns from an existing gene sequence. We also illustrated the use of a digest table that contains integers or digest values, as we referred to them earlier. Probable uses of integers include: indicating positions of occurrence, incorporating the *(l, d)* options as input on simple alteration of the algorithm in order to provide degenerative sequences as results. Usage of integer equivalent bit patterns for consecutive characters reduce the entire length to a bit sequence while continuing to indicate the length of repetition. This type of representation promotes compression as well.

References

1. Srinivas, V.R.: Bioinformatics - A Modern Approach. Prentice-Hall of India, Englewood Cliffs (2005)
2. Leung, H.C.M., Chin, F.Y.L.: An Efficient Algorithm For the Extended (l,d)-Motif Problem With Unknown Number of Binding Sites. In: Proceedings of the 5th IEEE Symposium on Bioinformatics and Bioengineering (BIBE 2005) (2005)
3. Jonassen, I., Collins, J.F., Higgins, D.: Finding Flexible Patterns in Unaligned Protein Sequences. Protein Science 4(8), 1587–1595 (1995)
4. Rigoutsos, I., Floratos, A.: Combinatorial Pattern Discovery in Biological Sequences: The TEIRESIAS Algorithm. Biofinformatics 14, 55–67 (1998)
5. Liu, X.S., Burtlag, L., Liu, J.S.: An Algorithm for Finding Protein-DNA Binding Sites with Applications to Chromatin Immuno Precipitation Microarray Experiments. Biotechnology 20, 835–839 (2002)
6. Neuwald, A.F., Liu, J.S., Lawrence, C.E.: Gibbs Motif Sampling: Detection of Bacterial Outer Membrane Protein Repeats. Protien Science 4, 1618–1632 (1995)
7. Hertz, G.Z., Stormo, G.D.: Identifying DNA and Protien Patterns with Statistically Significant Alignments of Multiple Sequences. Bioinformatics 15, 563–577 (1999)
8. Bailey, T.L., Elkan, C.: Unsupervised Learning of Multiple Motifs in Biopolymers using Expectation Maximization. Machine Learning 21, 51–80 (1995)
9. Liu, X., Burtlag, D.L., Liu, J.S.: Bioprospector: Discovering Conserved DNA Motifs in Upstream Ergulatory Regions of Co-expressed Genes. Pacific Symposium on Biocomputing 6, 127–138 (2001)
10. Needleman, S., Wunsch, C.: A General Method Applicable to the Search for Similarities in the Amino Acid Sequence of Two Proteins. Journal of Molecular Biology 48(3), 443–453 (2000)

Chapter 18
Merge Based Genetic Algorithm for Motif Discovery

Abstract. Motif discovery is an important problem in bio-informatics that involves the search for approximate matches. Various algorithms have been proposed, including exhaustive searches as well as heuristic searches that involve searching only a subset of all the possible solutions. One such often employed method is the genetic algorithm. A genetic algorithm based approach is employed in MDGA, using a single population. We build on that method using multiple populations, each evolving against different fitness landscapes. Individuals in each population compete for participation in the genetic events of crossover and mutation based on probabilities. Each of these fitness landscapes, is designed to solve a subset of the problem, thus optimizing a particular characteristic. Once evolution in each of these populations has saturated, they are merged according to a global fitness scheme and then evolved. This process continues till the final population also converges. Different options for implementation are also mentioned. We then proceed to compare our results with that of standard methods, on well known datasets and the results obtained are good.

18.1 Introduction

Bioinformatics and computational biology involve the use of techniques including applied mathematics, informatics, statistics, computer science, artificial intelligence, chemistry and biochemistry to solve biological problems usually on the molecular level. Bioinformatics seeks to solve problems of sequence alignment, protein structure prediction and gene prediction among many others. Computational biology is yet another related term.

Bioinformatics greatest claim to fame has been the much publicized compilation of the human genome sequence. Protein structure prediction which attempts to determine the 3 dimensional structure of proteins from their amino acid sequences is also a key and well known area. Detection and analysis of point mutations to detect cancer is a very important application. The large sequences that are generated from the sequencing efforts often contain noisy data. Sifting and analyzing this data is done by the use of informatics based approaches.

K.R. Venugopal, K.G. Srinivasa, L.M. Patnaik: Soft Comput. for Data Min. Appl., SCI 190, pp. 331–341.
springerlink.com

In molecular biology, a motif is a weakly conserved nucleotide or amino-acid sequence pattern that is widespread and has, or is conjectured to have, a biological significance [1]. Motif discovery is one of the holy grail problems of Bioinformatics research. Motif discovery seeks to find sequence patterns in multiple input patterns that is similar to a consensus motif of a given length with a certain permissible number of mutations. Mathematically, this is expressed as the (l,d) problem, where l denotes the length of the motif and d denotes the maximum number of mutations permissible within the candidate sequences to be considered to be similar to the consensus sequence. A d value of 0 indicates a perfect match.

The classic genetic algorithm (GA) models Charles Darwin's theory of evolution. A population of a species of organism, whose individuals represent solutions to the problem posed by their environment, as modeled by the fitness function, evolve to form better solutions. The weaker solutions, which are not very successful in solving the problem, are eliminated and become extinct. The stronger solutions then "mate" to combine the best qualities of both solutions. Also, existing solutions are mutated according to predefined probabilities. These crossover individuals and mutants then form the next generation of the species. This form of evolution continues until a suitably close solution is found. GAs are extremely good at scaling problems where the perceived gap between the solutions and the problem, as according to the fitness function, is large. However, as the solutions move towards optimality, the GA no longer has a "hill" to scale and as such, the rate of improvement slows down. Therefore, GAs are used in cases where an exact solution is not required, but where a merely good solution is sufficient.

Motivation for this study: Genetic algorithms present a non-exhaustive method for motif discovery. The search space in the motif discovery problem is extremely large and a good non-exhaustive method must intelligently choose the candidates to examine. Existing approaches using genetic algorithms have tried, via a variety of fitness landscapes, to select a set of candidate solutions. The selection of candidate solutions, is thus the key to the solution of the motif discovery problem. Our approach, attempts to maximize the examination of the solution space, by using multiple populations. Each population has its own fitness function, thereby allowing us to fine tune the search parameters for each population's quest.

Previous approaches, have so far concentrated on single population optimizations on the basic genetic algorithm. Those that have used multiple population models, have specified a relation or an interaction between them. Our proposed algorithm, seeks to isolate the individuals of one population from another and allows them to evolve unimpeded.

Each population grows on a different fitness landscape. The effect is that towards the end of the algorithm's execution, the final population has been maximized against each characteristic. Unlike other GAs however, if the algorithm halts the point of saturation is reached, the members of the population are not in the vicinity of the optimality. As the number of iterations in a GA increase, so does its "closeness" to the optimal value. In our algorithm, the number of iterations after the point of saturation has been crossed, defines the final population's "closeness" to the optimal value.

In order to solve the problem of convergence at local maxima, we propose the introduction of multiple geographically isolated populations. Each of these populations, uses a different fitness function. Hence, each population evolves towards its local maxima. At some point of saturation, all populations stops evolving towards the global optimum. We then seek to evaluate all the populations against a normalized function that determines the average fitness of each population. A strong population is then chosen and its individuals are merged into a weak population. This causes individuals which are very strong in one environment, to evolve in another environment, which presents yet another set of challenges. Hence the resulting population after sufficient evolution bears the characteristics of both initial populations. A population that is generated from the result of merging, may not be evaluated for fitness for a certain number of iterations. This gives the population a "window of opportunity" to evolve. This process is iterated, until only one population remains. This final new population, is then evolved until a sufficiently optimal solution is reached.

Previous applications of Genetic algorithms to Motif Discovery have used Multiple populations. Each of these populations has the same fitness function. However, these algorithms have a measured probability p_t which defines the probability on which individuals are transferred between populations. This has the effect of refining the population into which the new individuals are transferred. The extent of refinement, however is a matter of chance, since genetic algorithms are non exhaustive and stochastic in nature. The specificity of the fitness function, also plays a major role in the extent of refinement.

Geographical Isolation, occurs when a part of a species' population when separated from its parent population, over time, evolves to different characteristics. In nature, when geographical isolation does occur, the new environment presents a different set of problems for the population. Thus, the population has to solve the problems in order to avoid extinction. In the perspective of an algorithm, the population has a new "hill" to climb. This hill is formed due to a change in the fitness landscape.

Our approach, involves a known and predefined number of populations, which are bred, from the very start, against different fitness landscapes. After a specified period of individual evolution, these populations are merged, one at a time, with regular iterations. Different fitness landscapes are achieved by the use of different fitness functions. Each of these fitness functions evolves the population, in order to optimize a given characteristic or a set of characteristics. The end result of such evolutionary processes is a set of populations, the union of their sets of characteristics being the total set of characteristics desired.

These populations are then merged as according to a merging scheme. The end result of the algorithm, is a well balanced population, which has individuals, which having evolved in a competitive fitness landscape among competitive peers, provides a very desirable solution.

18.2 Related Work

Genetic algorithm has been previously applied to Motif Discovery to identify motifs in multiple unaligned DNA sequences [1], [2]. These methods differ in the fitness functions they use to evaluate the individuals of the population. Each of these methods uses a fitness function, which seeks to solve a specific problem. In some cases, these problems overlap and the final resulting populations have similar characteristics [3].

A variety of approaches have been applied to solve this problem. These can be broadly classified as deterministic and non-deterministic. Deterministic approaches conduct an exhaustive search of the solution space to realize the solution. On the other hand, non-deterministic approaches only search a subset of the solution space. This subset is determined via an algorithm that characterizes the approach. Commonly used non-deterministic approaches utilize statistical methods such as hidden Markov models, artificial intelligence methods such as neural networks and evolutionary computing methods such as genetic algorithms. Commonly used motif discovery tools are MEME [4], CONSENSUS [5], MotifSampler [6], BioProspector [7], Gibbs sampling [8] and AlignACE [9].

[10] and [11] have also proposed the use of multiple population models. They advocate the use of either a migration model or a specific distribution of the individuals. Very often, genetic algorithms tend to converge towards local optima rather than a global optimum. These solutions, give useful results when the solution space has only one global optimum. The fitness landscape, as described by the fitness function can be used to predict the occurrence of either. Some methods that are used to alleviate this problem include triggered hypermutation and random immigrants.

18.3 Algorithm

We now look into design of an algorithm, which uses multiple populations in order to achieve the required result. Our first work is to divide the given problem into a set of sub problems, each requiring a particular characteristic of the solution. This forms the basis for our algorithm. We then proceed to design fitness functions for each of these characteristics. Let us now briefly explore some of the design issues.

Design issues: The design of our proposed algorithm, requires the following issues to be considered.

- *The number of populations:* A more complex problem, involving very specific requirements, benefits from a larger number of populations. In such cases, each population can be tuned to work towards a specific characteristic that is required of the final solution.
- *The fitness functions for each population:* For each population required, a unique fitness function is required. If diversity of a property of the solution is sought, then more than one population may utilize the same fitness function. The fitness function for a population, must seek only to present a specific problem to the

individuals of the population. The set of characteristics that the different fitness functions seek to work with must be disjoint.

- *Normalized average fitness function:* Since this function decides when a population is "weak" or "strong", it must be defined with some care. The output of this function may be implementation dependent. Some alternatives for the implementation are:

  1. It can either be an unsigned score in a range $[min, max]$ where a population with score max is stronger than a population with score min.
  2. A ranking based system where a population with rank 1 is the strongest and a population with rank n is the weakest.
  3. The function itself may return a boolean value when two populations are passed as arguments. This result depicts the relative fitness of the populations.

- *Point of saturation:* After a certain number of iterations, the individuals in a GA stop evolving. This point, considered to be the point of saturation, is dependent on the fitness landscape of the fitness function. Considering the fitness landscapes of each population, this point of saturation may vary. The GA must run at least till this point is crossed. The evaluation of the normalized average fitness function must only be done on a population which has crossed its point of saturation. When all the populations have been merged into a single final population, this population must also be evolved until the point of saturation.
- *Merging scheme:* The order in which the populations are merged, must be decided. This can be implemented in the following ways:

  1. A ranking based scheme, which merges the second weakest population into the weakest population. This has the effect of evolving the "stronger" populations in a different landscape.
  2. An ordering scheme, which is derived from the problem statement itself. For instance, the evolution of a characteristic may be a pre-requisite for the evolution of another.

Proposed Algorithm: We now define some of the notations that we use to describe our proposed algorithm. We define the number of populations to be n. The fitness functions for each of these populations is given as $fitness[i]()$. The corresponding point of saturation for the population i is $saturation[i]$. The merging scheme is defined in the function $merge\_populations(pop_1, pop_2)$, where pop_2 and pop_2 are merged as according to the scheme. The normalized fitness function is written as $normalized\_fitness(pop)$. The algorithm for motif discovery using multiple geographically isolated populations, MGIGA is shown in Algorithm 1.

Explanation: We shall now explore the algorithm in some detail. The given problem statement, must first be carefully analyzed to derive specific characteristics that need to be optimized. Accordingly, once the desired characteristics of the solution have been decided upon, the number of populations for the algorithm must be decided. It may be required to use more than one population for a given

Algorithm 1. MGIGA

```
n ← Number of populations
for i = 0 to n do
    Design fitness[i]
    Decide saturation[i] based on fitness[i]
end for
Design merge_populations(pop1, pop2)
Design normalized_fitness(population)
while num_of_populations > 1 do
    if n > 1 then
        for i = 0 to num_of_populations do
            evolutions ← 0
            while evolutions < saturation[i] do
                evolve(population[i])
            end while
        end for
    end if
end while
for i ← 0 to n do
    Evaluate normalized_fitness(population[i])
end for
while num_of_populations! = 1 do
    Select pop1 and pop2.
    merge_populations(pop1, pop2)
end while
while fitness(final_population) < required_fitness do
    evolve(final_population);
end while
```

characteristic. It may also be required to use one population for a set of minor characteristics. In either circumstance, the fitness function for the population is to be designed appropriately.

The fitness functions, must work on mutually independent characteristics. If there is any overlap of characteristics, the populations would have individuals, with similar characteristics. The overlap, would be at the cost of exploring other candidate solutions and this exploration must be optimized in an exhaustive search. On the other hand, optimization of the same characteristic across different populations ensures that, the particular characteristic is studied thoroughly by the algorithm. This issue needs to be factored while designing the fitness functions for the populations.

Each fitness function, causes the population to saturate at a given point. This "saturation" is defined at the point, when the population no longer evolves appreciably with any further generations. This point of saturation, can be a fitness value which is reached upon after the sufficient evolution of the population, or the number of iterations. The point of saturation defines the stopping criteria for the algorithm and thus evolving a particular population of interest to the application.

The population merging function, defined by $merge\_populations(pop1, pop2)$, serves to move individuals, which have adapted well to one population to evolve in another landscape. This has the effect of improving the quality of mutation, without changing the mutation rate itself. For implementations, with fewer populations, this merging scheme could be predefined such as a trivial transfer of individuals from one population to another. For more complex implementations, the merging scheme should define the number of individuals to be transferred, as well as the order of merging. The order of merging defines the characteristics to be optimized and the order of optimization. In general, it would be more useful to transfer the individuals from stronger population to weaker population, as this has an effect of optimizing stronger individuals in a different landscape. In addition, the native population has a varied mutant quantum. The normalized fitness function, would have to be designed, keeping in mind the overall requirements of the solution. This function is used to evaluate individuals across populations and thus, must take into account all characteristics.

Once these functions have been setup as desired, the algorithm is executed. Each of the populations are evolved, without any interaction from the other populations. After all populations have been sufficiently evolved (once their points of saturation have been reached), the normalized average fitness function is called to evaluate the populations. The merging function is then used to merge individuals from different populations, in the order that is decided with the help of the normalized average fitness function. Each single merge is accompanied by iterative evolution, until the point of saturation is reached. The above process is iterated, until one single population is left. This final population is evolved, until a sufficiently fit solution is found. This final solution is optimized against all populations.

Implementation Strategies: The proposed algorithm may be implemented in parallel using a cluster or a grid. A Master node may be utilized to control the timing and flow of the implementation. Each worker node may be used to evolve one population. MPI may be used for purposes of communication between the master node and each of the worker nodes.

A character based chromosome may be used for simple implementation of the crossover and mutation operators. The crossover and mutation operators are based on set probabilities. To increase the diversity of a population, the probabilities of the operators may be changed when the iterations near the point of saturation.

18.4 Experimental Setup

The algorithm is implemented using the Python programming language, for its dynamic nature and ease of use. The genetic algorithm framework is implemented using the PyGene library. A single gene of an individual is made up of the starting positions of the candidate motifs, from the different sequences of the input. Selection is done using Roulette wheel selection procedure. This is internally implemented in the PyGene library.

Fitness Functions: For the first population, we used the following function. This function optimizes the search for weakly conserved motifs. The fitness of an individual X is given by $fitness(X)$. β is the bonus coefficient, that is used to give a bonus to a consensus motif that is found at least once in all the sequences. N gives the number of sequences. c_i gives the number of motif instances of sequence i having $0 \leq d_j \leq d$. γ is used to give distance based weight to a motif. Always $\gamma \geq 2$. Finally, $w_(j)$ defines the frequency based weight of motif j. A similar fitness function is used by [1]

$$fitness(X) = \beta \sum_{i=1}^{N} \sum_{j=1}^{c_i} \gamma^{(d-d_j)} w_(j))$$

The second fitness function we use, is given by

$$fitness(X) = \sum_{i=1}^{W} IC_i$$

W is the Motif width, defined by the number of columns in the motif. IC_i is the information content of column i. This information content is calculated as

$$IC = \sum obs_b \log_2 \frac{obs_b}{background_b}$$

where obs_b is the observed frequency of nucleotide b on a particular column and the $background_b$ is a background frequency of the same nucleotide. Thus the fitness of an individual is calculated as the sum of the information in each column of the motif alignment. This function is similar to the one used by [3].

Data sets used: To test the results of our algorithm, we ran it against the 15 target genes of transcription factor YDR026c in *Saccharomyces cerevisiae*, as well as the 6 sequences MCB transcriptional factors of the same and finally the LEU3 motifs as well. All sequence datasets are obtained from [12].

YDR026c Binding Sites: The dataset consists of 15 genes of YDR026c in yeast. Empirical results for the binding site of this sequence have not yet been confirmed. We also compare our results with those obtained from MEME and MDGA. Motif patterns detected by all three programs are very similar.

Motif program	Predicted Motif
MEME	TCCGGGTAAA
MDGA	TCCGGGTAAA
MGIGA	TACCGGGTAA

MCB Transcriptional Factor: Six sequences from positions -500 upstream to +50 downstream of the regulated genes of *Saccharomyces cerevisiae* are extracted and used as inputs. Our results are exactly the same as observed by MDGA. Thus the additional population is not needed in this case.

Embedded motif	MDGA	MGIGA
ACGCGT	ACGCGT	ACGCGT
ACGCGA	ACGCGT	ACGCGT
CCGCGT	ACGCGT	ACGCGT
TCGCGA	ACGCGT	ACGCGT
ACGCGT	ACGCGT	ACGCGT
ACGCGT	ACGCGT	ACGCGT

The consensus of the embedded motifs is WCGCGW. Notice, that the consensus of our predicted motifs is ACGCGT. This is the true consensus motif and reflects that obtained by MDGA.

LEU3 Transcriptional Factors: Two sequences from positions -500 upstream and +50 downstream of start sites of two regulated genes of *Saccharomyces cerevisiae* are extracted. The two regulated genes are LEU1 and LEU2, and have a consensus motif of CCGNNNNCGG. For the consensus CCGGAACCGG, and consensus CCGTAACCGG, the motifs predicted by MGIGA and by MDGA are identical.

Consensus Motif	MDGA	MGIGA
CCGGGACCGG	CCGGAACCGG	CCGGAACCGG
	CCGGGACCGG	CCGGGACCGG
CCGGAACCGG	CCGGAACCGG	CCGGAACCGG
	CCGGAACCGG	CCGTAACCGG
CCGTAACCGG	CCGTAACCGG	CCGTAACCGG

18.5 Performance Analysis

The running time of our algorithm, is significantly larger from that of other genetic algorithm based approaches, since multiple genetic algorithms must be run. Even after running the GAs in parallel, the merging scheme requires additional GAs to complete evolution. For the experiments shown, two GAs are run and merged after some evolutions. Thus the running time is the sum of the running time of both these GAs. Figure 18.1, shows the convergence rate of our GA after all merging has completed. The rate of convergence of MDGA and of the genetic algorithm proposed by [1] is much slower than that of MGIGA. This is because, at the final GA in our algorithm, individuals are already much fitter than the random individuals chosen by MDGA and [1] at the beginning.

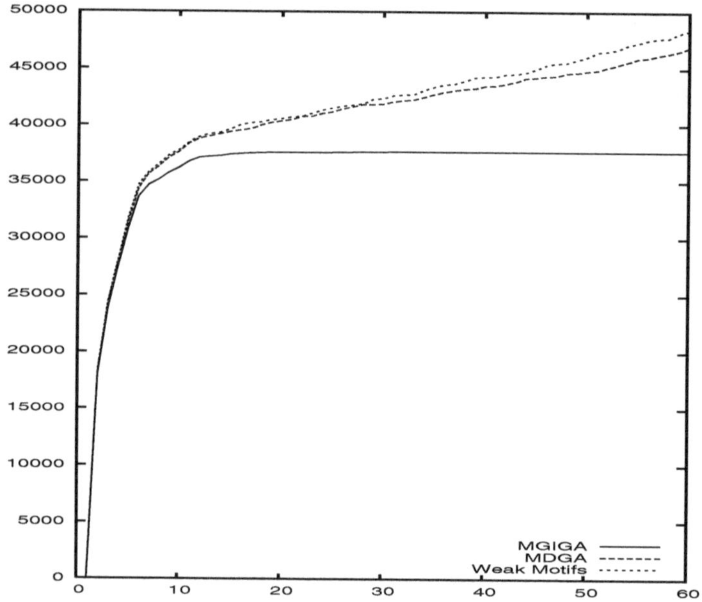

Fig. 18.1 Comparative convergence rate of the final GA after all merging has completed, MDGA and from [1]

18.6 Summary

We have proposed an algorithm that utilizes many populations in order to reach the global optimum. Each population represents a characteristic. Each population is optimized for the local maxima. Individuals from these populations are combined to form the population for the final evolution that reaches the global optimum. As applied to Motif Discovery, our algorithm is as effective as existing genetic algorithm based approaches. The scope of this work is only exploratory and thus more validation needs to be done. For shorter sequences, the overhead of the multiple genetic algorithms outweigh the benefits of scaling the local maxima. This algorithm may be more suited to longer sequences, when searching for motifs with two or more characteristics.

References

1. Paul, T.K., Iba, H.: Identification of Weak Motifs in Multiple Biological Sequences using Genetic Algorithm. In: Proceedings of GECCO 2006, Seattle, USA (2006)
2. Fogel, G.B., Weekes, D.G., Varga, G., Dow, E.R., Harlow, H.B., Onyia, J.E., Su, C.: Discovery of Sequence Motifs Related to Coexpression of Genes using Evolutionary Computation. Nucleic Acids Research 32(13), 3826–3835 (2004)
3. Che, D., Song, Y., Rasheed, K.: MDGA: Motif Discovery using a Genetic Algorithm. In: Proceedings of GECCO 2005, pp. 447–452 (2005)

4. Baile, T.L., Elkan, C.: Unsupervised Learning of Multiple Motifs in Biopolymers using Expectation Maximization. Machine Learning 21, 51–80 (1995)
5. Hertz, G.Z., Stormo, G.D.: Identifying DNA and Protein Patterns with Statistically Significant Alignment Sets of Multiple Sequences. Bioinformatics 15, 563–577 (1999)
6. Thijs, G., Marchal, K., Lescot, M., Rombauts, S., De Moore, B., Rouze, P., Moreau, Y.: A Gibbs Sampling Method to Detect Over-represented Motifs in the Upstream Regions of Coexpressed Genes. Journal of Computational Biology 9, 447–464 (2002)
7. Liu, X., Burtlag, D.L., Liu, J.S.: Bioprospector: Discovering Conserved DNA Motifs in Upstream Regulatory Regions of Co-expressed Genes. In: Pacific Symposium on Biocomputing, vol. 6, pp. 127–138 (2001)
8. Neuwald, A.F., Liu, J.S., Lawrence, C.E.: Gibbs Motif Sampling: Detection of Bacterial Outer Membrane Protein Repeats. Protein Science 4, 1618–1632 (1995)
9. Roth, F.P., Hughes, J.D., Estep, P.W., Church, G.M.: Finding DNA Regulatory Motifs within Unaligned Noncoding Sequences Clustered by Whole-Genome mRNA Quantization. Nature Biotechnology 16, 939–945 (1998)
10. Srinivasa, K.G., Sridharan, K., Shenoy, P.D., Venugopal, K.R., Patnaik, L.M.: A Dynamic Migration Model for Self Adaptive Genetic Algorithms. In: Gallagher, M., Hogan, J.P., Maire, F. (eds.) IDEAL 2005. LNCS, vol. 3578, pp. 555–562. Springer, Heidelberg (2005)
11. Srinivas, M., Patnaik, L.M.: Binomially Distributed Populations for Modelling GAs. In: Proceedings of Fifth International Conference in Genetic Algorithms, pp. 138–143. Morgan Kauffmann Publishers, San Francisco (1993)
12. Fraenkel lab downloads, `http://jura.wi.mit.edu/fraenkel/download/release_v24/fsafiles/`

FSC
www.fsc.org
MIX
Papier | Fördert gute Waldnutzung
FSC® C083411

Kolibri 360 GmbH
Ferdinand-Jühlke-Straße 7
99095 Erfurt, Deutschland
produktsicherheit@kolibri360.de